Electromechanical Energy Devices and Power Systems

Electromechanical Energy Devices and Power Systems

Zia A. Yamayee
Juan L. Bala, Jr.
Gonzaga University

John Wiley & Sons, Inc.
New York   Chichester   Brisbane   Toronto   Singapore

ACQUISITIONS EDITOR	Steven Elliot
MARKETING MANAGER	Susan Elbe
PRODUCTION EDITOR	Richard Blander
DESIGNER	Maddy Lesure
MANUFACTURING MANAGER	Andrea Price
ILLUSTRATION COORDINATOR	Jaime Perea

This book was typeset in Times Roman by Publication Services and
printed and bound by Courier Stoughton. The cover was printed by Phoenix Color Corp.

Library of Congress Cataloging in Publication Data:

Yamayee, Zia A.
   Electromechanical energy devices and power systems /
Zia A. Yamayee, Juan L. Bala, Jr.
      p.   cm.
   Includes bibliographical references and index.
   ISBN 0-471-57217-9 (alk. paper)
   1. Electric power systems.   2. Electric machinery.   I. Bala,
Juan, 1947–   . II. Title
TK1001.Y36   1994
621.319–dc20   93-5529
                                                                    CIP

Printed in the United States of America

10   9   8   7   6   5   4   3   2   1

Preface

This textbook is intended to serve as a book for a junior level, one-semester course in electric power engineering or energy conversion. Students using this book should have taken an electric circuits analysis course and have been introduced to electromagnetic fields. The student should have gained a working knowledge of complex algebra, sinusoidal analysis, phasor diagrams, phasor analysis, and basic power concepts.

As the curricula of electrical engineering programs became more and more overcrowded, many electrical engineering departments decided to stop requiring all electrical engineering students to take two power engineering courses and began (a) requiring a combined electromechanical energy conversion and power systems course, or (b) to drop the power system analysis portion altogether and require only an electromechanical energy conversion course.

In a recent article prepared by the Power Engineering Education Committee of the Power Engineering Society of IEEE, it was discovered that almost 15% of the electrical engineering programs across the United States and Canada require a combined course in energy conversion and power systems. The objectives in preparing this book were (a) to provide all electrical engineering students with an understanding of the principles of electromechanical energy conversion, as well as an overview of the power system, and (b) to provide students who are interested in power engineering sufficient understanding of machinery and power systems so they can take more advanced courses in the power area. The book has been written to satisfy the requirements of electrical engineering programs that offer a basic core course in electric power or energy conversion to serve all electrical engineering students. Our goal has been to write a book that could be used either for an electromechanical energy conversion course, or a course that combines the electromechanical energy conversion and power systems analysis into one course with primary emphasis on electromechanical energy conversion topics.

Features

This textbook offers a number of special features including:

- *Drill Problems* Drill exercises are presented within the different chapters to enable students to test their understanding of the subject matter just studied.
- *Homework Problems* Several homework problems are given at the end of the chapter with varying levels of difficulty—from simple problems requiring only simple manipulations of the various formulas to more difficult ones requiring in-depth analysis.
- *Illustrations and Diagrams* Several illustrations and circuit diagrams are presented in the text to help demonstrate and describe the theory being studied. The use of circuit diagrams and phasor diagrams proves very helpful in understanding the problem and its solution.
- *Numerical Examples* Illustrative and numerical examples are used extensively throughout the text to help the student fully understand the concepts and how the theory applies to the solution of application type problems.

Text Organization

A chapter-by-chapter overview of the textbook follows:

Chapter 1 This chapter introduces the student to different energy sources and various methods of electric energy conversion.

Chapter 2 The second chapter presents an overview of the electric power system and its components.

Chapter 3 This chapter reviews circuit and power concepts in electrical circuits.

Chapter 4 This chapter introduces the student to magnetic circuits and transformers, both single-phase and three-phase transformers.

Chapter 5 Chapter 5 presents the fundamentals of rotating machines. It provides a unified approach to all the electromechanical energy conversion machines discussed in this book.

Chapter 6 The operational characteristics of various types of DC machines are discussed.

Chapter 7 Chapter 7 presents a thorough coverage of both wound-rotor and salient-pole synchronous machines (generators and motors).

Chapter 8 Chapter 8 presents the theory and application of both three-phase and single-phase induction motors.

Chapter 9 This chapter covers (a) the basic parameters of the conductors that are used in transmission lines, (b) the various formulas for these parameters, and (c) the proper use of tables of conductor characteristics for obtaining the values of the parameters. Models of transmission lines are then presented for use in power system studies.

Chapter 10 This chapter presents the different power flow solution techniques. In other words, the focus of this chapter is the normal steady state operation of the power system.

Chapter 11 Chapter 11 covers the abnormal operating conditions of power systems including fault studies, system protection, and power system stability.

Appendices Six appendices provide supplementary information for material covered in the 11 chapters of this textbook. The final one, Appendix G, contains a glossary of key terms.

In writing this book, it was our goal to provide a degree of flexibility in meeting the needs of various engineering curricula as regards the conduct of the associated power courses. Listed below are suggestions that can serve as guidelines in the use of the text material.

A. One-semester course on electric machines and power systems:
Chapter 1
Chapter 2
Chapter 3 (if needed)
Chapter 4
Chapter 5
Chapter 6 (excluding Section 6.8)
Chapter 7 (excluding Section 7.5)
Chapter 8 (excluding Section 8.5)
Chapter 9
A brief coverage of Chapters 10 and 11

B. One-semester course on electric machines:
Chapter 1
Chapter 2
Chapter 3 (if needed)
Chapter 4
Chapter 5
Chapter 6
Chapter 7
Chapter 8

Supplements

Solutions Manual A solutions manual provides solutions to all drill and end-of-chapter problems. Answers to problems are not provided in the textbook.

Software For those interested, the authors have developed a Power Systems Simulation Package (PSSP), which is available separately from our publisher, John Wiley & Sons, Inc. PSSP has three major functions:

1. Power flow analysis of an electric power system
2. Short circuit analysis of an electric power system
3. Data preparation and editing

In closing, it should be stated that the intent has not been to go in great depth into electromechanical energy conversion or power system analysis, but to provide a thorough and understandable coverage of these two topics to undergraduate electrical engineering students. Our hope has been to expose the student to more than a familiarity with just conventional machines. The authors are of the opinion that although this is a worthwhile objective in itself, it is also important that electrical engineering graduates possess some knowledge of electric power systems. Time for these topics is made available by forgoing some of the in-depth study of conventional machines. The authors would appreciate the comments of the user of this book.

June, 1993   Zia A. Yamayee
                                                                    Juan L. Bala, Jr.

Acknowledgments

This book could not have been written without the contribution of many of the authors' students and colleagues, and of Wiley's editorial team. We thank all those who contributed to the development of this text.

The authors are thankful to a number of engineering students at Gonzaga University, including Andrew R. Santos, for drawing the initial sketches; Holger H. Peller, for solving some of the problems and reading the initial manuscript; and Jianmin Zhao, for solving a few of the problems. The authors also thank Juan Bala, III, for typing the initial manuscript.

The authors are indebted to the following individuals who reviewed and made constructive suggestions for the improvement of the textbook: Professor Peter Sauer, University of Illinois; Professor Ali Keyhani, Ohio State University; Professor Richard D. Schultz, University of Wisconsin at Platteville; Professor Keith Stanek, University of Missouri-Rolla; Professor Robert Laramore, Purdue University; Professor Thomas H. Ortmeyer, Clarkson University; Professor J. O. Ojo, Tennessee Technological University; and Professor Clifford H. Grigg, Rose-Hulman Institute of Technology. These reviewers' comments and suggestions have greatly improved the quality of the text.

The authors are grateful to their editor, Steven Elliot, of John Wiley & Sons, for his encouragement and dedication to this project. The authors also thank their assistant, Shirley Grant, for her contributions to the publication of this manuscript. Finally, the authors thank their respective families for their patience and encouragement during the course of completing this book.

ZAY
JLB

Contents

Electromechanical Energy Devices and Power Systems

One

Energy Resources and Electric Energy Conversion

1.1 INTRODUCTION

The ability of humans to develop sources of energy needed to accomplish useful work has played an essential role in the continual improvement in the standard of living of people around the globe. The use of energy can be seen in everyday devices such as home appliances and machinery. Energy consumption in homes and factories increased beyond the point where useful forms of energy could be produced at the locations at which energy was being used. Hence, centralized energy processing and generating stations, together with elaborate transmission and distribution systems, were developed and the electric power system emerged as a tool for converting and transmitting energy. Energy is converted from its basic source, such as coal or hydro, into electric energy at places usually remote from population centers. Then the electric energy is sent over transmission and distribution systems to various locations, where it is converted to light, heat, and mechanical energy.

The objective of this book is to investigate how energy conversion, particularly electromechanical energy conversion, and transmission and distribution of electric energy take place. The devices used in the operation of a power system are studied. The book is structured so that a power system is described element by element in each chapter.

The first section of this chapter gives a brief description of the various energy resources available to humans. These energy resources are usually not in directly usable form; they have to be converted into more useful forms. The conventional electric energy conversion techniques that are currently in use are described in the next section. Finally, alternative sources or nonconventional energy conversion techniques are presented in the last section.

1

1.2 ENERGY RESOURCES

Energy resources are the various materials that contain energy in usable quantities. These are present in any of the various energy forms that are transformable to other forms, including electrical, mechanical, chemical, and nuclear energy. The main energy forms include chemical, hydro, nuclear, and geothermal energy. Chemical energy includes such fuels as coal, oil, and natural gas. Energy resources are usually classified in two general categories: renewable resources and expendable resources. Renewable resources, such as water, wind, solar, and tide, are replaced continuously by nature. Expendable resources, such as oil and coal, are expended when used.

Energy may be classified as either primary energy, which is obtained directly from nature, or secondary energy, which is energy derived from primary energy. Primary energy sources include hydro, solar, wind, oil, natural gas, and geothermal. Electrical energy is a form of secondary energy. It is derived from primary energy by using thermal power plants to convert heat energy and hydroelectric plants to transform the energy of water under pressure.

Energy resources are being consumed at a high rate because of the growing requirements of industries and the increasing demands of people. It has been predicted that the world's supply of fossil fuels will be used up within a few hundred years.

The rate of consumption of oil has been rapidly growing. In many countries, oil has been the primary fuel used for electricity generation. It is estimated that 60% of the world's proven oil reserves are in the Middle East and only about 10% are in the United States. The fast growth of oil consumption is due mainly to its use as fuel for automobiles and airplanes and the fact that it is easier to recover and transform to other energy forms than solid fuels.

The distribution of natural gas reserves is not accurately known. Estimates of these reserves are approximate. In the United States, the use of natural gas has risen consistently because of its relatively low cost.

It has been estimated that the world's hydropower resources amount to about twice the current annual generation of all hydroelectric plants in the world. The available hydropower depends on water inflows, and estimates of the available hydro energy could be very approximate.

Geothermal energy is an almost limitless reserve. The temperature of the Earth increases with depth in the Earth's crust. For example, the temperature is about 1200°C at a depth of 30 miles. During earthquakes or volcanic eruptions, molten rocks may be able to escape to the surface in the form of lava. Some of the molten rocks do not reach the surface, but they produce gases and steam by heating water they encounter on their way to the surface. The heated water may escape through the surface as hot springs, and the steam can be used in geothermal plants.

Geothermal energy has been used to supply hot water for the city of Reykjavik, the capital of Iceland. In the northern part of California, geothermal power plants produce over 800 MW of electrical power.

It has been estimated that the heat content of the world's reserves of uranium is more than 300 times the heat content of the world's reserves of fossil fuels. However, the cost of mining for uranium is extremely high.

Using nuclear fission, current nuclear power plants convert only a small fraction of the energy present in the nuclear fuel. However, the unexpended fuel can be reprocessed for use in breeder reactors.

In nuclear fission, the nucleus of a heavy element such as uranium-235 is split. In nuclear fusion, on the other hand, nuclei of light elements are fused to form heavier nuclei. Either process is accompanied by the release of large amounts of heat.

Nuclear fusion is a more attractive energy source than fission because it is inherently safe, does not produce radioactive waste, does not present any danger of explosion, and cannot result in runaway meltdown, and the fuel (deuterium) used is readily available from the oceans. However, it is still not technically feasible to use nuclear fusion for commercial electric power generation.

The oceans regularly rise and fall in response to the relative positions of the Sun, Earth, and Moon. The water elevation is not the same at different places. The variation of water elevation can be used by electric power plants to produce electric energy. However, such power plants are extremely expensive, and the power output is quite variable because of the continually changing tides.

Wind energy has been used in windmills around the world in countries such as The Netherlands. Wind turbines have also been used for electrical energy production in many countries, including the United States, but these have been in the low power (few megawatt) levels.

The Sun is the ultimate source of energy because it has immense energy reserves. It has been estimated that the total amount of solar energy received by Earth in a year exceeds its total energy requirements. Solar energy is usually used for space heating and water heating. Although the use of solar energy is growing, it is still at a low level. Because the power density is low at the surface of the Earth, large collectors are necessary to accumulate small amounts of power.

Research has been under way to develop other methods of utilizing solar energy for electric power production. However, solar cells have not yet become a feasible alternative for bulk, or large-scale, power production.

1.3 CONVENTIONAL METHODS FOR ELECTRIC ENERGY CONVERSION

At present, two methods are commonly used for electric bulk-power generation. Both methods employ an electric generator that converts the mechanical energy of the prime mover to electrical energy. The main difference between the two methods is the source of the mechanical energy used to rotate the generator.

One method makes use of water under hydraulic pressure to provide the mechanical energy to rotate a hydraulic turbine whose shaft is coupled to the generator shaft. The efficiency of conversion of the energy available from the water to electric energy is quite high—up to 80%–90%.

The other method employs a boiler to convert the energy of the fuel (coal, oil, or nuclear fuel) into heat energy, which is used to transform water into high-temperature steam at very high pressure. The steam rotates the steam turbine in the same way as the water turns the hydraulic turbine. The efficiency of conversion of the energy available from the fuel to electric energy is much lower than in the first method, typically ranging from 30% to 40%.

The different types of electric bulk-power generating plants are described in this section. Other power plants employing variations of these methods, such as pumped-storage plants, are also briefly discussed.

1.3.1 Hydroelectric Plants

A schematic diagram of a hydroelectric power plant is shown in Fig. 1.1. The electric power generated by a hydroelectric plant depends on the amount of water flowing through the hydraulic turbine and on the water pressure. The plant capacity, expressed in kilowatts, is given by

$$P = 9.81QH\eta \qquad (1.1)$$

where

Q = rate of flow in m³/s

H = pressure head of the water in m

η = efficiency expressed as a fraction of 1

In order to increase the pressure head of the water, river dams or diversion canals are constructed. The hydraulic turbine converts the kinetic energy of the water coming from the reservoir into mechanical energy available at its rotating shaft. A hydraulic turbine may be classified as either an impulse turbine or a reaction turbine.

In a high-pressure impulse turbine, a convergent nozzle transforms the po-tential energy of the hydrostatic pressure into kinetic energy of moving water.

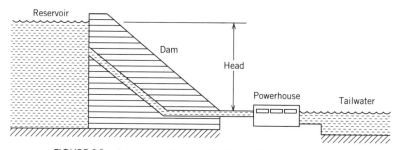

FIGURE 1.1   Schematic diagram of a hydroelectric power plant.

The pressure at the nozzle outlet is atmospheric. The amount of water entering the turbine is adjusted automatically to obtain the desired power output.

When the water flow rate Q is high and the pressure head H is rather low, the reaction turbine is preferred. In a reaction turbine, water is introduced to the runner blades through a convergent pipe section known as a constrictor, wherein part of the potential energy of the water is converted into kinetic energy. Further energy conversion is carried out at the turbine blades.

The turbine and generator of an electric power plant are located in a common shaft. This results in a common rotational speed n_s, which depends on the number of poles p of the generator rotor and the operating frequency f of the electric power system. Thus,

$$n_s = \frac{120f}{p} \qquad (1.2)$$

Hydroelectric plants are usually designed and built for multiple reasons and purposes in addition to electric energy production. These include river flood control, river navigation, irrigation of agricultural lands, water and electric energy supply to power-consuming industries using local raw material resources, and recreational purposes.

1.3.2   Pumped-Storage Plants

The electric power generated by the power plants must be exactly equal to the electric power consumed by various users at all times. The electrical demand of individual consumers may be described as random and irregular, so the total usage of electric power is nonuniform.

The variation of the total power needed by users with the time of day is described by the demand curve. This curve contains a number of peaks and valleys; that is, the total power required is high during certain hours and low during other hours. This indicates that some of the generators are operated below rated capacity, or even shut down, some of the time, which means that much money is invested in power equipment that is not being fully utilized.

Thermal power plants are operated most economically at a constant power level. They are not designed for quick and large changes in generation level. Start-ups and shutdowns of thermal plants take much time and effort and lead unnecessarily to rapid depreciation of the equipment. Therefore, "peak shaving" is and will be done by pumped-storage hydroelectric plants, which can be started up and brought to full power within a few minutes.

During the off-peak periods when the demand for electricity is at a minimum, the pumped-storage hydro plant is used as a pump to transfer water from the lower reservoir to the upper reservoir. At the same time, the electric machine operates as a synchronous condenser, resulting in an improved overall system power factor. During the periods of peak demand, the pumped-storage hydro

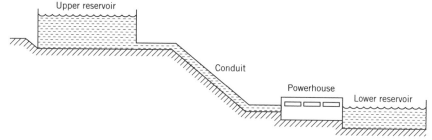

FIGURE 1.2   Schematic diagram of a pumped-storage power plant.

plant uses the water stored in the upper reservoir to drive the electric generator to produce electric power. A schematic diagram of a pumped-storage power plant is shown in Fig. 1.2.

Pumped-storage plants employ a reversible hydraulic turbine that performs the functions of both turbine and pump. This mechanism is illustrated in Fig. 1.3.

The prospects for the application of pumped-storage plants depend largely on their efficiency, that is, the ratio of the energy produced by the generator to the energy expended by the pump. Most modern pumped-storage plants have efficiencies of 70%–75%. Another advantage of pumped-storage plants is their low cost of construction, since there is no need to build a high dam across a river, long tunnels, and so forth.

1.3.3   Condensing Power Plants

A condensing power plant converts the energy of a chemical fuel into mechanical energy, which is then converted into electric energy by a generator. Heat engines are used to convert the energy of the steam or gas into mechanical energy that is available at the rotating shaft.

Heat engines may be classified according to the working substance used, which may be either steam or gas. They may also be classified according to the method used to convert heat energy to mechanical energy as either piston or rotary. The different types of condensing power plants are given in Table 1.1.

The internal combustion engine is not usually used in power plants, except for low-power applications. At modern thermal power plants, steam turbines are generally employed. To increase the efficiency of heat engines, the temperature and pressure of the working substance are raised to very high levels. Modern steam power units operate at a steam temperature of 600°C at a pressure of

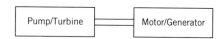

FIGURE 1.3   Dual-function pumped-storage power plant.

Table 1.1   Types of Condensing Power Plants

Principle of Operation	Work Substance: Steam	Work Substance: Gas
Piston (engine)	Steam engine	Internal combustion
Rotary	Steam turbine	Gas turbine

30 MPa, where 1 MPa $= 10^6$ N/m^2. After the working substance (steam) exits from the turbine, it is cooled by water to reduce its temperature down to 30°–40°C, which is followed by a sharp decrease in steam pressure. A schematic diagram of a thermal power plant is shown in Fig. 1.4.

Steam is produced in the steam generator, or boiler. The water supplied to the boiler is of high purity. Water flows through small-diameter steel pipes designed to withstand high steam pressures. The chemical fuel in the form of pulverized coal, gas, or atomized oil is burned at 1500°–2000°C in the furnace. Large amounts of heated air are introduced by fans in order to improve fuel combustion. The heat produced raises the temperature of water, converts the water into steam, and increases the temperature and pressure to operating levels. Hot gases are removed from the steam generator by induced-draft fans and brought out to the atmosphere through the stack.

Two types of steam generators are employed: drum-type and once-through boilers. The drum-type boiler contains a steel drum that has steam and water in its upper and lower parts, respectively. When the steam leaves the drum, it is heated again in the steam superheater to raise the temperature further.

The once-through boiler has no drum. Water flows through the boiler tubes, is converted into steam, is heated further in a steam superheater, and is then introduced into the turbine. Once-through boilers require fast response and precise control of the feedwater, which must be of high quality. Once-through boilers are cheaper than drum-type boilers and are capable of operating at higher pressures than the drum type.

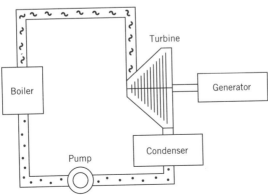

FIGURE 1.4   Schematic diagram of a thermal power plant.

The steam produced by the boiler, at a temperature of about 600°C and a pressure of 30 MPa, is passed to nozzles that transform the internal energy of the steam into kinetic energy. After the steam leaves the nozzles, it enters the turbine rotor blades.

The exhaust steam from the turbine is brought to the condenser, which is used to cool and condense the steam. Cooling water enters the condenser at 10°–15°C and leaves at 20°–25°C. The steam flows over the tubes from the top downward, condenses, exits at the bottom, and is returned as water to the boiler.

1.3.4   Geothermal Power Plants

Geothermal power plants use steam extracted from the Earth. Beneath the Earth's surface, the temperature increases with depth, reaching 1000°–1200°C at a depth of 6–30 miles. The presence of hot springs is a good indication that fairly high temperatures exist right under the surface, and these places are good candidate sites for geothermal power development.

In geothermal power plants, the hot geothermal steam is used just like the steam produced by the boiler in a condensing power plant. Unlike the steam produced in the boilers of thermal power plants, however, the steam derived from the Earth contains various impurities. This dirty steam can cause corrosion and scaling in the different parts of the power plant, and the exhaust steam and gases contribute to pollution of the surrounding atmosphere. Thus, there is a need to clean the steam before introducing it into the turbine. For this purpose, specially designed corrosion-resistant equipment and steam cleaners, called scrubbers, are employed and a rigorous maintenance program is implemented.

A schematic diagram of a geothermal power plant is shown in Fig. 1.5. The hot geothermal steam is passed through a low-pressure turbine because the steam pressures available are much lower than those at modern coal or nuclear

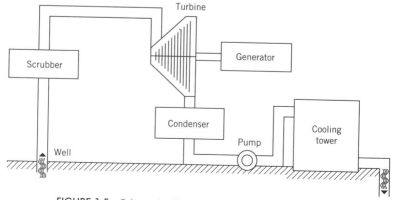

FIGURE 1.5   Schematic diagram of a geothermal power plant.

thermal power plants. As the steam comes out of the turbine, it is brought to a condenser. The condensed steam is then taken to a cooling tower, where it is sprayed downward against the flow of air blown at the water spray by a fan. The greater part of the water is evaporated into the atmosphere by the cooling tower, and the remaining water that reaches the bottom of the tower is returned to the ground through dry wells.

1.3.5   Gas-Turbine Power Plants

Gas-turbine power plants are normally designed to operate during peak conditions and thus are called peaking power plants. The mixture resulting from the combustion of the fuel and hot air is used to rotate the turbine. The gas turbine converts the heat energy of the gases into mechanical energy for driving the electric generator.

Gas turbines are designed and are operated in much the same way as steam turbines. They have approximately the same efficiency as internal combustion engines. Gas turbines require less space than steam turbines or internal combustion engines.

Modern gas-turbine power plants usually employ liquid fuel, but they may also use gaseous fuel, natural gas, or gas produced artificially by gasification of a solid fuel. In the gas turbine, liquid or gaseous fuel and air are brought into the combustion chamber, where hot combustion gases are produced at high pressure. These gases are discharged against the rotor blades of the turbine. The turbine rotates the electric generator.

1.3.6   Combined Steam- and Gas-Turbine Power Plants

A gas-turbine generating unit may be present together with a steam-turbine unit in a single power plant. In this combination power plant, the gas-turbine unit and the steam unit share the heat produced by burning the fuel. Part of the heat energy is used to produce steam and raise it to the proper temperature and pressure to drive the steam turbine. The hot gases produced during burning of the fuel are used to rotate the gas turbine. Upon leaving the turbine, the exhaust gases are used to heat the feedwater of the steam generator.

The overall efficiency of a combined power plant is significantly higher than that of a single gas-turbine power plant or a conventional steam power plant.

1.3.7   Nuclear Power Plants

In a nuclear power plant, the heat energy released by nuclear fission is used to produce the steam that rotates the turbine that drives the electric generator. The major difference between a nuclear power plant and a conventional thermal power plant is the method of producing heat energy. In the former, heat is produced in a reactor by fission of the nuclear fuel; in the latter, heat is produced in a boiler by the combustion of chemical fuel. In both cases, the heat produced is used to convert water into steam, which is fed into the steam turbine.

The nuclear reactor is the principal component of a nuclear power plant, and it is shown in Fig. 1.6. In the nuclear reactor, nuclei of uranium are split when bombarded by neutrons. The resulting nuclear fragments, neutrons, and other products of the process scatter with extremely high velocities, and they trigger a chain reaction in which more nuclei are split, continuing the process. This is accompanied by the release of enormous amounts of energy in the form of heat.

The nuclear fuel in the form of rods is placed into fuel tubes in the reactor core. Nuclear reaction takes place in the rods, and a large amount of heat is released in the process. The reaction is controlled by raising and lowering the control rods in the reactor core. A coolant flows through the reactor in order to remove the heat produced. The coolant used may be light water, heavy water, steam, or other fluid. The coolant flows over the surface of the fuel rods, where it is heated up. As the coolant exits, it brings out the heat. A large part of the energy released during nuclear reaction is expended in heating the nuclear fuel. Since heat is transferred to the coolant by convection, the velocity of the coolant must be high (3–7 m/s) to increase the output heat.

1.4 ALTERNATIVE METHODS FOR ELECTRIC ENERGY CONVERSION

With the conventional methods used to convert various forms of energy into electrical energy, chemical fuels are burned in furnaces, thereby dissipating their limited reserves. The maximum efficiency of thermal power plants is about 40%. This low conversion efficiency means that a large part of the heat

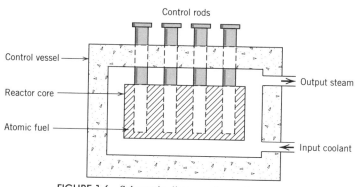

FIGURE 1.6 Schematic diagram of a nuclear reactor.

generated is lost to the environment, contributing to the thermal pollution of nearby bodies of water. In addition, the inefficient combustion of fuel results in large amounts of waste by-products being released to the atmosphere. These inherent disadvantages of thermal power plants, however, do not imply that they should be discontinued. In the future, they will continue to be one of the primary types of electric power plants.

The development of rivers for hydroelectric power production will continue in the future because it is an advantageous, although cost-intensive, method of converting a renewable form of energy.

The conventional methods of producing electric power involve the construction of huge dams and the inefficient use of chemical fuels. In the future, the continually rising demand for cheap energy and the increasing requirements of other industries for the raw material resources will necessitate the replacement of existing traditional methods of energy conversion by alternative techniques. These alternatives will be energy conversion methods that can directly convert heat, nuclear, and chemical energy into electricity.

The methods for converting various forms of energy directly into electricity are based on known physical phenomena. However, these methods of direct energy conversion are still not able to compete with the conventional energy conversion techniques used in the power industry on a large scale.

Direct conversion techniques have been used for electricity generation as low-capacity, self-contained power sources. They are used in inaccessible and remote areas not serviced by the electric utility. Galvanic cells and storage batteries produce electric energy from the chemical reaction between different materials. The principles of operation of thermal converters, photoelectric batteries, and thermionic converters are based on various physical and natural phenomena. At present, these direct-conversion plants operate at a lower efficiency than existing commercial electric power plants.

1.4.1   Magnetohydrodynamic Plants

Magnetohydrodynamic (MHD) generators are devices that convert heat directly into electricity. This direct conversion makes use of fuel resources more efficiently. The theoretical efficiency of heat engines depends on the maximum and minimum temperatures of the working substance. Whereas the operating temperature of a steam turbine is about 750°C, which results in an efficiency of energy conversion of about 40%, the operating temperatures of an MHD generator can be as high as 2700°–3000°C, which could result in much higher efficiency.

The theory of operation of MHD generators is based on Faraday's law of electromagnetic induction, which states that an electromotive force (emf) is induced in a conductor moving in a magnetic field. The emf is induced in any conductor regardless of its physical state, that is, whether it is solid, liquid, or gaseous. The interaction of an electrically conducting liquid or gas with a magnetic field is called magnetohydrodynamics.

A schematic diagram of an MHD generator is shown in Fig. 1.7. The ionized gas is passed between metal plates located in a strong magnetic field. An emf is induced, which causes an electric current to flow between electrodes inside the generator duct and in an external circuit. The flow of ionized gas, or what is more commonly called plasma, is opposed by electromagnetic forces resulting from the interaction of the current and the magnetic flux. The work done against the retarding forces provides the mechanism by which energy is converted from one form to another.

Traditionally, the three known states of matter have been solid, liquid, and gas. Gas has been considered electrically neutral. When a gas is heated, however, the outer electrons are knocked out as a result of multiple head-on collisions between the atoms. With its nuclei stripped of all the electrons, the gas is in a fourth state of matter referred to as high-temperature plasma. This state of matter is not normally found on Earth because very high temperatures and pressures are required to bring a substance to this state. At 3000°C, some gases turn to low-temperature plasmas composed of free atoms, ions, and electrons. A low-temperature plasma is highly electrically conducting.

1.4.2   Thermal Converters

The thermal converter is the device most often used for direct heat-to-electricity conversion. However, thermal converters are characterized by a relatively low power output.

The main advantages of thermal converters are that they have no moving parts, do not require high pressures, can use any heat source, and can operate for a long time. They are used extensively as power sources on spacecraft, missiles, beacons, and so forth.

Present thermal converters have outputs varying from several watts to a few kilowatts. The efficiency of these converters is up to about 10%. Although higher efficiencies will probably be achieved in the future, present thermal

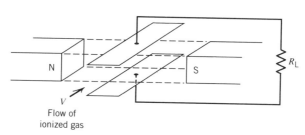

FIGURE 1.7   Schematic diagram of an MHD generator.

converters are still not competitive in cost and efficiency with conventional high-capacity electric power plants.

1.4.3   Thermionic Converters

A schematic representation of a simple thermionic converter is shown in Fig. 1.8. Electrons are emitted from the cathode when it is heated. Part of the energy supplied to the cathode is transferred by electrons to the anode and is delivered to the external circuit in the form of electric current.

In thermionic converters used for power generation, the cathode can be heated by nuclear reaction. The efficiency of the first converters of this type is about 15%, and current estimates are that this figure can be raised to 40%.

1.4.4   Electrochemical Cells

In an electrochemical cell, or fuel cell, chemical energy is converted directly to electricity. The fuel and oxidizer are stored externally and are supplied as needed. The electrodes and the electrolyte are not consumed in the energy conversion process. However, commercially attractive and cheap fuel cells using natural fuels and oxygen are still unavailable at present.

In a galvanic cell, an emf is produced by metal ions passing into a solution on interaction with the molecules and ions of the latter. When a zinc electrode is immersed in a solution of zinc sulfate ($ZnSO_4$), as shown in Fig. 1.9, the molecules of water tend to settle around the positive ions of metallic zinc, which then go into the solution of zinc sulfate.

The passage of positive ions into solution raises the electrode to a higher negative potential, which inhibits the ions from going into solution. At a certain point, the two oppositely directed flows of ions—from the electrode to the solution and back—become equal, that is, come to dynamic equilibrium.

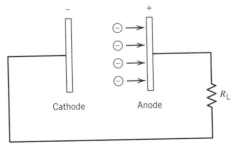

FIGURE 1.8   Schematic diagram of a thermionic converter.

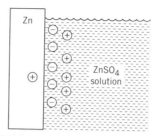

FIGURE 1.9   Zinc electrode in ZnSO₄ solution.

An important application of galvanic cells is in storage batteries. A current from an external source is passed through the electrodes of a storage battery for some time during the charging cycle so that at a later time, current can be delivered to an external load. However, the power industry has limited use of storage batteries because of the scarcity of reactive chemical fuels and their low power capabilities.

Fuel cells are quiet and efficient in operation, and they do not release any harmful air pollutants. With progress in fuel cell design and electrochemistry, the applications of fuel cells in the future may include electric power generation and power supply for automobiles.

1.4.5   Solar Power Plants

The principle of operation of a solar cell is based on the phenomenon of photoelectricity; that is, when light is incident on a body, electrons are liberated. Although photoelectricity has long been known, it has been used only lately, primarily because of advances in semiconductor technology.

When an n-type semiconductor is brought into contact with a p-type semi-conductor, a contact potential difference is established at the interface by the diffusion of electrons. When the p-type semiconductor is exposed to light, its electrons absorb photons of light and pass into the n-type semiconductor. Thus electric current is produced in a closed circuit.

At present, the most advanced devices of this type are silicon solar cells. Silicon solar cells may be up to 15% efficient. However, the difficulty of manufacturing semiconductors and their high cost restrict the use of silicon solar cells to special applications, such as in satellites.

1.4.6   Tidal Power Plants

Tidal energy has been used for a long time in various devices, particularly mills. An advantage of tidal power plants over run-of-river hydroelectric plants with no pondage or reservoir is that the performance of the former is determined by cosmic phenomena rather than by weather conditions. The main disadvantage of tidal power plants is their irregular operation. Daily variations in the available tidal energy prevent the tidal power from being used regularly in power systems during periods of peak demand.

The turbine can be run either for power generation or for pumping. When the generator is shut down, the seawater is allowed to flow directly to or from a tidal basin. The turbine operates as a pump to transfer the seawater to the basin, thus raising the operating head of water for power generation.

Use of a reversible variable-pitch turbine allows the operation of a tidal power plant to be varied and adjusted in accordance with the load demand curve. During off-peak hours, when the load demand is low, the turbine pumps water from the sea, increasing the water level in the basin. During the peak demand hours, the turbine drives the generator, converting the stored energy into electricity.

REFERENCES

1. Angrist, Stanley W. *Direct Energy Conversion*. 4th ed. Allyn & Bacon, Boston, 1982.
2. Committee on Nuclear and Alternative Energy Systems, National Research Council. *Energy in Transition 1985–2010*. National Academy of Sciences, Washington, D.C., 1979.
3. Elgerd, Olle I. *Electric Energy Systems Theory: An Introduction*. 2nd ed. McGraw-Hill, New York, 1982.
4. Healey, Timothy J. *Energy, Electric Power, and Man*. Boyd & Fraser, San Francisco, 1974.
5. Matsch, Leander W., and J. Derald Morgan. *Electromagnetic and Electromechanical Machines*. 3rd ed. Harper & Row, New York, 1986.
6. Nasar, Syed A. *Electric Energy Conversion and Transmission*. Macmillan, New York, 1985.
7. Venikov, V. A., and E. V. Putyatin. *Introduction to Energy Technology*. Mir, Moscow, 1981.
8. Wildi, Theodore. *Electrical Machines, Drives, and Power Systems*. 2nd ed. Prentice Hall, Englewood Cliffs, N.J., 1991.

Two

Power System Components and Analysis

2.1   INTRODUCTION

One of the primary contributors to the advances and improvements in human life-style over the centuries has been the ability to use and control energy. To discover new sources of energy, to obtain an essentially inexhaustible supply of energy for the future, to make energy available wherever needed, and to convert energy from one form to another and use it without creating pollution are among the greatest challenges facing our world. The electric power system is one of the tools for converting and transporting energy that is playing an important role in meeting this challenge. Highly trained engineers are needed to develop and implement the advances in science and technology to solve the problems of the electric power industry and to ensure a high degree of system reliability along with the utmost regard for the protection of our ecology.

An electric power system consists of three principal divisions: the generating system, the transmission system, and the distribution system. Transmission lines are the connecting links between the generating stations and the distribution systems and lead to other power systems over interconnections. A distribution system connects all the individual loads to the transmission lines at substations that perform voltage transformation and switching functions.

Energy is converted at the generating station from one of its basic forms, such as fossil fuels, hydro, and nuclear, into electric energy. This electric energy is then sent through a transmission system to loads at various places, where it is usually converted into other useful forms of energy. Thus, electric energy is used primarily to transmit the energy from one source, such as heat from burning coal, at one location to another location to do work, such as running a compressor on a refrigerator. This book discusses how these

16

energy conversions and electric power transmissions take place and describes the various devices used in the operation of the electric power system.

2.2   POWER SYSTEM STRUCTURE

The physical plant associated with a power system can be divided into generation (G), transmission (T), and distribution (D) facilities as shown in Fig. 2.1. The generating system provides the system with electric energy.

The transmission and subtransmission systems are meshed networks; that is, there is more than one path from one point to another. This multiple-path structure increases the reliability of the transmission system. The transmission network is a high-voltage network designed to carry power over long distances from generators to load points. The subtransmission network is a low-voltage network whose purpose is to transport power over shorter distances from bulk power substations to distribution substations. The transmission system, which is usually 138 to 765 kilovolts (kV), and the subtransmission system, which is usually 34 to 138 kV, consist of

1. Insulated wires or cables for transmission of power
2. Transformers for converting from one voltage level to another
3. Protective devices, such as circuit breakers, relays, communication and control systems
4. Physical structures for containing the foregoing, such as transmission towers and substations

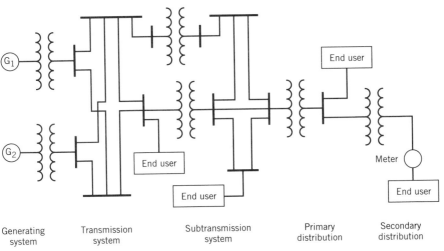

| Generating system | Transmission system | Subtransmission system | Primary distribution | Secondary distribution |

FIGURE 2.1   Sample power system structure.

The distribution of electric power includes that part of an electric power system below the subtransmission level, that is, the distribution substation, primary distribution lines or feeders, distribution transformers, secondary distribution circuits, and customers' connections and meters. The substation contains a transformer to bring the voltage down from subtransmission to distribution levels. Distribution voltages are typically 4 to 34 kV. Each substation transformer serves one or more feeders.

The primary distribution system extends from the distribution substations to the distribution transformers. A distinction is made between main feeders, which are connected to the substation, and lateral feeders, which are connected to the main feeders. These main feeders and laterals are illustrated in Fig. 2.2.

Each feeder is equipped with a circuit breaker or recloser to protect itself and the substation transformer against damage by short-circuit currents. Beyond the substation, the primary feeder consists of underground cables or overhead lines to transport the power and associated devices to control and protect the feeder. These include switches, capacitors, fuses, voltage regulators, sectionalizers, reclosers, and step-down distribution transformers. Many transformers are connected to the primary feeder for stepping voltage levels down to

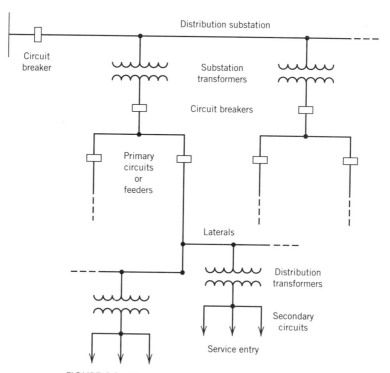

FIGURE 2.2   Electric distribution system components.

customer levels, which are 240, 208, or 120 V. The distribution transformers serve the secondary distribution system, which has small conductors connecting 1 to 10 residential customers to each distribution transformer.

Depending on the size of their power demand, customers may be connected to the transmission system, subtransmission system, primary distribution, or secondary distribution. Each customer is connected to the power system through a meter.

The primary distribution circuits in central business districts of large urban areas consist of underground cables that are used to interconnect the distribution transformers in an electric network. With this exception, the primary system is most often radial; that is, it constitutes a tree. For additional security, it is quite frequently loop-radial: the main feeder loops through the load area and comes back to the substation. The two ends of the loop are usually connected to the substation by two separate circuit breakers. A loop-radial configuration provides a backup capability. Opening selected sectionalizing switches results in a radial configuration that is used for normal operation. When a failure occurs in any section served by a radialized feeder, that section is isolated and the other part of the loop can be used to supply the unaffected customers downstream of the faulted section.

2.2.1   Electric Energy Production

Most of the electric power in the United States is generated in steam-turbine plants. Water power accounts for less than 20% of the total, and that percentage will drop because most of the available sources of water power have been developed. Gas turbines are used to a minor extent for short periods when a system is carrying peak load.

Coal is the most widely used fuel for the steam plants. Nuclear plants fueled by uranium account for a continually increasing share of the load, but their construction is slow and uncertain because of the difficulty of financing the higher costs of construction, increasing safety requirements, public opposition to the operation of nuclear plants, and delays in licensing.

The supply of uranium is limited, but the fast breeder reactor, which is now prohibited in the United States, has greatly extended the total energy available from uranium in Europe. Nuclear fusion is the great hope for the future, but a controllable fusion process on a commercial scale is not expected to become feasible until well after the year 2000.

There is some use of geothermal energy in the form of steam derived from the ground in the United States and other countries. Solar energy, still mainly in the form of direct heating of water for residential use, should eventually become practical through research on photovoltaic cells. Great progress has been made in increasing the efficiency and reducing the cost of these cells, but there is still a long way to go. Electric generators driven by windmills are operating in a number of places, and they supply small amounts of energy to power systems. Efforts to extract power from the changing tides and from waves are under way. An indirect form of solar energy is alcohol obtained from grain and mixed with gasoline to make an acceptable fuel for automobiles. Synthetic gas made from garbage and sewage is another form of indirect solar energy.

2.2.2   Transmission and Distribution

The voltage of large generators is usually in the range 13.8 to 24 kV. Large modern generators, however, are built for voltages ranging from 18 to 24 kV. No standard for generator voltages has been adopted.

Generator voltage is stepped up to transmission levels in the range 115 to 765 kV. The standard high voltages (HVs) are 115, 138, and 230 kV. Extra-high voltages (EHVs) are 345, 500, and 765 kV. Research is being conducted on lines in the ultra-high-voltage (UHV) levels of 1000 to 1500 kV. The advantage of higher levels of transmission voltage is apparent when consideration is given to the transmission capability in megavolt-amperes (MVAs) of a line. The capability of transmission lines varies with the square of the voltage. Capability is also dependent on the thermal limits of the conductor, on the allowable voltage drop, on reliability, and on the stability requirements for maintaining synchronism among the machines of the system. Most of these factors are dependent on line length.

High-voltage transmission usually employs overhead lines supported by steel, cement, or wood structures. At present, the application of underground transmission is negligible in terms of the total length of transmission lines. This is primarily because of the much higher investment, as well as repair and maintenance, costs of underground compared to overhead transmission. The use of underground cables is mostly confined to heavily populated urban areas or wide bodies of water.

The first step down of voltage from transmission levels is at the bulk-power substation, where the reduction is to a range of 34.5 to 138 kV, depending on the transmission voltage. Some industrial customers may be supplied at these voltage levels. The next step down in voltage is at the distribution substation, where the voltage on the transmission lines leaving the substation ranges from 4 to 34.5 kV and is commonly between 11 and 15 kV. This is the primary distribution system. A very popular voltage at this level is 12 kV (line-to-line). This voltage is usually described as 12-kV Y/7.2-kV Δ. A lower primary system voltage that is less widely used is 4160-V Y/2400-V Δ. Most industrial loads are fed from the primary system, which also supplies the distribution transformers providing secondary voltages over single-phase, three-wire circuits for residential use. Here the voltage is 240 V between two wires and 120 V between each of these and the third wire, which is grounded. Other secondary circuits are three-phase, four-wire systems rated 208-V Y/120-V Δ or 480-V Y/277-V Δ.

2.2.3   Electrical Load Characteristics

The ultimate objective of any power system is to deliver electrical energy to the consumer safely, reliably, economically, and with good quality. Operation of the power system requires that proper attention be given to the safety not only of the utility personnel but also of the general public.

At the consumer load centers, electrical energy is converted to other more desirable and useful forms of energy. This implies that the supply of electricity should be available where, when, and in whatever amount the consumer requires.

The quality of the supplied electrical energy is partially dependent on energy usage by the consumers. When there is high demand on the limited capabilities of the power system, voltage may deviate from its acceptable levels. Switching of large machinery could cause fluctuation of the voltage as well as the frequency. Unusually high and prolonged demands may lead to overloaded equipment, which may cause tripping of protective devices to prevent further damage to the equipment.

Finally, the power system must be operated such that the overall costs of producing electricity, including all attendant losses in the generation and delivery systems, are minimized. The most economical conditions derive not only from proper operating procedures but also from efficient system planning and design.

Electrical loads are commonly grouped into four categories: residential, commercial, industrial, and other. These are sometimes subdivided into subgroups depending on their usage levels, for example, residential A, B, or C. The residential loads are private homes and apartments. They include lighting, cooking, comfort heating and cooling, refrigerators, water heaters, washers and dryers, and many more different appliances.

Commercial loads include office buildings, department stores, grocery stores, and shops. Industrial loads consist of factories, manufacturing plants, steel, lumber, paper, mining, textile, and other industrial factories. In both commercial and industrial groups, loads include lighting, comfort heating and cooling, and various types of office equipment. In addition, industrial loads contain various types and sizes of motors, fans, presses, furnaces, and so on.

Electrical load refers to the amount of electrical energy or electrical power consumed by a particular device or by a whole community. It is also referred to as electrical demand. At the individual consumer level, the electrical demand is quite unpredictable. However, as the demands of the various users are accumulated and added at a feeder or a substation, they begin to exhibit a definite pattern.

A typical plot of the electrical load is shown in Fig. 2.3, and it is referred to as a *load curve*. The period of interest is normally 1 day; thus, it is called a daily load curve. The daily load curve shows the kilowatt demand from midnight to noon to midnight.

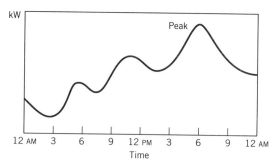

FIGURE 2.3   A typical load curve.

The daily load curves for Monday through Friday are similar in shape and maximum values. Weekend load curves are generally different, particularly for industrial and commercial customers because of shutdown of operations on Saturdays and Sundays.

Load curves may be constructed for feeders, substations, generating plants, or for the whole system. Load curves are also drawn for different periods or seasons of the year. Thus, the system peak load for summer, or winter, may be read as the highest ordinate from the corresponding load curve.

The daily load curves may be accumulated for a whole year and presented in another curve, called the annual *load duration curve* (*LDC*). A typical load duration curve is shown in Fig. 2.4. The LDC shows the 8760 hourly loads during the whole year, although not in the order in which they occurred. Rather, the LDC shows the number of hours during which the load exceeded a certain kilowatt demand. If the area under the LDC curve is calculated and divided by the total number of hours, the average demand is determined.

2.3   POWER SYSTEM ANALYSIS PROBLEMS

Power system studies can be grouped into two types: steady-state analysis for normal system conditions and transient analysis for abnormal conditions.

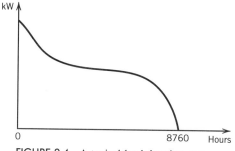

FIGURE 2.4   A typical load duration curve.

Economic dispatch and power (load) flow fall in the steady category. Transients include fault analysis for symmetrical and unsymmetrical faults, electrical transients, and stability analysis.

2.3.1 Economic Dispatch

The power industry may seem to lack competition. This idea arises because each power company operates in a geographic area not served by other companies. Competition is present, however, in attracting new industries to an area. Favorable electric rates are a compelling factor in the location of an industry, although this factor is much less important in times when costs are rising rapidly and rates charged for power are uncertain than in periods of stable economic conditions. Regulation of rates by state public utility commissions places constant pressure on companies to achieve maximum economy and earn a reasonable profit in the face of advancing costs of production.

Economic dispatch is the process of apportioning the total load on a system among the various generating plants to achieve the greatest economy of operation. The power plants are controlled by a computer as the load changes so that the total power demand is allocated for the most economical operation.

2.3.2 Power Flow

The electric power flow problem is perhaps the most studied and documented problem in power engineering. A power flow study is the determination of the voltage, current, power, and power factor or reactive power at various points in an electric network under existing or contemplated conditions of normal operation. These studies are essential in planning the future development of the system because its satisfactory operation depends on knowing the effects of interconnections with other power systems, new loads, new generating stations, and new transmission lines even before they are installed.

Digital computers provide the solutions of power flow studies of complex systems. For instance, some computer programs handle more than 1500 buses or nodes, 2500 transmission lines, 500 transformers, and so forth. Complete analysis results are printed quickly and economically.

2.3.3 Fault Calculations

A *fault* in a circuit is any failure that interferes with the normal flow of current. Most faults on transmission lines of 115 kV and higher are caused by lightning, which results in the flashover of insulators. The high voltage between a phase conductor and the grounded supporting tower causes ionization, which provides a path to ground for the charge induced by the lightning stroke. Once an ionized path to ground is established, the resultant low impedance to ground allows flow of current from the conductor to ground and through the ground to the grounded neutral of a transformer or generator, thus completing the circuit. Line-to-line faults not involving ground are less common.

Opening circuit breakers to isolate the faulted portion of the line from the rest of the system interrupts the flow of current in the ionized path and allows deionization to take place. After an interval of about 20 cycles, circuit breakers can usually be reclosed without reestablishing the arc. Experience in the operation of transmission lines has shown that ultra-high-speed reclosing breakers successfully reclose after most faults. Of the cases in which reclosure is not successful, an appreciable number are caused by permanent faults where reclosure would be impossible regardless of the interval between opening and reclosing. Permanent faults are caused by lines being on the ground, by insulator strings breaking because of ice loads, by permanent damage to towers, and by lightning arrester failures.

Experience has shown that between 70% and 80% of transmission line faults are single line-to-ground faults, which arise from the flashover of only one line to the tower and ground. The smallest number of faults, roughly 5%, involve all three phases and are called three-phase faults. Other types of transmission line faults are line-to-line faults, which do not involve ground, and double line-to-ground faults. Except for the three-phase fault, all other faults are unsymmetrical, which cause an imbalance between the voltages and currents in the three phases.

The short-circuit current that flows in different parts of a power system immediately after the occurrence of a fault differs from the current that flows a few cycles later, just before the circuit breakers are called on to open the line on both sides of the fault. Both of these currents differ from the current that would flow under steady-state conditions, if the fault were to remain and was not isolated from the rest of the system by the operation of the circuit breakers. Two of the factors on which the proper selection of circuit breakers depends are the current flowing immediately after the fault occurs and the current that the breaker must interrupt to isolate the fault.

Fault calculations involve determining these currents for various types of faults at various locations in the system. The data obtained from fault calculations also serve to determine the settings of relays that control the circuit breakers.

2.3.4   System Protection

Two possible abnormal conditions may occur while operating a power system. One is equipment or transmission line overloading, in which the current rating of the system element is exceeded. The second abnormal operating condition is the occurrence of a short circuit or fault, which could be due to insulation breakdown or phase conductors coming into contact with other phase conductors or with ground.

The amount of short-circuit current that would flow depends on the type of fault, its location, and how long the fault persists before it is cleared or removed. It is also affected by the size and configuration of the power system, the presence of fault-current limiters, and the speed and effectiveness of protective and switching devices.

The value of the fault current and the speed at which the fault or the faulted element is isolated determine the extent of damage to equipment and undesirable voltage and frequency dips. Such deviations from normal conditions cause additional system losses and could lead to loss of synchronism and complete system breakdown.

Therefore, there is a need for an automatic protection scheme that can quickly identify the faulted element and disconnect it from the system. In the case of overloads where there is no imminent danger, such protection gives an alarm signal to alert utility personnel for corrective action to remedy the problem.

Protection schemes are classified as either primary (main) protection or backup protection. Primary protection provides rapid and selective clearing of faults within the primary zone of protection. Backup protection provides the protection needed whenever the primary protection fails or is under maintenance.

Primary relaying is provided for each section or major piece of equipment, including generators, switchgears, transmission lines, transformers, and distribution feeders. The individual zones of protection overlap, so no possible section or area is left unprotected.

The backup relay is normally slower acting than the primary relay to allow the latter to perform its job. It is set to be energized at a higher level and its time setting is longer.

An important feature of any protection scheme is its selectivity. This pertains to its ability to search for the particular point or element of the system where the fault occurred and isolate the fault by tripping the nearest circuit breaker(s). This ensures minimum disruption of electric service to the customers. Selectivity is accomplished by coordinating the operating currents and time settings of the protective relays at adjacent and nearby sections or stations. With the complexity and size of modern power systems, this could prove to be a challenging task for the protection engineer.

2.3.5 Transmission Line Transients

The electric transients that occur on a power system are generally caused by a sudden change in the operating condition or configuration of the system. These transient overvoltages are caused either by lightning striking transmission lines or by switching operations. Lightning is always potentially harmful to equipment, but switching operations are also potentially damaging.

Overhead lines are usually protected from lightning by one or more wires, called *ground wires*, which are located above the phase conductors. The ground wires are connected to ground through the transmission towers. In most cases, lightning hits the ground wires instead of the phase conductors.

When a ground wire or a phase conductor is hit by lightning, a current is produced, half of which flows in one direction and the other half in the other. The crest value of current depends on the intensity of the lightning stroke, with typical values of 10,000 A and upward. When a phase conductor is hit by lightning, the damage to equipment at the line terminals is caused by the overvoltages resulting from the currents that travel along the line. The voltage and current waves propagate along the transmission line at a velocity near the speed of light. These voltages are typically above a million volts. When lightning hits a ground wire, high-voltage surges are also produced on the transmission line by electromagnetic induction.

Equipment at the line terminals is protected by surge arresters, also called lightning arresters. The arrester is connected from line to ground. When the voltage across the surge arrester becomes greater than a specified value, the arrester becomes conducting; thus, it serves to limit the voltage across its terminals to the specified value. The surge arrester becomes nonconducting again when the voltage drops below the specified value.

2.3.6   Stability Analysis

An electric power system is a dynamic nonlinear system. The dynamics are due to change in demand, generation, line switching, lightning surges, and faults. These dynamics are often classified by the speed of occurrence: the high-speed phenomena (less than 5 cycles = 5/60 s) include switching and lightning surges; the intermediate-speed occurrences (less than 100 cycles) are primarily electromechanical transients of the synchronous machine rotors. Slower phenomena, such as changes in load and generation, are virtually steady-state phenomena. The models needed to study these dynamics vary in detail, depending on the speed of occurrence.

The problem of stability is concerned with maintaining the synchronous operation of the generators and motors of the system. Stability studies are classified according to whether they involve steady-state or transient conditions. There is a definite limit to the amount of power that an AC generator can deliver and to the load that a synchronous motor can carry. Instability results from attempting to increase the mechanical input to a generator, or the mechanical load on a motor, beyond this definite amount of power, called the stability limit. A limiting value of power is reached even if the change is made gradually. Loss of synchronism may be due to suddenly applied loads, occurrence of faults, loss of field excitation of a generator, or equipment switching. Depending on whether instability is reached by a sudden and large change or a gradual change in system conditions, the limiting value of power is called the transient stability limit or steady-state stability limit, respectively.

2.4   THE COMPUTER CONNECTION

Many people believe that the only connection between power systems and computers is that computers must be plugged into an electrical outlet to operate. True, electricity does indeed power computers, either directly for desktop units or indirectly for large computers through uninterruptible power supplies that are DC batteries charged by the power system. There is far more, however, to the connection between computers and power systems.

The power industry is the third largest user of computers in the United States, next to the federal government and financial institutions. Computers are used heavily by electric utilities for customer billing, employee payrolls, and record keeping. They are also used to plan, design, control, and operate the power system.

Electric power systems are enormously large and complex. In the United States and Canada, three synchronously operated networks function independently of one another. The largest one ranges from Florida to Montana and Texas to the Hudson Bay. Power is generally produced in large central power-generating stations, sent over high-voltage transmission lines, and distributed over low-voltage lines to each customer.

There are, for example, over 600 gigawatts (600×10^9 W) of installed generating capacity in the United States. The typical cost of new generating capacity is $4000 per kilowatt. Power plants are connected to load centers through 275,000 miles of overhead and underground transmission lines operating at 115 through 765 kV. Underground transmission, typically costing $1.2 million per mile, is 5–15 times as costly as overhead transmission. Several factors make the power system a complex engineering problem:

- Power must be available the instant it is required because no economical method of storing electricity exists. This means that forecasting of the system loads must be sophisticated and system control fast.

- Customers expect 100% reliability. The high cost of supplying power prevents building a significant amount of redundancy into the power system, contrary to how other large complex engineering systems are built.

- Fossil-fueled power plants take some 10 years to build, and nuclear power plants take even longer. Hence, electric utility planners must make decisions on system expansions up to 20 years into the future.

The size of the industry and its complexities are important reasons why computers are used by the power industry. Another important reason is the enormous amount of money at stake. Revenues of the largest U.S. utilities in the mid-eighties were well over $100 billion, more than $40 billion was spent on fuel alone, and another $35 billion was spent on new equipment.

These staggering numbers make it obvious that reducing costs even a fraction of a percent would result in substantial savings. In an effort to control costs, the electric utilities use computers at all levels of the system. The system is too complex and stakes are too high to plan and operate a utility system without computers.

2.4.1   Typical Applications

Process control–type computers are used in power plants to assist plant operators with monitoring and control functions. These computers collect and process data from all over the plant. They alert the operator when various plant parameters such as temperatures, pressures, and flows are outside design operating values. They do performance calculations that tell operators how efficiently the plant and its individual components are operating.

In system control centers, operators (system dispatchers) use computers to assist them in selecting the least expensive mix of generators from the many possibilities. They are constantly monitoring and controlling the complex transmission network. They must be prepared for any emergency, such as storms or equipment failure. System changes for maintenance must be scheduled and controlled to avoid outages. Computers also help train both new and experienced system control dispatchers and power plant operators.

In load control centers, computers monitor lower-voltage transmission and major distribution circuits during periods of both normal and abnormal operation, such as during storms and system disturbances in which customers experience outages.

System planners use computers in the planning and design of future power system networks. Their requirements challenge the complex simulation techniques offered by mathematicians and computer scientists. Often the size and complexity of power system problems are beyond state-of-the-art capabilities.

Computer applications extend to electrical system components as well. Relays, for example, are used as sensors to control circuit breakers and remove shorted components from the system. At one time relays were electromechanical devices. They were, in turn, surpassed by solid-state devices. State-of-the-art relays now use microprocessor technology.

Microprocessors are also becoming commonplace in power plants to collect and process data for transfer to the plant computer. High reliability is expected from these microprocessors, even in the hostile physical and electromagnetic atmospheres of power plants and substations.

2.4.2   System Planning and Operation

Power system planning often extends up to 20 years from the present. Computers are invaluable tools in the efforts required to develop plans that are technically and economically feasible and environmentally and socially acceptable. Planning complications arise because of the enormous size of the systems studied, as well as the countless possible options and situations.

At this time, more than 40 application programs are available for planning the power system. Some examples of how these individual tools assist the planning effort follow.

The power flow (or load flow) program is used as the basis of steady-state analysis. Typical network study sizes range from 1000 to 4000 buses or nodes. A power flow program is typically written in Fortran and contains 10,000 to 40,000 Fortran statements using sparse matrix methods to reduce computer memory requirements. Large mainframe computers can perform a power flow computation in 10 to 30 s. Recently, personal computers have performed smaller power flow studies.

Another computer simulation tool is used extensively to analyze system response during the first 3 to 10 s after a major disturbance, such as loss of generating unit or transmission line. This transient stability program is substantially larger than the power flow program because it solves sets of both algebraic nonlinear equations and the tens of thousands of differential equations that describe system transient response. The stability program numerically integrates the differential equations at time steps of about 0.05 s. The number of equations and the small time step needed illustrate the challenge in computer applications of power system analysis.

Computer optimization is playing a larger role than ever before in planning tomorrow's power systems. Examples of the application of computer optimization include scheduling the operation of generating units while maintaining security constraints on individual transmission lines, coordinating the use of hydroelectric and fossil-fueled generating units, siting and scheduling of reactive power sources, planning generating system and transmission system expansions, and scheduling of fuel use.

Like the power system planner, the system dispatcher has more than 40 computer programs to assist in job performance. But in many ways the dispatchers are ahead of the planners in their use of computers.

The dispatchers are responsible for monitoring and controlling the power system. In addition to "keeping the lights on," dispatchers must operate the power system, that is, choose patterns of generation and energy transmission so that loss of any single system element such as a generator or transmission line will not lead to a blackout. When the power system is secure, the dispatcher can pay attention to operating the power system as efficiently as possible. A dedicated set of computer hardware and software assists the dispatcher.

The computer system used to monitor and control the power system has a large database management system at its heart. It is designed for redundant operation with automatic fail-over schemes. The hundreds of buses in the power system are scanned every 2 to 4 s to determine current operating conditions. Because it is impractical to monitor every single point, the newer system control centers use state estimation techniques to check for missing or bad data. From the available data, another computer program develops a model of the current operating condition.

Automatic generation control provides on-line scheduling of the power system generating units. Control signals are sent out to the plants every 6 to 10 s. Economic dispatch is a separate computer program that establishes set points for each generating unit that is on line and under automatic control. The unit commitment program determines which generating units should be operated on a daily basis; the program includes minimum run times and minimum down times for reducing maintenance requirements. Both economic dispatch and unit commitment are optimization programs.

An additional dispatcher responsibility is to purchase power from and sell power to other electric utilities. Buy and sell decisions are based on short-term load forecasts and knowledge of which generating units are operating and their operating costs.

Control centers make substantial use of computer graphics to display information to dispatchers. But so much information is available that a large application problem is determining exactly what information to display. Thus, the control center involves not only a computation and communication challenge but also the human problem of presenting information effectively.

2.4.3   The Computing Environment

Most power system simulation and analysis today is performed using either large mainframe computers or super-minicomputers. "Smart" terminals are often used as workstations. Computations are generally performed using an interactive data setup followed by batch solution execution, with interactive output analysis completing the computer run. Graphics are widely used but not yet to their full capability.

Personal computer applications are just beginning. These generally rely on commercial software packages, such as spreadsheets and word processing. Some engineering programs are becoming available, most of which have been developed by the utilities.

Database management systems are not yet in wide use by power system planners as a means of interfacing the various application computer programs. Almost all application programs are written in Fortran, although PL/1, Basic, C, and Pascal have been used.

The computing environment in the future will be significantly different. Desktop workstations in the form of personal computers will have the capability of present computer mainframes. Artificial intelligence methods, particularly expert systems, will become widely used. Program consolidation and increased use of company-wide databases using modern, relational database management systems will become common. These tools will allow studies to be performed with less human intervention. Office automation techniques will also become an integral part of the available tools.

2.5   THE ROLE OF THE POWER ENGINEER

This introduction shows that many challenges are waiting for the future electric power engineer. One of the objectives of this book is to help students choosing the area of power engineering as their career to prepare for further study and eventual employment in the industry. The power industry is different from most industries in that it is regulated by state and federal governments. So, not only does industry touch virtually everyone through the service it provides, but everyone has the ability to affect the industry through the regulatory bodies. Another objective of this book is to educate electrical engineers who do not choose power engineering as a profession in some of the basic principles of power system operation. In this way they can exercise their input into the industry from an informed point of view.

With revenues of over $100 billion, the electric utility industry is one of the nation's largest. Computer applications in the industry today are indeed extensive, but the coming years promise even greater challenges in such areas as computer modeling, process control, digital communications, mathematics, and software engineering. Thus, for electric utilities to make the best use of computers, they will require young engineers with diverse backgrounds. Your role in the industry can eventually become highly significant.

REFERENCES

1. Blackburn, J. L., ed. *Applied Protective Relaying.* Westinghouse Electric Corporation, Newark, N.J., 1976.

2. Del Toro, Vincent. *Electric Power Systems.* Prentice Hall, Englewood Cliffs, N.J., 1992.

3. Glover, J. Duncan, and Mulukutla Sarma. *Power System Analysis and Design.* PWS, Boston, 1987.

4. Gönen, Turan. *Electric Power Transmission System Engineering Analysis and Design.* Wiley, New York, 1988a.

5. ——. *Modern Power System Analysis.* Wiley, New York, 1988b.

6. Gross, Charles A. *Power System Analysis*. 2nd ed. Wiley, New York, 1986.

7. Heydt, G. T. *Computer Analysis Methods for Power Systems*. Macmillan, New York, 1986.

8. Kimbark, Edward W. *Power System Stability: Synchronous Machines*. Dover, New York, 1956.

9. Kirchmayer, Leon K. *Economic Operation of Power Systems*. Wiley, New York, 1958.

10. Rustebakke, Homer M., ed. *Electric Utility Systems and Practices*. 4th ed. Wiley, New York, 1983.

11. Stagg, Glenn A., and A. H. El-Abiad. *Computer Methods in Power System Analysis*. McGraw-Hill, New York, 1968.

12. Stevenson, William D., Jr. *Elements of Power System Analysis*. 4th ed. McGraw-Hill, New York, 1982.

Three

Basic AC Circuit Concepts

3.1 INTRODUCTION

The analysis of electric power systems involves the study of the performance of the system under both normal and abnormal conditions. The power system engineer regularly performs such analyses, which consider both single-phase and three-phase circuits. The power engineer, therefore, must be competent in steady-state AC circuit analytical techniques.

In this chapter, the notations used and a review of basic circuit analysis are presented first, followed by per-unit representation of the electrical quantities such as voltage, current, power, and impedance.

3.2 NOTATIONS

Single-subscript notation is discussed first. Then double-subscript notation is introduced to eliminate the need for the polarity markings for voltages and directions for currents.

3.2.1 Single-Subscript Notation

Consider the AC circuit shown in Fig. 3.1. The circle represents a voltage source with emf $\mathbf{E_g}$ and terminal voltage $\mathbf{V_t}$. The voltage across the load is designated as $\mathbf{V_L}$. The internal impedance of the source is Z_g, the impedance of the feeder is Z_{fdr}, and the impedance of the load is Z_L.

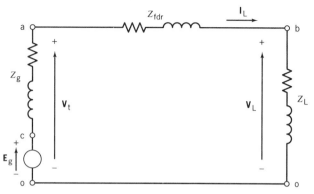

FIGURE 3.1   A simple AC circuit.

On the diagram, $+$ and $-$ polarity marks are assigned to each of the various voltages. The polarity marks specify that the voltage is positive when the terminal marked $+$ is at a higher potential than the terminal marked $-$. In an AC circuit, this is the case during half of a cycle. During the next half-cycle, the voltage is negative because the terminal with the $+$ marking is actually at a lower potential.

Alternatively, the voltages may be specified using arrows. In this case, the tip of the arrow points to the terminal corresponding to the $+$ marking, whereas the tail corresponds to the $-$ marking.

Similarly, an arrow is used to designate the positive direction of the flow of current. The current is considered negative if the actual direction of current flow is opposite to the arrow direction. For the given circuit, the phasor current can be calculated as follows:

$$\mathbf{I}_L = \frac{\mathbf{V}_t - \mathbf{V}_L}{Z_{fdr}} \tag{3.1}$$

The terminal voltage of the source is

$$\mathbf{V}_t = \mathbf{E}_g - \mathbf{I}_L Z_g \tag{3.2}$$

On the circuit diagram, the different nodes have been identified by letters. The voltages at these nodes may be referred to the reference node o, such that $\mathbf{V}_a$ is positive when node a is at a higher potential than reference node o. Thus,

$$\mathbf{V}_a = \mathbf{V}_t, \qquad \mathbf{V}_b = \mathbf{V}_L, \qquad \mathbf{V}_c = \mathbf{E}_g \tag{3.3}$$

Note that voltage and current phasors are set in boldface. Impedance is a complex quantity, but is not a phasor, and so is not set in boldface.

3.2.2 Double-Subscript Notation

The double-subscript notation eliminates the need for both polarity markings for voltages and direction arrows for currents. It is even more useful for representing voltages and currents in three-phase circuits, resulting in greater clarity and less confusion.

The voltage phasor with double subscripts represents the voltage across the two nodes identified by the two subscripts. The voltage is positive during the half-cycle in which the node named by the first subscript is at a higher potential than the node named by the second subscript.

The current phasor with the double subscript represents the current flowing between two nodes of a circuit. The current is considered positive when it flows in the direction from the node identified by the first subscript toward the node identified by the second subscript.

The voltage across the line impedance in Fig. 3.1 can be expressed in double-subscript notation as

$$\mathbf{V}_{ab} = \mathbf{I}_{ab} Z_{fdr} \tag{3.4}$$

Using double-subscript notation, Eqs. 3.1 and 3.2 can be rewritten as Eqs. 3.5 and 3.6.

$$\mathbf{I}_{ab} = \frac{\mathbf{V}_{ao} - \mathbf{V}_{bo}}{Z_{fdr}} \tag{3.5}$$

$$\mathbf{V}_{ao} = \mathbf{E}_{co} - \mathbf{I}_{ab} Z_g \tag{3.6}$$

Because node o is the reference node, it may be dropped from the subscripts without loss of clarity. Thus,

$$\mathbf{I}_{ab} = \frac{\mathbf{V}_a - \mathbf{V}_b}{Z_{fdr}} \tag{3.7}$$

$$\mathbf{V}_a = \mathbf{E}_c - \mathbf{I}_{ab} Z_g \tag{3.8}$$

3.3 SINGLE-PHASE AC CIRCUITS

In this section, basic concepts for the analysis of single-phase circuits are presented. The effective value of sinusoidal voltages and currents and definitions of power and complex power are discussed.

3.3.1 Effective or Root-Mean-Square Value

Consider the sinusoidal voltage $v(t) = V_m \cos(\omega t + \theta)$ whose peak value, or amplitude, is V_m. For this sinusoid, the period is $T = 2\pi/\omega$.

The *effective* or *root-mean-square (rms) value* is the value of the sinusoidal voltage that, when connected across a resistor, delivers the same amount of electric energy to the resistor in T seconds that a constant (DC) voltage would. The rms value is found by using the formula

$$V = \sqrt{\frac{1}{T} \int_0^T v^2(t)\,dt} \tag{3.9}$$

Substituting the expressions for the voltage and its period yields

$$V = \sqrt{\frac{V_m^2}{2\pi/\omega} \int_0^{2\pi/\omega} \cos^2(\omega t + \theta)\,dt} = \frac{V_m}{\sqrt{2}} \tag{3.10}$$

EXAMPLE 3.1

The voltage across a certain impedance load is given by $v(t) = 170 \cos \omega t$, and the current flowing through the load is $i(t) = 14.14 \cos(\omega t - 30°)$. Find the rms values of the voltage and current. Find the expressions for the voltage and current phasors.

Solution

a. The rms values of the voltage and current are

$$V = 170/\sqrt{2} = 120 \text{ V}$$

$$I = 14.14/\sqrt{2} = 10 \text{ A}$$

b. The voltage and current phasors are

$$\mathbf{V} = 120 \underline{/0°} \text{ V}$$

$$\mathbf{I} = 10 \underline{/-30°} \text{ A}$$

3.3.2   Power in a Single-Phase Circuit

Consider the impedance load consisting of a resistance R and an inductive reactance X_L connected in series as shown in Fig. 3.2.

Let the current flowing through the resistance and inductive reactance be expressed as

$$i(t) = \sqrt{2}I \cos \omega t \qquad (3.11)$$

The applied voltage $v(t)$ is equal to the sum of the voltage drops across the resistance and reactance; that is,

$$v(t) = \sqrt{2}V_R \cos \omega t + \sqrt{2}V_X \cos(\omega t + 90°) \qquad (3.12)$$

where $V_R = IR$ and $V_X = IX_L$. Combining the two cosine functions yields

$$v(t) = \sqrt{2}V \cos(\omega t + \theta) \qquad (3.13)$$

where

$$V = \sqrt{V_R^2 + V_X^2} = I \sqrt{R^2 + X_L^2} = IZ \qquad (3.14)$$

$$\theta = \tan^{-1}\left(\frac{V_X}{V_R}\right) = \tan^{-1}\left(\frac{X_L}{R}\right) = \arg(Z) \qquad (3.15)$$

The instantaneous power $p(t)$ delivered to the load may be expressed as follows:

$$
\begin{aligned}
p(t) = v(t)i(t) &= [\sqrt{2}V \cos(\omega t + \theta)][\sqrt{2}I \cos \omega t] \\
&= 2VI \cos(\omega t + \theta)\cos \omega t \qquad (3.16)
\end{aligned}
$$

The power $p(t) = v(t)i(t)$ absorbed by the load is positive when both $v(t)$ and $i(t)$ have the same sign (either both positive or both negative). This power becomes negative when $v(t)$ and $i(t)$ are of opposite sign.

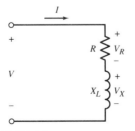

FIGURE 3.2   Series $R - X$ load.

In the case of a pure resistance load, the voltage and current are in phase and the power is always positive. For a load consisting of a pure inductance or pure capacitance, the power has alternately positive and negative, but equal, portions and its average value is zero.

Using the trigonometric identity $\cos A \cos B = \frac{1}{2}[\cos(A - B) + \cos(A + B)]$ (see Appendix F), Eq. 3.16 reduces to

$$p(t) = VI \cos \theta + VI \cos(2\omega t + \theta) \tag{3.17}$$

The first term on the right-hand side of Eq. 3.17 has a constant value. The second term is a sinusoid of twice the frequency, and its average is zero. Thus, the average of the whole expression, that is, the *average power P* absorbed by the load, is

$$P = VI \cos \theta \tag{3.18}$$

The quantity P is also called *real power*, or active power, and is measured in watts (W), kW, or MW.

The product of the rms values of voltage and current, VI, is referred to as the *apparent power S*. It is measured in volt-amperes (VA), kVA, or MVA.

The cosine of the phase angle θ between the voltage and the current is called the *power factor*. It may be calculated from Eq. 3.18 as follows:

$$\text{Power factor} = \cos \theta = \frac{P}{VI} \tag{3.19}$$

An inductive circuit is said to have a *lagging power factor* because the current lags the voltage. On the other hand, a capacitive circuit is said to have a *leading power factor* because the current leads the voltage.

The instantaneous power may be expressed as the sum of two sinusoids of twice the frequency. Thus, Eq. 3.17 is written as

$$p(t) = VI \cos \theta(1 + \cos 2\omega t) + VI \sin \theta \cos(2\omega t + 90°) \tag{3.20}$$

The first term on the right-hand side is a sinusoid displaced upward by an amount equal to the average power P.

The second term on the right-hand side of Eq. 3.20 is called the instantaneous reactive power. It has an average value of zero. Its peak value is designated as Q and is simply called *reactive power*. It is equal to

$$Q = VI \sin \theta \tag{3.21}$$

Reactive power is measured in volt-ampere reactive (VAR), kVAR, or MVAR.

From Eqs. 3.18 and 3.21, it is seen that the apparent power $S = VI$ may be computed from P and Q. Thus,

$$\sqrt{P^2 + Q^2} = \sqrt{V^2I^2 \cos^2 \theta + V^2I^2 \sin^2 \theta} = VI = S \qquad (3.22)$$

The instantaneous power to the load of Fig. 3.2 may also be found as the product of Eqs. 3.11 and 3.12. Thus,

$$p(t) = 2I^2R \cos^2 \omega t + 2I^2X_L \cos(\omega t + 90°) \cos \omega t \qquad (3.23)$$

Using the trigonometric identity cited previously, Eq. 3.23 reduces to

$$p(t) = I^2R(1 + \cos 2\omega t) + I^2X_L \cos(2\omega t + 90°) \qquad (3.24)$$

Comparing Eq. 3.24 with Eq. 3.20 yields the following:

$$P = VI \cos \theta = I^2R \qquad (3.25)$$
$$Q = VI \sin \theta = I^2X_L \qquad (3.26)$$
$$S = VI = \sqrt{P^2 + Q^2} = I^2Z \qquad (3.27)$$

Then the following are derived:

$$\text{Power factor} = \cos \theta$$
$$= \cos \left[\tan^{-1} \left(\frac{Q}{P} \right) \right]$$
$$= \cos \left[\tan^{-1} \left(\frac{X_L}{R} \right) \right] \qquad (3.28)$$

3.3.3   Complex Power

Consider an arbitrary load. The voltage impressed across the load is $\mathbf{V} = V \angle \theta_V$, and the current flowing through it is $\mathbf{I} = I \angle \theta_I$. The *complex power* $\mathbf{S}$ is equal to the product of the voltage times the complex conjugate of the current; that is,

$$\mathbf{S} = \mathbf{V}\mathbf{I}^* = VI \angle \theta_V - \theta_I = VI \angle \theta \qquad (3.29)$$

where $\theta = \theta_V - \theta_I$. Thus, it is seen that the magnitude of the complex power is the apparent power VI and the angle of the complex power is the phase angle difference between the voltage and the current, or the *power factor angle* θ.

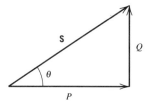

FIGURE 3.3   Power triangle.

The complex power may be expressed in rectangular form by using Euler's formula as follows:

$$\mathbf{S} = VI(\cos\theta + j\sin\theta)$$
$$= VI\cos\theta + jVI\sin\theta$$
$$= P + jQ \qquad (3.30)$$

The reactive power Q is positive when $\theta_V > \theta_I$ or $\theta > 0$, which means that the current is lagging the voltage as in inductive loads. Conversely, Q is negative when $\theta_V < \theta_I$; that is, the current leads the voltage, as for the case of capacitive loads.

A power triangle for an inductive or lagging power factor load is shown in Fig. 3.3.

EXAMPLE 3.2

A generator supplies a load through a feeder whose impedance is $Z_{\text{fdr}} = 1 + j2\ \Omega$. The load impedance is $Z_L = 8 + j6\ \Omega$. The voltage across the load is 120 V. Find the real power and reactive power supplied by the generator. Take the load voltage $\mathbf{V}_L$ as the reference phasor.

Solution   A schematic representation of the system and a diagram showing the various phasors are given in Fig. 3.4.

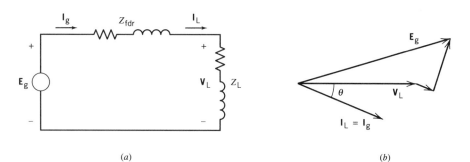

(a)   (b)

FIGURE 3.4   (a) Power system; (b) phasor diagram of Example 3.2.

The impedances may be expressed in polar form as follows:

$$Z_{\text{fdr}} = 1 + j2 = 2.24 \underline{/63.4°} \ \Omega$$

$$Z_{\text{L}} = 8 + j6 = 10.0 \underline{/36.9°} \ \Omega$$

The voltage across the load is taken as reference phasor; thus,

$$\mathbf{V}_{\text{L}} = 120 \underline{/0°} \ \text{V}$$

The load current is computed as follows:

$$\mathbf{I}_{\text{L}} = \mathbf{V}_{\text{L}}/\mathbf{Z}_{\text{L}} = \frac{120 \underline{/0°}}{10 \underline{/36.9°}} = 12 \underline{/-36.9°} \ \text{A} = \mathbf{I}_{\text{g}}$$

That is, the generator current is the same as the load current.

The generator voltage is found by writing a Kirchhoff's voltage equation around the loop.

$$\mathbf{E}_{\text{g}} = \mathbf{V}_{\text{L}} + \mathbf{I}_{\text{L}} Z_{\text{fdr}}$$
$$= 120 \underline{/0°} + (12 \underline{/-36.9°})(2.24 \underline{/63.4°}) = 144.5 \underline{/4.8°} \ \text{V}$$

The complex power is given by

$$\mathbf{S}_{\text{g}} = \mathbf{E}_{\text{g}} \mathbf{I}_{\text{g}}^{*} = \mathbf{E}_{\text{g}} \mathbf{I}_{\text{L}}^{*} = (144.5 \underline{/4.8°})(12 \underline{/-36.9°})^{*}$$
$$= 1734 \underline{/41.7°} = 1295 + j1154 \ \text{VA}$$

Therefore, the real power and reactive power are

$$P = 1295 \ \text{W} \quad \text{and} \quad Q = 1154 \ \text{VAR}$$

3.3.4   Direction of Power Flow

The convention used for positive power flow is described with the help of Fig. 3.5 and Eqs. 3.29 and 3.30. Figure 3.5a applies to a generator, and Fig. 3.5b is for a load.

For a generator, the electric current is assumed to flow out of the positive terminal in the direction of the voltage rise. Depending on the phase angle $\theta = \theta_V - \theta_I$, the values of P and Q may be positive or negative. If P is positive, the generator delivers positive real power. However, if P is negative,

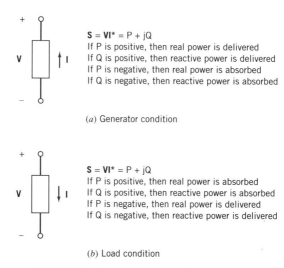

$S = VI^* = P + jQ$
If P is positive, then real power is delivered
If Q is positive, then reactive power is delivered
If P is negative, then real power is absorbed
If Q is negative, then reactive power is absorbed

(*a*) Generator condition

$S = VI^* = P + jQ$
If P is positive, then real power is absorbed
If Q is positive, then reactive power is absorbed
If P is negative, then real power is delivered
If Q is negative, then reactive power is delivered

(*b*) Load condition

FIGURE 3.5 Convention for positive power flow.

the generator delivers negative real power; in other words, it absorbs positive real power. Similarly, if Q is positive, the generator delivers positive reactive power. However, if Q is negative, the generator delivers negative reactive power, or, alternatively, it absorbs positive reactive power.

For an electrical load, the current is assumed to enter the positive terminal in the direction of the voltage drop. Depending on the phase angle $\theta = \theta_V - \theta_I$, the values of P and Q may be positive or negative. If P is positive, the load absorbs positive real power. However, if P is negative, the load absorbs negative real power, or it delivers positive real power. Similarly, if Q is positive, the load absorbs positive reactive power. However, if Q is negative, the load absorbs negative reactive power, or it delivers positive reactive power.

DRILL PROBLEMS

D3.1 The instantaneous voltage $v(t)$ across a series-connected impedance and the instantaneous current $i(t)$ entering the positive terminal of the circuit element are given by the following expressions:

$$v(t) = 110\cos(\omega t + 65°) \text{ V}$$
$$i(t) = 15\sin(\omega t - 20°) \text{ A}$$

Determine:

a. The maximum or peak value of the voltage and current
b. The rms value of the voltage and current
c. The phasor expressions for the voltage and current

D3.2 Determine the rms value of each of the following currents.

   a. $12 \sin 4t$

   b. $3 \cos 2t + 4 \sin 2t$

   c. $10 + 5 \sin 3t$

D3.3 Given that $v(t) = \sqrt{2}V \cos(\omega t + \alpha)$, $i(t) = \sqrt{2}I \cos(\omega t + \beta)$, and $\omega = 2\pi/T$, show that the average power is given by

$$P = \frac{1}{T} \int_0^T v(t)i(t)dt = VI \cos(\alpha - \beta)$$

D3.4 For the voltage and current given in Drill Problem D3.1, find

   a. The expression for instantaneous power

   b. The average or real power, and state whether this power is absorbed or supplied by the impedance

   c. The reactive power, and state whether absorbed or supplied

   d. The power factor, and state whether lagging or leading

D3.5 An electrical load has an impedance of $10 \underline{/30°} \ \Omega$.

   a. Compute the equivalent series resistance and reactance. State whether the reactance is inductive or capacitive.

   b. The load is connected to a 60-Hz, 120-V source. Draw a phasor diagram showing the current, voltage across the resistance, voltage across the reactance, and voltage of the source as reference.

D3.6 A circuit consists of two impedances, $Z_1 = 50 \underline{/45°} \ \Omega$ and $Z_2 = 25 \underline{/30°} \ \Omega$, connected in parallel. They are supplied by a source whose voltage is $\mathbf{V} = 50 \underline{/0°}$ volts. Determine the following:

   a. Current drawn by each impedance

   b. Complex power absorbed by each impedance

   c. Total current

   d. Complex power supplied by the source

3.4 BALANCED THREE-PHASE AC CIRCUITS

In the United States, as in most parts of the world, electric bulk-power generation, transmission, and distribution are usually accomplished with three-phase systems. Although electric lights and small motors are single phase, they are assigned equally to the three phases of the distribution system so that the phases effectively form a balanced set.

A three-phase generator is shown in Fig. 3.6. The three voltage sources have equal magnitude, and their phase differences are each 120°. Each phase has a series resistance R_a and inductive reactance X_s. In the accompanying phasor diagram, $\mathbf{E}_{an}$ is seen leading $\mathbf{E}_{bn}$ by 120°, and $\mathbf{E}_{bn}$ is leading $\mathbf{E}_{cn}$ by 120°. Hence, the phase sequence is said to be positive sequence, or *an-bn-cn sequence*, or simply *abc sequence*.

3.4.1   Wye-Connected Load

A wye-connected load is shown in Fig. 3.7. Each phase impedance $Z_Y \underline{/\theta}$ consists of a resistance R_Y and an inductive reactance X_Y. The *balanced set* of voltages applied to the terminals of the load is of abc phase sequence and is given by

$$\mathbf{V}_{an} = V\underline{/0°}, \qquad \mathbf{V}_{bn} = V\underline{/-120°}, \qquad \mathbf{V}_{cn} = V\underline{/-240°} \qquad (3.31)$$

The line-to-line voltages are computed as follows:

$$\mathbf{V}_{ab} = \mathbf{V}_{an} - \mathbf{V}_{bn} = \sqrt{3}\,\mathbf{V}_{an}\underline{/30°} = \sqrt{3}\,V\underline{/30°} \qquad (3.32)$$

$$\mathbf{V}_{bc} = \mathbf{V}_{bn} - \mathbf{V}_{cn} = \sqrt{3}\,\mathbf{V}_{bn}\underline{/30°} = \sqrt{3}\,V\underline{/-90°} \qquad (3.33)$$

$$\mathbf{V}_{ca} = \mathbf{V}_{cn} - \mathbf{V}_{an} = \sqrt{3}\,\mathbf{V}_{cn}\underline{/30°} = \sqrt{3}\,V\underline{/-210°} \qquad (3.34)$$

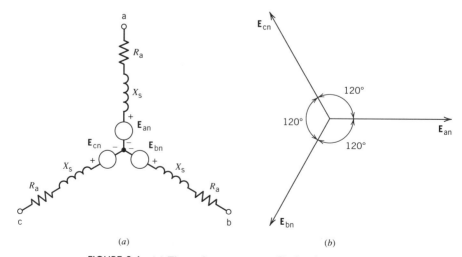

(*a*)   (*b*)

FIGURE 3.6   (*a*) Three-phase generator; (*b*) abc phase sequence.

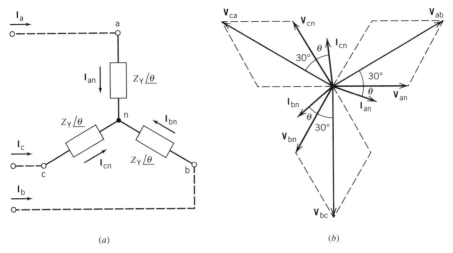

(a)   (b)

FIGURE 3.7 (a) Wye-connected load; (b) phasor diagram.

Thus, the line voltage has a magnitude equal to $\sqrt{3}$ times the magnitude of the phase voltage and it leads the corresponding phase voltage by $30°$; that is,

$$\mathbf{V}_L = \sqrt{3}\mathbf{V}_P \underline{/30°} \tag{3.35}$$

It may also be observed that the line currents are identical to the corresponding phase currents:

$$\mathbf{I}_L = \mathbf{I}_P \tag{3.36}$$

These currents are calculated as follows:

$$\mathbf{I}_a = \mathbf{I}_{an} = \frac{\mathbf{V}_{an}}{Z_Y \underline{/\theta}} = \frac{V}{Z_Y} \underline{/-\theta} = I \underline{/-\theta} \tag{3.37}$$

$$\mathbf{I}_b = \mathbf{I}_{bn} = \frac{\mathbf{V}_{bn}}{Z_Y \underline{/\theta}} = \frac{V}{Z_Y} \underline{/-\theta - 120°} = I \underline{/-\theta - 120°} \tag{3.38}$$

$$\mathbf{I}_c = \mathbf{I}_{cn} = \frac{\mathbf{V}_{cn}}{Z_Y \underline{/\theta}} = \frac{V}{Z_Y} \underline{/-\theta - 240°} = I \underline{/-\theta - 240°} \tag{3.39}$$

It may be noted that the currents form a balanced set of phasors and the sum of the currents is zero:

$$\mathbf{I}_a + \mathbf{I}_b + \mathbf{I}_c = 0 \tag{3.40}$$

3.4.2   Delta-Connected Load

A delta-connected load is shown in Fig. 3.8. Each phase impedance $Z_\Delta \underline{/\theta}$ consists of a resistance R_Δ and an inductive reactance X_Δ. It is assumed that balanced abc phase sequence voltages are applied and $\mathbf{V}_{ab} = V \underline{/0°}$.
The phase currents will be

$$\mathbf{I}_{ab} = \frac{\mathbf{V}_{ab}}{Z_\Delta \underline{/\theta}} = \frac{V}{Z_\Delta} \underline{/-\theta} = I \underline{/-\theta} \qquad (3.41)$$

$$\mathbf{I}_{bc} = \frac{\mathbf{V}_{bc}}{Z_\Delta \underline{/\theta}} = \frac{V}{Z_\Delta} \underline{/-\theta - 120°} = I \underline{/-\theta - 120°} \qquad (3.42)$$

$$\mathbf{I}_{ca} = \frac{\mathbf{V}_{ca}}{Z_\Delta \underline{/\theta}} = \frac{V}{Z_\Delta} \underline{/-\theta - 240°} = I \underline{/-\theta - 240°} \qquad (3.43)$$

The line currents are calculated as follows:

$$\mathbf{I}_a = \mathbf{I}_{ab} - \mathbf{I}_{ca} = \sqrt{3}\mathbf{I}_{ab}\underline{/-30°} = \sqrt{3}I\underline{/-\theta - 30°} \qquad (3.44)$$

$$\mathbf{I}_b = \mathbf{I}_{bc} - \mathbf{I}_{ab} = \sqrt{3}\mathbf{I}_{bc}\underline{/-30°} = \sqrt{3}I\underline{/-\theta - 150°} \qquad (3.45)$$

$$\mathbf{I}_c = \mathbf{I}_{ca} - \mathbf{I}_{bc} = \sqrt{3}\mathbf{I}_{ca}\underline{/-30°} = \sqrt{3}I\underline{/-\theta - 270°} \qquad (3.46)$$

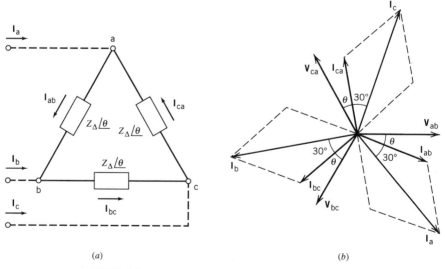

(a)   (b)

FIGURE 3.8   (a) Delta-connected load; (b) phasor diagram.

It may be observed that the line-to-line voltages are identical to the phase voltages for delta-connected loads:

$$V_L = V_P \qquad (3.47)$$

The delta-connected load may be transformed into an equivalent wye-connected load so that the terminal behavior of the two configurations will be identical; that is, corresponding line-to-line voltages and line currents will be the same. A derivation of this Δ-to-Y transformation is given in Ref. 5. When the load is balanced, the impedance per phase of the wye-connected load will be one-third of the impedance per phase of the delta-connected load:

$$Z_Y = \tfrac{1}{3}Z_\Delta \qquad (3.48)$$

3.4.3   Analysis of Balanced Three-Phase Systems

If a three-phase system is balanced, it may be analyzed by using a single-phase equivalent circuit. Since the sum of the phase currents is equal to zero, a neutral wire may be connected between the source neutral and the load neutral. This neutral wire would not affect voltages or currents in the circuit.

Figure 3.9 shows a wye-connected generator supplying a wye-connected load through a three-phase feeder. It can be shown that the voltages and currents belonging to a particular phase are identical to corresponding voltages and currents in the other phases except for 120° shifts in their respective phase angles. Therefore, a single circuit consisting of one phase and a neutral wire may be analyzed and the results applied to the other phases by including the corresponding 120° phase shift. This procedure will be illustrated in the following sample system.

When the three-phase source (or load) is delta connected, it is customary to transform it into an equivalent wye-connected source (or load) before applying the procedure. Subsequently, the results are reconverted into their delta equivalents.

EXAMPLE 3.3

A three-phase power system consists of a wye-connected ideal generator connected to a wye-connected load through a three-phase feeder. The load has an impedance $Z_L \underline{/\theta} = 20 \underline{/30°}$ Ω /phase, and the feeder has an impedance $Z_{fdr} = 1.5 \underline{/75°}$ Ω /phase. The terminal voltage of the load is 4.16 kV. Determine (a) the terminal voltage of the generator and (b) the line current supplied by the generator.

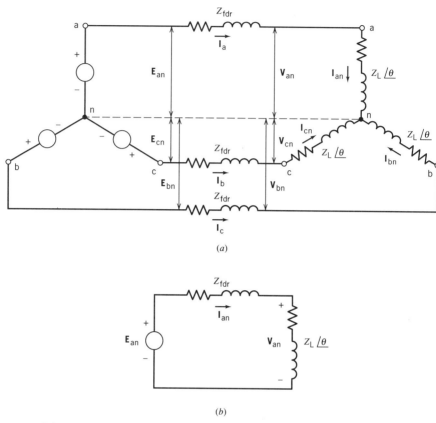

(a)

(b)

FIGURE 3.9 (a) Three-phase power system; (b) single-phase equivalent circuit.

Solution With both generator and load wye connected, the single-phase analysis is used in conjunction with the single-phase equivalent circuit shown in Fig. 3.10.

The phase a voltage at the load is chosen as reference phasor. It is given by

$$\mathbf{V}_{an} = \left(4160/\sqrt{3}\right)\underline{/0^\circ} = 2400\,\underline{/0^\circ}\ \text{V} \qquad \text{(line-to-neutral)}$$

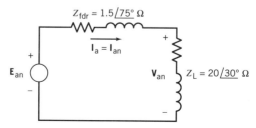

FIGURE 3.10 Per-phase equivalent circuit for Example 3.3.

The phase a current, which is identical to the line a current, is found as follows:

$$I_a = I_{an} = \frac{V_{an}}{Z_L \angle \theta} = \frac{2400 \angle 0°}{20 \angle 30°} = 120 \angle -30° \text{ A}$$

a. The phase a voltage of the generator is found as follows:

$$E_{an} = V_{an} + I_a Z_{fdr} = 2400 \angle 0° + (120 \angle -30°)(1.5 \angle 75°)$$
$$= 2527 + j127 = 2530 \angle 3° \text{ V}$$

Consequently, the phase voltages of the load and the generator are computed as follows:

$$V_{an} = 2400 \angle 0° \text{ V} \qquad E_{an} = 2530 \angle 3° \text{ V}$$
$$V_{bn} = 2400 \angle -120° \qquad E_{bn} = 2530 \angle -117°$$
$$V_{cn} = 2400 \angle 120° \qquad E_{cn} = 2530 \angle 123°$$

The line-to-line voltages are

$$V_{ab} = 4160 \angle 30° \text{ V} \qquad E_{ab} = 4382 \angle 33° \text{ V}$$
$$V_{bc} = 4160 \angle -90° \qquad E_{bc} = 4382 \angle -87°$$
$$V_{ca} = 4160 \angle 150° \qquad E_{ca} = 4382 \angle 153°$$

b. The load current and generator current are equal and are given by

$$I_a = I_{an} = 120 \angle -30° \text{ A}$$
$$I_b = I_{bn} = 120 \angle -150°$$
$$I_c = I_{cn} = 120 \angle 90°$$

3.4.4   Power in Balanced Three-Phase Systems

The total average power absorbed by a three-phase balanced load, or delivered by a three-phase generator, is equal to the sum of the powers in each phase. The voltages and currents in each phase are equal; that is,

$$V_P = V_{an} = V_{bn} = V_{cn} \qquad (3.49)$$

$$I_P = I_{an} = I_{bn} = I_{cn} \qquad (3.50)$$

Therefore, the total three-phase power is

$$P_T = 3P_P = 3V_P I_P \cos \theta_P \tag{3.51}$$

where θ_P is the phase angle between the voltage and the current. Similarly, the total three-phase reactive power is

$$Q_T = 3Q_P = 3V_P I_P \sin \theta_P \tag{3.52}$$

Also the total three-phase apparent power is given by

$$S_T = 3S_P = \sqrt{P_T^2 + Q_T^2} = 3\sqrt{P_P^2 + Q_P^2} = 3V_P I_P \tag{3.53}$$

For a three-phase wye-connected generator, or wye-connected load, $I_L = I_P$, and $V_L = \sqrt{3}V_P$. Thus the real, reactive, and apparent powers may be expressed as follows:

$$P_T = 3\left(\frac{V_L}{\sqrt{3}}\right)I_L \cos \theta_P = \sqrt{3}V_L I_L \cos \theta_P \tag{3.54}$$

$$Q_T = 3\left(\frac{V_L}{\sqrt{3}}\right)I_L \sin \theta_P = \sqrt{3}V_L I_L \sin \theta_P \tag{3.55}$$

$$S_T = 3\left(\frac{V_L}{\sqrt{3}}\right)I_L = \sqrt{3}V_L I_L \tag{3.56}$$

For a three-phase delta-connected generator, or delta-connected load, $I_L = \sqrt{3}I_P$ and $V_L = V_P$. In terms of the line quantities, the power expressions may be written as

$$P_T = 3V_L\left(\frac{I_L}{\sqrt{3}}\right)\cos \theta_P = \sqrt{3}V_L I_L \cos \theta_P \tag{3.57}$$

$$Q_T = 3V_L\left(\frac{I_L}{\sqrt{3}}\right)\sin \theta_P = \sqrt{3}V_L I_L \sin \theta_P \tag{3.58}$$

$$S_T = 3V_L\left(\frac{I_L}{\sqrt{3}}\right) = \sqrt{3}V_L I_L \tag{3.59}$$

Ratings of three-phase equipment, such as generators, motors, transformers, and transmission lines, are normally given as total or three-phase real power in MW, or as total apparent power in MVA, and as line-to-line voltage in kV.

EXAMPLE 3.4

A three-phase motor draws 20 kVA at 0.707 lagging power factor from a 220-V source. It is desired to improve the power factor to 0.90 lagging by connecting a capacitor bank across the terminals of the motor.

    a. Calculate the line current before and after the addition of the capacitor bank.

    b. Determine the required kVA rating of the capacitor bank.

Solution   The real and reactive powers of the load are

$$P_M = (20)(0.707) = 14.14 \text{ kW}$$
$$Q_M = 20\sin(\cos^{-1}0.707) = 14.14 \text{ kVAR}$$

    a. The line current of the motor is

$$I_M = \frac{S_T}{\sqrt{3}V_L} = \frac{20,000}{\sqrt{3}\,220} = 52.5 \text{ A}$$

For a power factor $PF_{corr} = 0.90$, the new value of line current is

$$I_{corr} = \frac{P_M}{\sqrt{3}V_L PF_{corr}} = \frac{14,140}{\sqrt{3}(220)(0.9)} = 41.2 \text{ A}$$

    b. The corrected value of reactive power is

$$Q_{corr} = P_M\tan(\cos^{-1}0.90) = 14.14\tan 25.8° = 6.85 \text{ kVAR}$$

The kVA rating of the capacitor bank required to bring the power factor from 0.707 to 0.90 lagging is found as

$$Q_{cap} = Q_{corr} - Q_M = 6.85 - 14.14 = -7.29 \text{ kVAR}$$

A power triangle depicting power factor correction by using capacitors is shown in Fig. 3.11.

3.4.5   Instantaneous Power in Balanced Three-Phase Systems

For a balanced three-phase load, the total instantaneous power is equal to the sum of the individual powers of the three phases. Thus,

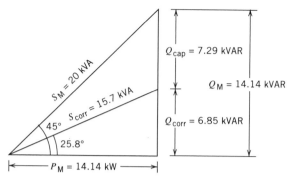

FIGURE 3.11   Power triangle of Example 3.4.

$$p_T = p_a + p_b + p_c \tag{3.60}$$

In terms of the instantaneous phase voltages and currents, the total power is

$$p_T = v_{an}i_a + v_{bn}i_b + v_{cn}i_c \tag{3.61}$$

Assuming a positive phase sequence with v_{an} taken as the reference, the total power may be expressed as follows:

$$
\begin{aligned}
p_T = \; & V_m I_m \cos \omega t \cos(\omega t - \theta_P) \\
& + V_m I_m \cos(\omega t - 120°) \cos(\omega t - \theta_P - 120°) \\
& + V_m I_m \cos(\omega t - 240°) \cos(\omega t - \theta_P - 240°)
\end{aligned} \tag{3.62}
$$

where V_m and I_m are the peak values of the phase voltages and currents, respectively, and θ_P is the phase angle by which the current lags the voltage in each phase. By using the trigonometric identity on the product of two cosine functions that was cited previously in Section 3.3.2, Eq. 3.62 can be written as

$$
\begin{aligned}
p_T = \; & \tfrac{3}{2} V_m I_m \cos \theta_P + \tfrac{1}{2} V_m I_m [\cos(2\omega t - \theta_P) \\
& + \cos(2\omega t - \theta_P + 120°) + \cos(2\omega t - \theta_P - 120°)]
\end{aligned} \tag{3.63}
$$

Since the second term on the right-hand side of Eq. 3.63 is identically equal to zero, it reduces to

$$p_T = \tfrac{3}{2} V_m I_m \cos \theta_P \tag{3.64}$$

This demonstrates an important property of a balanced three-phase system: the total instantaneous power is time invariant.

3.4.6 Three-Phase Power Measurements

In three-phase power systems, two wattmeters can be used to measure total power. The connection and phasor diagrams are shown in Fig. 3.12 for an assumed abc phase sequence and lagging power factor.

The wattmeter readings are given by

$$W_1 = V_{ab}I_a \cos \angle(V_{ab}, I_a) = V_L I_L \cos(30° + \theta) \qquad (3.65)$$

$$W_2 = V_{cb}I_c \cos \angle(V_{cb}, I_c) = V_L I_L \cos(30° - \theta) \qquad (3.66)$$

The sum of the two wattmeter readings gives the total three-phase power:

$$P_T = W_1 + W_2 = V_L I_L[\cos(30° + \theta) + \cos(30° - \theta)]$$

$$= \sqrt{3} V_L I_L \cos\theta \qquad (3.67)$$

The difference of the two readings is

$$W_2 - W_1 = V_L I_L[\cos(30° - \theta) - \cos(30° + \theta)]$$

$$= V_L I_L \sin\theta \qquad (3.68)$$

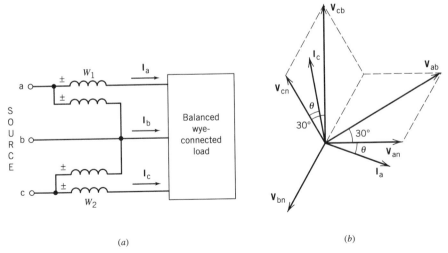

(a)   (b)

FIGURE 3.12 Two-wattmeter method: (a) connection diagram; (b) phasor diagram.

which is $1/\sqrt{3}$ times the total three-phase reactive power. Thus, Q_T can be found from

$$Q_T = \sqrt{3}(W_2 - W_1) \tag{3.69}$$

The power factor angle can also be found from

$$\theta = \tan^{-1}\left(\frac{Q_T}{P_T}\right) = \tan^{-1}\left[\frac{\sqrt{3}(W_2 - W_1)}{W_2 + W_1}\right] \tag{3.70}$$

DRILL PROBLEMS

D3.7 A balanced, three-phase load is connected to a 2200-V feeder. The load draws a line current of 60 A at a power factor of 0.90 lagging. Calculate the real, reactive, and apparent power absorbed by the load.

D3.8 A delta-connected load consists of three identical impedances of 8 + $j6$ Ω each and is supplied from a three-phase, 200-V source. Calculate

a. The phase current and the line current
b. The power factor
c. The real, reactive, and apparent power taken by the load

D3.9 Repeat Problem D3.8 when the load impedances are wye connected.

D3.10 Two balanced wye-connected loads of 8 + $j5$ Ω/phase and 6 − $j2$ Ω/phase are supplied by a three-phase source at a line-to-line voltage of 440 V. Find

a. The line current drawn by each load
b. The total line current supplied by the source
c. The real, reactive, and complex power absorbed by each load
d. The real, reactive, and complex power delivered by the source

3.5 PER-UNIT ANALYSIS

The per-unit method of power system analysis offers distinct advantages over the use of actual amperes, volts, and ohms. It eliminates the need for conversion of the voltages, currents, and impedances across every transformer in the circuit; thus, there is less chance of computational errors. Second, the need to transform from three-phase to single-phase equivalents, and vice versa, is avoided with the use of per-unit quantities; hence, there is less confusion in handling and manipulating the various parameters in three-phase systems.

In the per-unit system, any electrical quantity may be expressed in per unit as the ratio of the actual quantity and the chosen base value for that quantity. This quotient is presented either in decimal form or as a percentage. However, care must be taken when handling percent quantities. Although the product of two quantities in per unit is expressed in per unit itself, the product of two quantities in percent must be divided by 100 to obtain the correct result in percent.

Four base electrical quantities must be considered: power, voltage, current, and impedance bases. Just like the actual quantities, these bases satisfy the electrical laws.

In single-phase systems, the relationships among the base quantities are

$$S_{\text{base}} = V_{\text{base}} I_{\text{base}} \qquad (3.71)$$

$$V_{\text{base}} = I_{\text{base}} Z_{\text{base}} \qquad (3.72)$$

With only two equations relating the four base quantities, it is necessary to specify two base values. The power and voltage bases are usually chosen equal to the rated values, or nominal values, and the other two are computed from the foregoing electrical relationships as follows:

$$I_{\text{base}} = \frac{S_{\text{base}}}{V_{\text{base}}} \qquad (3.73)$$

$$Z_{\text{base}} = \frac{V_{\text{base}}}{I_{\text{base}}} = \frac{(V_{\text{base}})^2}{S_{\text{base}}} = \frac{(kV_{\text{base}})^2}{MVA_{\text{base}}} \qquad (3.74)$$

The specified power base is applicable to all parts of the power system. The voltage base varies across a transformer, and so do the current base and impedance base.

The per-unit electrical quantities are calculated as follows:

$$\mathbf{S}_u = \frac{P + jQ}{S_{\text{base}}} = P_u + jQ_u \qquad (3.75)$$

$$\mathbf{V}_u = \frac{\mathbf{V}}{V_{\text{base}}} \qquad (3.76)$$

$$\mathbf{I}_u = \frac{\mathbf{I}}{I_{\text{base}}} \qquad (3.77)$$

$$Z_u = \frac{Z}{Z_{\text{base}}} \qquad (3.78)$$

The complex power into a lossless transformer is equal to the complex power out. On the other hand, actual voltages, currents, and impedances change across the transformer. However, because base voltage, base current, and base impedance also change across the transformer, the per-unit values are the same on both sides of the transformer. Furthermore, since the voltage base is usually chosen to be either the nominal voltage or the rated voltage, the per-unit value of voltage is almost always near unity.

EXAMPLE 3.5

Solve Example 3.2 using per-unit representation. Choose $S_{base} = 1500$ VA and $V_{base} = 120$ V.

Solution The base current is calculated as

$$I_{base} = S_{base}/V_{base} = 1500/120 = 12.5 \text{ A}$$

The base impedance is

$$Z_{base} = (V_{base})^2/S_{base} = (120)^2/1500 = 9.6 \text{ }\Omega$$

The per-unit values are computed as follows:

$$\mathbf{V}_L = \frac{120 \angle 0°}{120} = 1.0 \angle 0° \text{ pu V}$$

$$Z_L = \frac{10 \angle 36.9°}{9.6} = 1.04 \angle 36.9° \text{ pu } \Omega$$

$$\mathbf{I}_L = \frac{1.0 \angle 0°}{1.04 \angle 36.9°} = 0.96 \angle -36.9° \text{ pu A}$$

$$Z_{fdr} = \frac{2.24 \angle 63.4°}{9.6} = 0.233 \angle 63.4° \text{ pu } \Omega$$

The generator voltage in per unit is computed as follows:

$$\begin{aligned}
\mathbf{E}_g &= \mathbf{V}_L + \mathbf{I}_L Z_{fdr} \\
&= 1.0 \angle 0° + (0.96 \angle -36.9°)(0.233 \angle 63.4°) \\
&= 1.204 \angle 4.8° \text{ pu} \\
&= (1.204 \angle 4.8°)(120) = 144.5 \angle 4.8° \text{ V}
\end{aligned}$$

The complex power is then calculated as follows:

$$\mathbf{S_g} = \mathbf{E_g I_g^*} = \mathbf{E_g I_L^*}$$
$$= (1.204 \underline{/4.8°})(0.96 \underline{/-36.9°})^* = 1.156 \underline{/41.7°}$$
$$= 0.863 + j0.769 \text{ pu}$$

Therefore,

$$P_g = (0.863)(1500) = 1295 \text{ W}$$
$$Q_g = (0.769)(1500) = 1154 \text{ VAR}$$

For three-phase systems, the base power is total three-phase power and the base voltage is line-to-line voltage. With a wye connection assumed, the base line current is equal to the base phase current. The base impedance is per phase. The three-phase base quantities are related to the single-phase bases as follows:

$$\text{Base power:} \quad S_{T,base} = 3S_{P,base} \quad (3.79)$$

$$\text{Base voltage:} \quad V_{L,base} = \sqrt{3}V_{P,base} \quad (3.80)$$

$$\text{Base current:} \quad I_{L,base} = \frac{S_{T,base}}{\sqrt{3}V_{L,base}} = \frac{S_{P,base}}{V_{P,base}} = I_{P,base} \quad (3.81)$$

$$\text{Base impedance:} \quad Z_{base} = \frac{(V_{L,base})^2}{S_{T,base}} = \frac{(V_{P,base})^2}{S_{P,base}} \quad (3.82)$$

The impedance characteristic of an electrical equipment or device is usually expressed as a percentage based on its ratings. When such a device is connected in a power system in which the selected base values are different from the machine ratings, the per-unit quantities have to be expressed in terms of the system bases. The per-unit value of impedance may be converted to the new bases as follows:

$$Z_{u,new} = Z_{u,old} \left(\frac{S_{base,new}}{S_{base,old}} \right) \left(\frac{V_{base,old}}{V_{base,new}} \right)^2 \quad (3.83)$$

EXAMPLE 3.6

A three-phase, 60-Hz, 30-MVA, 13.8-kV, wye-connected synchronous generator has an armature resistance $R_a = 2 \, \Omega$ per phase and a synchronous reactance $X_s = 10 \, \Omega$ per phase.

a. Express the machine impedance in per unit based on the machine ratings.

b. Using the results of part (a), find the per-unit impedance based on a new $S_{base} = 50$ MVA and $V_{base} = 34.5$ kV.

Solution

a.

$$Z_{base} = (13.8)^2/30 = 6.35 \ \Omega$$

$$Z_s = R_a + jX_s = 2 + j10 = 10.2 \underline{/78.7°} \ \Omega$$

$$Z_{su} = Z_s/Z_{base} = (2 + j10)/6.35 = 0.315 + j1.575$$

$$= 1.606 \underline{/78.7°} \text{ pu } \Omega$$

b.

$$Z_{su,new} = 1.606 \underline{/78.7°}(50/30)(13.8/34.5)^2$$

$$= 0.428 \underline{/78.7°} = 0.084 + j0.420 \text{ pu } \Omega$$

REFERENCES

1. Bobrow, Leonard S. *Elementary Linear Circuit Analysis*. Holt, Rinehart & Winston, New York, 1987.

2. Glover, J. Duncan, and Mulukutla Sarma. *Power System Analysis and Design*. PWS, Boston, 1987.

3. Gross, Charles A. *Power System Analysis*. 2nd ed. Wiley, New York, 1986.

4. Gungor, Behic R. *Power Systems*. Harcourt Brace Jovanovich, New York, 1988.

5. Nilsson, James W. *Electric Circuits*. 3rd ed. Addison-Wesley, New York, 1990.

6. Stevenson, William D., Jr. *Elements of Power System Analysis*. 4th ed. McGraw-Hill, New York, 1982.

PROBLEMS

3.1 Given the complex numbers $A_1 = 5 \underline{/30°}$ and $A_2 = -3 + j4$, calculate the following, giving the answers in both rectangular and polar forms.

a. $A_1 + A_2$

b. $A_1 A_2$

c. $A_1/(A_2)^*$

d. $(A_2)^2$

e. $A_1[1 + (A_2)^2]$

3.2 For a given electrical circuit, the expressions for voltage and current as functions of time are given as follows:

$$v(t) = 283 \sin \omega t \text{ V}$$

$$i(t) = 35 \cos(\omega t + 25°) \text{ A}$$

a. Find the rms values of the voltage and current.

b. Find the phasor expressions for the voltage and current in both polar and rectangular form.

c. State whether the circuit is inductive or capacitive.

3.3 The electrical network shown in Fig. 3.13 has a voltage source $V = 100 \underline{/0°}$, and the values of the impedances are as follows:

$$Z_1 = 8 - j6 \ \Omega, \qquad Z_2 = 3 - j4 \ \Omega, \qquad Z_3 = 5 + j5 \ \Omega$$

Determine (a) the real power absorbed by each impedance and (b) the reactive power taken by each impedance.

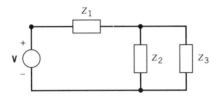

FIGURE 3.13 Electrical network of Problem 3.3.

3.4 A single-phase source supplies a load consisting of a resistor $R = 20 \ \Omega$ and a capacitive reactance $X_C = 10 \ \Omega$, which are connected in parallel. The instantaneous voltage of the source is given by

$$v(t) = 120 \cos(\omega t + 45°) \text{ V}$$

Find the following:

a. Phasor voltage V of the source

b. Phasor current I supplied by the source

c. Instantaneous current $i(t)$ supplied by the source

3.5 Repeat Problem 3.4 if the resistor and capacitor are connected in series.

3.6 For the parallel $R - C$ load of Problem 3.4, determine

a. Instantaneous power absorbed by the resistor

b. Instantaneous power absorbed by the capacitor

c. Real power absorbed by the resistor

d. Reactive power absorbed by the capacitor

e. Power factor of the combined load

3.7 Repeat Problem 3.6 if the resistor and capacitor are connected in series.

3.8 Consider a single-phase load with an applied voltage $v(t)$ and load current $i(t)$ specified as follows:

$$v(t) = 220 \cos(\omega t + 20°) \text{ V}$$
$$i(t) = 40 \cos(\omega t - 30°) \text{ A}$$

a. Find the real and reactive power absorbed by the load.

b. Draw the power triangle.

c. Find the power factor, and state whether it is lagging or leading.

d. Calculate the reactance in ohms of capacitors to be connected in parallel with the load in order to improve the power factor to 0.9 lagging.

3.9 A series circuit has an impedance of $25\,\underline{/53.1°}$ and is connected to a single-phase 220-V source.

a. Find the resistance and reactance of the load.

b. Find the real and reactive power absorbed by the load.

c. Find the power factor of the circuit, and state whether lagging or leading.

3.10 A single-phase source has a terminal voltage $\mathbf{V} = 120\,\underline{/-15°}$. It supplies a current of $15\,\underline{/45°}$ to an electrical load.

a. Find the complex power supplied by the source.

b. Determine the real power. State whether the source is delivering or absorbing.

c. Determine the reactive power, and state whether the source is delivering or absorbing.

3.11 Two ideal voltage sources are connected to each other through a feeder with impedance $Z = 1.5 + j6$ Ω as shown in Fig. 3.14. Let $\mathbf{E}_1 = 120\,\underline{/0°}$ V and $\mathbf{E}_2 = 110\,\underline{/45°}$ V.

a. Determine the real power of each machine, and state whether the machine is supplying or absorbing real power.

b. Determine the reactive power of each machine, and state whether each machine is delivering or receiving reactive power.

c. Determine the real and reactive power of the impedance, and state whether supplied or consumed.

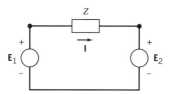

FIGURE 3.14 Electric circuit of Problem 3.11.

3.12 Repeat Problem 3.11 if the feeder impedance between the voltage sources of Fig. 3.14 is $Z = 1.5 - j6$ Ω.

3.13 A single-phase electrical load draws 10 MW at 0.6 power factor lagging.

a. Find the real and reactive power absorbed by the load.

b. Draw the power triangle.

c. Determine the kVAR of a capacitor to be connected across the load to raise the power factor to 0.95.

3.14 A capacitor is connected across the series impedance of Problem 3.9. This capacitor supplies 1000 VARs.

a. Find the real and reactive power supplied by the source.

b. Find the resultant power factor.

3.15 An industrial plant consists of several induction motors. The plant absorbs 300 kW at 0.6 PF lagging from the substation bus.

a. Compute the required kVAR rating of the capacitor connected across the load to raise the power factor to 0.9 lagging.

b. A 200-hp, 90% efficiency, synchronous motor is operated from the same bus at rated conditions and 0.8 power factor leading. Calculate the resulting power factor.

3.16 A 230-V source supplies two loads in parallel. One draws 5 kVA at a lagging power factor of 0.80, and the other draws 3 kW at a lagging power factor of 0.90. Find the source current.

3.17 A 440-V, 30-hp, three-phase motor operates at full load, 88% efficiency, and 65% power factor lagging.

a. Find the current drawn by the motor.

b. Find the real and reactive power absorbed by the motor.

3.18 A three-phase, 50-kVA, 600-V, 60-Hz generator operates at rated terminal voltage and supplies a line current of 48 A per phase at a 0.8 lagging power factor to a balanced three-phase load. Determine the real, reactive, and apparent power.

3.19 A 345-kV, three-phase transmission line delivers 500 MVA, 0.866 power factor lagging, to a three-phase load connected to its receiving-end terminals. Assume that the load is Δ connected and the voltage at the receiving end is 345 kV.

a. Find the complex load impedance per phase.

b. Calculate the line and phase currents.

c. Find the real and reactive power per phase.

d. Find the total real and reactive power.

3.20 Repeat Problem 3.19 assuming that the load is wye connected.

3.21 A three-phase load draws 120 kW at a power factor of 0.85 lagging from a 440-V bus. In parallel with this load is a three-phase capacitor bank that is rated 50 kVAR. Find (a) the total line current and (b) the resultant power factor.

3.22 A three-phase motor draws 40 kVA at 0.65 power factor lagging from a 230-V source. A capacitor bank is connected across the motor terminals to make the combined power factor 0.95 lagging.

a. Determine the required kVAR rating of the capacitor bank.

b. Determine the line current before and after the capacitors are added.

3.23 Two balanced wye-connected loads are connected in parallel with each other. The first draws 15 kVA at 0.8 PF lagging, and the second requires 20 kW at 0.9 PF leading. The two loads are supplied by a balanced three-phase, wye-connected, 2400-V source.

a. Determine the phasor current drawn by each load.

b. Find the real and reactive power absorbed by each load.

c. Compute the phasor current supplied by the source.

d. Calculate the total real and reactive power drawn by the combined load.

e. What is the overall power factor?

3.24 The motor of Problem 3.17 is connected to a substation bus through a three-phase feeder with an impedance $0.5 + j1.5$ Ω per phase. Find the line-to-line voltage at the bus if the voltage at the motor terminals is 440 V.

3.25 A three-phase substation bus supplies two wye-connected loads that are connected in parallel through a three-phase feeder with an impedance of $0.5 + j2.0$ Ω per phase. Load 1 draws 50 kW at 0.866 lagging power factor, and load 2 draws 36 kVA at 0.9 leading power factor. The line-to-line voltage at the loads is 460 V. Find the following:

a. Impedance of each load per phase

b. Total line current flowing through the feeder

c. Line-to-line voltage at the substation bus

d. Total real and reactive power supplied by the bus

3.26 A delta-connected load consists of three identical impedances $Z_\Delta = 45 \underline{/60°}$ Ω per phase. It is connected to a three-phase, 208-V source by a three-phase feeder with conductor impedance $Z_{fdr} = (1.2 + j1.6)$ Ω per phase.

a. Calculate the line-to-line voltage at the load terminals.

b. A delta-connected capacitor bank with a reactance of 60 Ω per phase is connected in parallel with the load at its terminals. Find the resulting line-to-line voltage at the load terminals.

3.27 The total power being absorbed by a balanced three-phase load is measured using the two-wattmeter method. The phase sequence is abc. The current coils of the wattmeters are connected in lines a and b. Let the readings of the two meters be P_a and P_b, respectively. The line voltage is 2400 V, and the load is 30 kVA.

a. Show a wiring diagram.

b. Calculate P_a and P_b if the power factor is 1.0.

c. Calculate P_a and P_b if the power factor is 0.2 lagging.

d. Calculate P_a and P_b if the power factor is 0.5 leading.

e. Sketch the phasor diagram showing all voltages and currents for each case.

3.28 Using 100 MVA and 115 kV as base values, express 110 kV, 75 MVA, 375 A, and 26.5 Ω in per-unit and percent values.

3.29 The per-unit impedance of an electric load is 0.5. The base power is 200 kVA, and the base voltage is 12 kV.

    a. Find the per-unit impedance of the load if 400 kVA and 24 kV are selected as base values.

    b. Find the ohmic value of the impedance.

3.30 A 350-MVA, 13.8-kV AC generator has a synchronous reactance of 1.20 per unit. The generator is connected to a circuit for which the specified bases are 100 MVA, 13.2 kV.

    a. Find the per-unit value of the generator synchronous reactance on the specified bases.

    b. Find the ohmic value of the synchronous reactance.

3.31 A single-phase source is connected to an electrical load. The load draws a 0.6 pu current at 1.10 pu voltage while taking a real power of 0.4 pu at a lagging power factor. Choose a base voltage of 8 kV and a base current of 125 A. Calculate the following:

    a. Real power in kW

    b. Reactive power in kVAR

    c. Power factor

    d. The ohmic values of the resistance and the reactance of the load

Four

Magnetic Circuits
and Transformers

4.1 INTRODUCTION

A transformer is a device used to convert AC electric energy at one voltage and current level to AC electric energy at another voltage and current level. This conversion takes place in an electromagnetic system consisting of two or more windings supported by a ferromagnetic structure.

Other forms of electromechanical energy conversion take place in rotating machines, both AC and DC machines, and other devices such as transducers, solenoids, and relays. In these machines and devices, electromagnetic fields that are confined in magnetic structures also play an important role in the conversion process. In the first few sections of this chapter, some basic concepts of electromagnetic theory are reviewed, typical magnetic circuits are analyzed, and other parameters, including flux linkages and inductances, are defined. These parameters are used in the discussion of the theory of energy conversion in rotating machines and transformers.

In the succeeding sections of this chapter, the principle of operation of a transformer is discussed, equivalent circuits for the transformer at steady state are developed, and the operating performance of the transformer is analyzed. Also, electrical tests for determining the parameters of the transformer are described. Finally, three-phase interconnections of transformers for three-phase voltage transformations are discussed.

4.2 MAGNETIC CIRCUITS

Ferromagnetic materials constitute a large portion of any electrical machine, including transformers. Hence, the study and design of electrical machinery includes the analysis of the magnetic circuits involved in these machines.

64

To be able to apply electric circuit concepts to these magnetic circuits, the following assumptions are made:

a. The frequencies involved — 60 Hz or less — and the sizes (dimensions) of the magnetic structures are such that the displacement term in Maxwell's equation based on *Ampère's law* can be neglected. Thus,

$$\oint \mathbf{H} \cdot \mathbf{dl} = I \qquad (4.1)$$

b. A three-dimensional magnetic field is reduced to a one-dimensional circuit equivalent, that is, magnetic circuit. Equation 4.1 states that the line integral of the *magnetic field* **H** over a closed path is equal to the net current enclosed by the path. **H** is expressed in amperes per meter when the current is given in amperes.

A *magnetic circuit* consists of a magnetic structure built mainly of high-permeability magnetic material. Thus, magnetic flux is confined to the paths presented by the high-permeability material, just as electric current is confined to the paths presented by the high-conductivity conductors of the electric circuit. A simple magnetic circuit is shown in Fig. 4.1. The source of magnetic flux, the *magnetomotive force (mmf)*, is the electric current flowing in the N-turn winding. Applying Eq. 4.1 to a closed path yields

$$\oint \mathbf{H_c} \cdot \mathbf{dl} = NI \qquad (4.2)$$

The magnetic field $\mathbf{H_c}$ is approximately constant, and its direction is the same as that of the magnetic flux ϕ, as shown in the figure. Hence, the integral in Eq. 4.2 reduces to $H_c l_c$. Therefore, the magnetomotive force (mmf), designated as

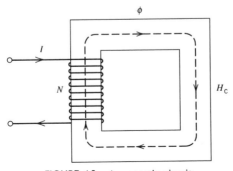

FIGURE 4.1   A magnetic circuit.

F, is given by

$$F = NI = H_c l_c \tag{4.3}$$

where l_c is the average length of the magnetic path or core.

The core is usually made of ferromagnetic material. The *magnetic flux density* B (expressed in tesla or weber/m^2) in the core is related to the magnetic field H according to the saturation curve, or B-H curve, of Fig. 4.2. The slope of this curve is designated as μ, the permeability of the material in henries per meter (H/m). Therefore, the relationship between **B** and **H** may be expressed as

$$\mathbf{B} = \mu\mathbf{H} \tag{4.4}$$

It may be seen from Fig. 4.2 that the slope of the B-H curve depends on the operating value of magnetic flux density. However, if the range of operating values is confined below the knee of the curve, the relationship between B and H can be approximated as linear with reasonable accuracy. Therefore, in the following discussions, it is assumed that the permeability of the magnetic materials used for core is a constant.

The permeability of a magnetic material is usually given relative to the permeability μ_0 of free space. Thus,

$$\mu = \mu_r \mu_0 \tag{4.5}$$

where μ_r is the relative permeability. In SI units, the permeability of free space $\mu_0 = 4\pi \times 10^{-7}$ H/m.

The *magnetic flux* ϕ (expressed in webers) through a given surface is found as follows:

$$\phi = \int_S \mathbf{B} \cdot d\mathbf{S} \tag{4.6}$$

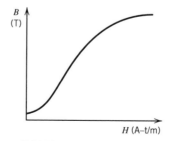

FIGURE 4.2   A B-H curve.

Table 4.1 Analogy Between Magnetic and Electric Circuits

Electric Circuit	Magnetic Circuit
i = current (A)	ϕ = flux (Wb)
V = emf (V)	F = mmf (A-t)
R = resistance (Ω)	R = reluctance (A-t/Wb)
σ = conductivity (S/m)	μ = permeability (H/m)

The direction of the differential area is along the perpendicular to the cross-sectional area A_c of the core itself. Since the flux density B_c has the same direction as $\mathbf{dS}$ and is approximately uniform over A_c, Eq. 4.6 reduces to

$$\phi = B_c A_c = \mu H_c A_c = \mu \left(\frac{NI}{l_c} \right) A_c \qquad (4.7)$$

Rearranging Eq. 4.7 gives

$$R\phi = NI = F \qquad (4.8)$$

where $R = l_c/(\mu A_c)$ = *reluctance* of the magnetic circuit in A-t/Wb.

Equation 4.8 is analogous to Ohm's law for resistive circuits. The analogies between other magnetic and electric circuit quantities are presented in Table 4.1.

EXAMPLE 4.1

The magnetic circuit shown in Fig. 4.1 has the following dimensions: $A_c = 16$ cm², $l_c = 40$ cm, and $N = 350$ turns. The relative permeability of the core is $\mu_r = 50,000$. For a magnetic flux density of 1.5 T in the core, determine

a. The flux ϕ

b. The total flux linkage $\lambda = N\phi$

c. The required current through the coil

Solution

a. An electric circuit analog is drawn for the magnetic circuit. This is shown in Fig. 4.3.

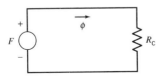

FIGURE 4.3 Electric circuit analog for Example 4.1.

The magnetic flux is given by

$$\phi = BA = (1.5)(16 \times 10^{-4}) = 2.4 \text{ mWb}$$

b. The total flux linkage is

$$\lambda = N\phi = (350)(2.4 \times 10^{-3}) = 0.84 \text{ Wb-t}$$

c. The reluctance of the magnetic circuit is computed as

$$R_c = l_c/(\mu_r \mu_0 A_c)$$

$$= \frac{40 \times 10^{-2}}{(50,000)(4\pi \times 10^{-7})(16 \times 10^{-4})} = 3979 \text{ A-t/Wb}$$

By using Eq. 4.8, the current is computed as follows:

$$I = \frac{R_c \phi}{N} = \frac{(3979)(2.4 \times 10^{-3})}{350} = 27.3 \text{ mA}$$

A magnetic core of this type may be used in a single-phase transformer.

EXAMPLE 4.2

An air gap of length $g = 0.5$ mm is cut in the right leg of the magnetic circuit of Fig. 4.1. If a current of 1.2 A flows through the coil, calculate

  a. The flux ϕ
  b. The total flux linkage $\lambda = N\phi$
  c. The magnetic flux density B

Solution

  a. The electric circuit analog is shown in Fig. 4.4.

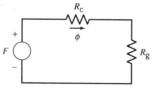

FIGURE 4.4   Electric circuit analog for Example 4.2.

The mmf is given by

$$F = NI = (350)(1.2) = 420 \text{ A-t}$$

The reluctance of the core is computed as follows:

$$R_c = \frac{l_c - g}{\mu_c A_c} \cong \frac{l_c}{\mu_c A_c} = 3979 \text{ A-t/Wb}$$

Assuming that fringing in the air gap is negligible, that is, $A_g = A_c$, the reluctance of the air gap is computed as follows:

$$R_g = \frac{l_g}{\mu_0 A_c}$$

$$= \frac{0.5 \times 10^{-3}}{(4\pi \times 10^{-7})(16 \times 10^{-4})} = 248{,}680 \text{ A-t/Wb}$$

For this magnetic circuit, the total reluctance of the flux path is

$$R_t = R_c + R_g = 3979 + 248{,}680 = 252{,}659 \text{ A-t/Wb}$$

The magnetic flux is calculated as

$$\phi = F/R_t = 420/252{,}659 = 1.66 \times 10^{-3} \text{ Wb}$$

b. The flux linkage is computed as follows:

$$\lambda = N\phi = (350)(1.66 \times 10^{-3}) = 0.58 \text{ Wb-t}$$

c. The magnetic flux density is

$$B = \phi/A = (1.66 \times 10^{-3})/(16 \times 10^{-4}) \cong 1.04 \text{ T}$$

A similar magnetic circuit is used in electrical machines, electric meters, and protective relays.

DRILL PROBLEMS

D4.1 The magnetic circuit shown in Fig. 4.1 has an air gap cut in the right leg of the core. The air gap is 0.1 mm long. The coil is connected to a voltage source, and the current drawn is adjusted so that the magnetic flux density in the air gap is 1.5 T. Assume that flux fringing in the air gap is negligible. Use the dimensions and the relative permeability of the magnetic core specified in Example 4.1.

   a. Find the value of the current.

   b. Calculate the magnetic flux.

   c. Determine the flux linkage of the coil.

D4.2 A magnetic core is built in the form of a circular ring having a mean radius of 10 cm. A coil containing 150 turns is wound uniformly throughout the length of the core. The coil is connected to a voltage source, and it draws a current of 15 A.

   a. Determine the mmf of the coil.

   b. Calculate the magnetic field intensity in the core.

D4.3 The circular ring of Drill Problem D4.2 has a mean cross-sectional area of 25 cm^2. The relative permeability of the material of the ring is 1500. Calculate

   a. The magnetic flux in the core

   b. The magnetic flux density in the core

   c. The flux linkage of the coil

   d. The reluctance of the core

D4.4 A magnetic core has a circular cross-sectional area of 2.0 in^2, a mean path length of 10 in, and an air-gap length of 0.125 in. A 350-turn coil is wound around the magnetic core, and a current of 5 A is supplied to the coil. Assume that the relative permeability of the core is infinite and fringing of flux in the air gap is negligible.

   a. Calculate the reluctance of the magnetic circuit.

   b. Find the magnetic flux density in the air gap.

D4.5 Repeat Drill Problem D4.4 assuming that the core has a relative permeability $\mu_r = 5000$.

4.3   FARADAY'S LAW

In 1820 Oersted observed that a compass needle is deflected by a current-carrying conductor. In 1831 Faraday discovered the principles of induced electromotive force (emf) on which the design and operation of generators, motors, and transformers are based. Faraday's experiments and findings can be described as follows:

$F = N L$

A time-varying magnetic field induces an electromotive force that produces a current in a closed circuit. This current flows in a direction such that it produces a magnetic field that tends to oppose the changing magnetic flux of the original time-varying field.

Mathematically, *Faraday's law* can be stated as follows:

$$\text{emf} = \frac{d\lambda}{dt} \tag{4.9}$$

where λ is the total flux linkage of the closed path. A nonzero value of $d\lambda/dt$ may result from any of the following conditions:

a. A time-varying flux linking a stationary path
b. Relative motion between a steady flux and a closed path
c. A combination of the first two

If the closed path consists of a winding with N turns and we assume that each turn links the same flux ϕ, then the induced emf is given by

$$e = \text{emf} = \frac{d\lambda}{dt} = N\frac{d\phi}{dt} \tag{4.10}$$

$\rightarrow$ *electromotive force*

where $\lambda = N\phi$ = total flux linkage.

EXAMPLE 4.3

The flux density **B** is normal to the plane of the rectangular loop and directed outward as shown in Fig. 4.5, and it is equal to

$$\mathbf{B} = B_0 \cos \omega t \, \mathbf{u}$$

where
B_0 = maximum value of flux density in tesla
ω = constant radian frequency
t = time in seconds
$\mathbf{u}$ = unit vector normal to the loop

a. Find an expression for the induced emf e_{xy}, between terminals x and y.
b. If a resistor R is connected between terminals x and y, determine the magnitude and direction of current in the resistor at $t = 0$ and at $t = \pi/(2\omega)$ s.

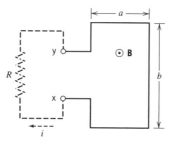

FIGURE 4.5   Rectangular loop.

Solution

a. If flux increases outward, point x will have a higher potential than point y. The magnetic flux is given by

$$\phi = \int \mathbf{B} \cdot \mathbf{dS} = Bab = B_0ab \cos \omega t$$

where ab = area of the loop.

The induced voltage is found as

$$e_{xy} = N\frac{d\phi}{dt}$$

$$= N\frac{d}{dt}(B_0ab \cos \omega t)$$

$$= -B_0ab\omega \sin \omega t$$

$$= B_0ab\omega \cos (\omega t + \pi/2)$$

since the loop contains $N = 1$ turn.

b. The current $i(t)$ that will flow through a resistor R connected across terminals x and y is given by

$$i(t) = e_{xy}(t)/R$$

At $t = 0$:

$$i(0) = \frac{B_0ab\omega}{R} \cos\left(\frac{\pi}{2}\right) = 0$$

At $t = \pi/(2\omega)$:

$$i\left(\frac{\pi}{2\omega}\right) = \frac{B_0ab\omega}{R} \cos\left[\omega\left(\frac{\pi}{2\omega}\right) + \frac{\pi}{2}\right] = -\frac{B_0ab\omega}{R}$$

Faraday's law is used to derive the expression for the induced emf in magnetic circuits. Such magnetic circuits have cores that are made of magnetic materials. Hence, a few remarks are given here on the losses associated with magnetic materials.

There are two sources of losses associated with magnetic materials. The first loss is called eddy current loss. Because of the time variation of flux in the core, eddy current loops are induced in the core. Since the core is made of ferromagnetic materials that contain resistances, current flow results in power loss in the form of heat. To reduce the effects of eddy currents, magnetic structures are usually built of laminations, or thin sheets, that are insulated from each other by a thin coat of insulating varnish.

The second loss is called hysteresis loss. This loss is the energy required to move the magnetic dipoles in the material. This energy is also dissipated as heat. To reduce hysteresis loss, core material is usually made of good-quality electrical steel having a narrow hysteresis characteristic loop. These losses due to hysteresis and eddy currents are collectively known as *core losses* or *iron losses*.

4.4 INDUCTANCE AND MAGNETIC ENERGY

Consider the magnetic circuit shown in Fig. 4.6. If the coil is connected to a voltage source of voltage v, a current i will flow. The current produces magnetic flux ϕ, as shown in the figure. The total *flux linkage* λ of the coil containing N turns is given by

$$\lambda = N\phi \tag{4.11}$$

If leakage flux is neglected, the magnetic flux is equal to the mmf of the coil divided by the reluctance of the flux path:

$$\phi = \frac{Ni}{R} = \frac{Ni}{l/(\mu A)} \tag{4.12}$$

where
- l = mean length of the flux path
- A = cross-sectional area of the core
- μ = permeability of the core

The *self-inductance L* of the coil is defined as follows:

$$L = \frac{\text{total flux linkage of the coil}}{\text{current producing the flux}} = \frac{\lambda}{i} \tag{4.13}$$

Substituting Eq. 4.12 into Eq. 4.11 and dividing by the current i yields the coil inductance:

$$L = \frac{\lambda}{i} = \frac{N^2 \mu A}{l} \tag{4.14}$$

When the voltage applied to the winding shown in Fig. 4.6 is a time-varying voltage, the current that flows and the magnetic flux produced are also time varying. Therefore, by Faraday's law, an emf is induced across the coil. This induced voltage is given by

$$e = \frac{d\lambda}{dt} = \frac{d(Li)}{dt} = L\frac{di}{dt} \tag{4.15}$$

The energy delivered to the inductor over the time interval from t_0 to t_1 is calculated as the integral of the power. If it is assumed that at time $t_0 = 0$, $i_0 = 0$, and at time t_1, $i = i_1$, then the energy is found as follows:

$$W = \int_{t_0}^{t_1} p\,dt = \int_{t_0}^{t_1} ie\,dt = \int_{t_0}^{t_1} i\left(\frac{d\lambda}{dt}\right) dt \tag{4.16}$$

$$W = \int_{\lambda_0}^{\lambda_1} i\,d\lambda = \frac{1}{L}\int_{\lambda_0}^{\lambda_1} \lambda\,d\lambda = \left(\frac{1}{2L}\right)\lambda_1^2 \tag{4.17}$$

$$W = L\int_{i_0}^{i_1} i\,di = \frac{1}{2}Li_1^2 \tag{4.18}$$

where $\lambda_0 = i_0 = 0$ at $t_0 = 0$ and $\lambda_1 = Li_1$ at time t_1.

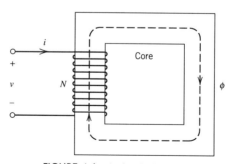

FIGURE 4.6   A simple inductor.

Next, suppose a second coil is added on the right leg of the magnetic circuit as shown in Fig. 4.7. If the first coil is excited ($i_1 \neq 0$) and the second coil is left unenergized ($i_2 = 0$), there is flux linking coil 2. The source of this flux is the current i_1 in coil 1. Then, similar to Eq. 4.13, the *mutual inductance* L_{21} is defined as

$$L_{21} = \frac{\text{total flux linking coil 2}}{\text{current flowing through coil 1}} = \frac{\lambda_{21}}{i_1} \qquad (4.19)$$

Assuming leakage flux is negligible, the total flux linkage in coil 2 may be expressed as follows:

$$\lambda_{21} = N_2\phi_{21} = N_2\phi \qquad (4.20)$$

where

$$\phi = \frac{N_1 i_1}{R} = \frac{N_1 i_1}{l/(\mu A)}$$

Substituting the expression for ϕ into Eq. 4.20 and dividing the result by i_1, the mutual inductance L_{21} is obtained.

$$L_{21} = N_2 N_1 \left(\frac{\mu A}{l}\right) \qquad (4.21)$$

On the other hand, if the second coil is excited ($i_2 \neq 0$) and the first coil is unenergized ($i_1 = 0$), the flux produced by coil 2 will link coil 1. Therefore, the mutual inductance L_{12} is defined as

$$L_{12} = \frac{\text{total flux linking coil 1}}{\text{current flowing through coil 2}} = \frac{\lambda_{12}}{i_2} \qquad (4.22)$$

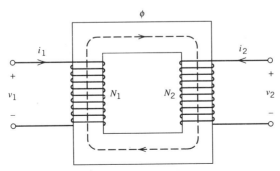

FIGURE 4.7   Mutual inductance.

Again assuming that leakage flux is negligible, the total flux linkage is

$$\lambda_{12} = N_1\phi_{12} = N_1\phi \qquad (4.23)$$

where

$$\phi = \frac{N_2 i_2}{R} = \frac{N_2 i_2}{l/(\mu A)}$$

Substituting the expression for ϕ into Eq. 4.23 and dividing the result by i_2, the mutual inductance L_{12} is obtained.

$$L_{12} = N_1 N_2 \left(\frac{\mu A}{l}\right) \qquad (4.24)$$

It may be observed from Eqs. 4.21 and 4.24 that $L_{12} = L_{21}$.

EXAMPLE 4.4

The magnetic circuit shown in Fig. 4.7 has the following dimensions: $A_c = 12$ cm^2 and $l_c = 50$ cm. The magnetic core has a relative permeability $\mu_r = 20,000$. The first coil has $N_1 = 500$ turns and the second has $N_2 = 1000$ turns.

a. The first coil is supplied with a current $i_1 = 10$ A, while the second is left unenergized (open circuited). Calculate the self-inductance L_{11} of coil 1 and the mutual inductance L_{21} between the two coils.

b. The first coil is de-energized ($i_1 = 0$), while the second is connected to a source from which it draws a current $i_2 = 8$ A. Calculate the self-inductance L_{22} of coil 2 and the mutual inductance L_{12} between the two coils.

Solution

a. The reluctance of the magnetic circuit is

$$R_c = \frac{l_c}{\mu_r \mu_0 A_c}$$

$$= \frac{50 \times 10^{-2}}{(20,000)(4\pi \times 10^{-7})(12 \times 10^{-4})} = 16.58 \times 10^3$$

The magnetic flux ϕ_1 due to the current in coil 1 is found as follows:

$$\phi_1 = \frac{N_1 i_1}{R_c} = \frac{(500)(10)}{16.58 \times 10^3} = 0.30 \text{ Wb}$$

The flux linkages of the two coils are given by

$$\lambda_{11} = N_1\phi_1 = (500)(0.30) = 150 \text{ Wb-t}$$
$$\lambda_{21} = N_2\phi_1 = (1000)(0.30) = 300 \text{ Wb-t}$$

Therefore, the self-inductance of coil 1 is

$$L_{11} = \lambda_{11}/i_1 = 150/10 = 15 \text{ H}$$

The mutual inductance is given by

$$L_{21} = \lambda_{21}/i_1 = 300/10 = 30 \text{ H}$$

b. The magnetic flux ϕ_2 due to the current in coil 2 is found as follows:

$$\phi_2 = \frac{N_2 i_2}{R_c} = \frac{(1000)(8)}{(16.58 \times 10^3)} = 0.48 \text{ Wb}$$

The flux linkages of the two coils are given by

$$\lambda_{12} = N_1\phi_2 = (500)(0.48) = 240 \text{ Wb-t}$$
$$\lambda_{22} = N_2\phi_2 = (1000)(0.48) = 480 \text{ Wb-t}$$

Therefore, the self-inductance of coil 2 is

$$L_{22} = \lambda_{22}/i_2 = 480/8 = 60 \text{ H}$$

The mutual inductance is given by

$$L_{12} = \lambda_{12}/i_2 = 240/8 = 30 \text{ H}$$

DRILL PROBLEMS

D4.6 Find the inductance of the coil in the magnetic circuit of Problem D4.2.

D4.7 Calculate the inductance of the coil of the magnetic circuit of Problem D4.4.

D4.8 In the magnetic circuit of Example 4.4, an air gap of length 0.5 mm is cut in the lower leg of the core. Find the self-inductances and mutual inductance of the two coils.

4.5 TRANSFORMERS

The transformer is an indispensable component of power systems. It is one of the main reasons for the widespread use of AC power systems. It makes possible electric power generation at the most economical voltage, transmission and distribution at the most economical voltage levels, and power utilization at the most suitable voltage. The transformer is also used in measurements of high voltages (potential transformers) and large currents (current transformers). Other uses of transformers include impedance matching, insulating one circuit from another, or insulating DC circuits from AC circuits.

A *single-phase transformer* basically consists of two or more windings coupled by a magnetic core. When one of the windings (primary) is connected to an AC voltage source, a time-varying flux is produced in the core. This flux is confined within the magnetic core, and it links the second winding. Therefore, a voltage is induced in the second winding (secondary). When an electrical load such as a resistor is connected to the secondary winding, a secondary current flows.

A single-phase transformer is illustrated in Fig. 4.8. The primary winding has N_1 turns and the secondary has N_2 turns. The voltages and currents are expressed in phasor form. In the following analysis, the capacitances of the transformer windings are not included; they do become important at higher frequencies.

4.5.1 Ideal Transformer

An *ideal transformer* is characterized by the following:

1. There is zero leakage flux. This implies that the fluxes produced by the primary and secondary currents are confined within the core. This assumption was also made in magnetic circuit analysis (see Section 4.2).

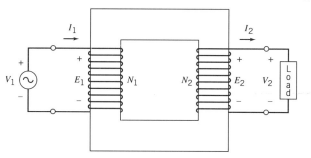

FIGURE 4.8 A transformer circuit.

2. The windings have no resistances. Therefore, the applied voltage v_1 equals the induced primary voltage e_1; that is, $v_1 = e_1$. Similarly, $v_2 = e_2$.

3. The core has infinite permeability. This implies that the reluctance of the core is zero. Hence, negligible current is required to set up the magnetic flux.

4. The magnetic core is lossless. Hysteresis and eddy current losses are, therefore, negligible.

Let the mutual flux linking both windings be sinusoidal, that is,

$$\phi_m = \Phi_p \sin \omega t \qquad (4.25)$$

Then, according to Faraday's law of electromagnetic induction, the induced emfs may be expressed as

$$e_1 = \frac{d\lambda_1}{dt} = N_1 \frac{d\phi_m}{dt} = \omega \Phi_p N_1 \cos \omega t \qquad (4.26)$$

$$e_2 = \frac{d\lambda_2}{dt} = N_2 \frac{d\phi_m}{dt} = \omega \Phi_p N_2 \cos \omega t \qquad (4.27)$$

The rms values of the induced voltages are

$$E_1 = \frac{1}{\sqrt{2}} \omega \Phi_p N_1 = 4.44 f \Phi_p N_1 \qquad (4.28)$$

$$E_2 = \frac{1}{\sqrt{2}} \omega \Phi_p N_2 = 4.44 f \Phi_p N_2 \qquad (4.29)$$

where $f = \omega/(2\pi)$ cycles per second or hertz.

The polarities of the induced voltages are given by Lenz's law; that is, the emfs produce currents that tend to oppose the flux change. The ratio of the induced voltages is

$$\frac{E_1}{E_2} = \frac{N_1}{N_2} = a \qquad (4.30)$$

where a is called the *turns ratio*. Since the transformer is ideal, the induced voltages are equal to their corresponding terminal voltages; that is, $E_1 = V_1$ and $E_2 = V_2$. Hence,

$$\frac{V_1}{V_2} = \frac{E_1}{E_2} = a \qquad (4.31)$$

The assumption that the magnetic circuit of the ideal transformer is lossless implies that the mmfs produced by the windings balance or cancel each other; that is, primary mmf equals secondary mmf. In terms of the winding currents, this may be expressed as

$$N_1 I_1 = N_2 I_2 \tag{4.32}$$

Equation 4.32 shows that the winding currents are in phase and that their magnitudes are related by

$$\frac{I_1}{I_2} = \frac{N_2}{N_1} = \frac{1}{a} \tag{4.33}$$

The primary voltage and current may be expressed in terms of their secondary counterparts as follows:

$$V_1 = a V_2 \tag{4.34}$$

$$I_1 = \left(\frac{1}{a}\right) I_2 \tag{4.35}$$

Multiplying Eqs. 4.34 and 4.35 together yields

$$V_1 I_1 = V_2 I_2 \tag{4.36}$$

Equation 4.36 states the *power invariance law* across the ideal transformer. In other words, the power input to the transformer is equal to its power output.

Dividing Eq. 4.34 by Eq. 4.35 gives the secondary impedance referred to the primary side.

$$\frac{V_1}{I_1} = \frac{a^2 V_2}{I_2} \tag{4.37}$$

$$Z_1 = a^2 Z_2 \tag{4.38}$$

The equivalent circuit of an ideal transformer is illustrated in Fig. 4.9, with all quantities referred to the same side.

EXAMPLE 4.5

A 60-Hz ideal transformer is rated 220/110 V. An inductive load $Z_2 = 10 + j10\ \Omega$ is connected across the low-voltage side at rated secondary voltage. Calculate the following:

a. Primary and secondary currents

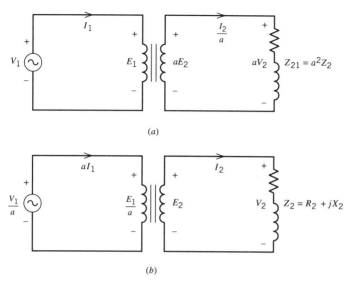

(a)

(b)

FIGURE 4.9 Equivalent circuit of an ideal transformer: (a) all quantities referred to primary side; (b) all quantities referred to secondary side.

b. Load impedance referred to the primary

c. Power supplied by the source

Solution

a. The turns ratio is

$$a = V_1/V_2 = 220/110 = 2$$

The primary and secondary currents are found as follows:

$$\mathbf{I}_2 = \frac{\mathbf{V}_2}{Z_2} = \frac{110\angle 0°}{(10 + j10)} = 7.78\angle{-45°}\ \text{A}$$

$$\mathbf{I}_1 = \left(\frac{1}{a}\right)\mathbf{I}_2 = (1/2)(7.78\angle{-45°}) = 3.89\angle{-45°}\ \text{A}$$

b. The load impedance referred to the primary side is

$$Z_1 = a^2Z_2 = (2)^2(10 + j10) = 40 + j40\ \Omega$$

c. The power supplied by the source is computed as follows:

$$P_1 = P_2 = V_2I_2\cos\theta = (110)(7.78)\cos 45° = 605\ \text{W}$$

$$P_1 = P_2 = I_2^2R_2 = (7.87)^2(10) = 605\ \text{W}$$

4.5.2 Nonideal or Actual Transformer

In the previous section, certain assumptions were made in characterizing an ideal transformer. These assumptions are no longer applicable when analyzing the performance of an actual transformer.

A *nonideal transformer* is illustrated in Fig. 4.10. It can be described as having resistances in its windings. Not all of the flux produced by one winding will link the other winding because of flux leakage. The core of the actual transformer is not perfectly permeable; it has a finite permeability. Thus, it requires a finite mmf for its magnetization. Because the flux in the core is alternating, there are hysteresis and eddy current losses, collectively called *core losses* or *iron losses*.

In deriving the equivalent circuit for the two-winding transformer of Fig. 4.10, the characteristics of an actual transformer described earlier need to be modeled.

Consider the primary circuit. A voltage equation around the loop may be written as

$$v_1 = R_1 i_1 + \frac{d\lambda_1}{dt} = R_1 i_1 + N_1 \frac{d\phi_1}{dt} \qquad (4.39)$$

where

R_1 = resistance of primary winding

N_1 = number of turns for primary winding

The primary winding flux ϕ_1 may be expressed as the sum of the mutual flux ϕ_m and the primary leakage ϕ_{l1}:

$$\phi_1 = \phi_m + \phi_{l1} \qquad (4.40)$$

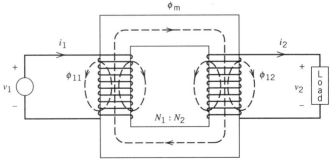

FIGURE 4.10   An actual transformer.

Thus, Eq. 4.39 reduces to

$$v_1 = R_1 i_1 + N_1 \frac{d\phi_{l1}}{dt} + N_1 \frac{d\phi_m}{dt} \qquad (4.41)$$

Since the leakage flux ϕ_{l1} is a linear function of the primary current i_1, the second term on the right-hand side of Eq. 4.41 may be expressed in terms of the inductance of the primary winding. Thus,

$$v_1 = R_1 i_1 + L_1 \frac{d i_1}{dt} + N_1 \frac{d\phi_m}{dt} \qquad (4.42)$$

The secondary circuit is considered next. From Fig. 4.10, the voltage equation in the secondary may be written as follows:

$$v_2 = -R_2 i_2 + \frac{d\lambda_2}{dt} = -R_2 i_2 + N_2 \frac{d\phi_2}{dt} \qquad (4.43)$$

From the flux directions, the secondary flux may be represented by the difference between the mutual flux and the secondary leakage flux:

$$\phi_2 = \phi_m - \phi_{l2} \qquad (4.44)$$

Substituting Eq. 4.44 into 4.43 yields

$$v_2 = -R_2 i_2 - N_2 \frac{d\phi_{l2}}{dt} + N_2 \frac{d\phi_m}{dt} \qquad (4.45)$$

Similarly, the leakage flux ϕ_{l2} is a linear function of the secondary current i_2. Thus, Eq. 4.45 may be written using the inductance of the secondary winding as

$$v_2 = -R_2 i_2 - L_2 \frac{d i_2}{dt} + N_2 \frac{d\phi_m}{dt} \qquad (4.46)$$

In Eqs. 4.42 and 4.46, the last terms represent the induced voltages across the primary and secondary windings, respectively; that is,

$$e_1 = N_1 \frac{d\phi_m}{dt} \qquad (4.47)$$

$$e_2 = N_2 \frac{d\phi_m}{dt} \qquad (4.48)$$

Dividing Eq. 4.47 by Eq. 4.48 yields the voltage ratio:

$$\frac{e_1}{e_2} = \frac{N_1}{N_2} = a \qquad (4.49)$$

Figure 4.11 shows the equivalent circuit of the two-winding transformer of Fig. 4.10. The circuit elements that are used to model the core magnetization and the core losses can be added to either the primary side or the secondary side. In Fig. 4.11, the inductor L_{m1} representing the core magnetization and the resistor R_{c1} representing the core losses (hysteresis and eddy current losses) have been connected in parallel and located in the primary side of the transformer equivalent circuit.

The core-related circuit elements R_{c1} and L_{m1} are usually determined at rated voltage and are referred to the primary side in Fig. 4.11. They are assumed to remain essentially constant when the transformer operates at, or near, rated conditions.

In phasor form, the transformer equivalent circuit takes the form shown in Fig. 4.12. The reactances are derived by multiplying the inductances by the radian frequency $\omega = 2\pi f$, where f is the frequency. The turns ratio $a = N_1/N_2$ is approximately equal to the voltage ratio V_1/V_2, the ratio of the rated primary voltage to the rated secondary voltage provided by the manufacturer.

A phasor diagram for a lagging power factor (inductive) load connected across the secondary of the transformer of Fig. 4.12 is shown in Fig. 4.13. The notation used is as follows:

$\mathbf{E}_1$ = primary induced voltage

$\mathbf{E}_2$ = secondary induced voltage

$\mathbf{V}_1$ = primary terminal voltage

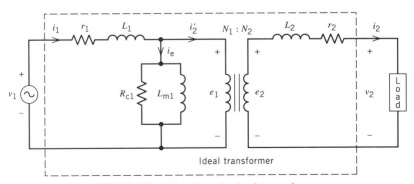

FIGURE 4.11   Equivalent circuit of a transformer.

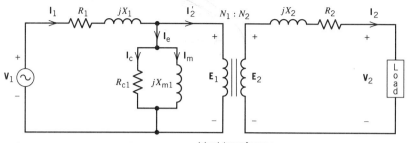

FIGURE 4.12 Transformer equivalent circuit in phasor form.

V_2 = secondary terminal voltage

I_1 = primary current

I_2 = secondary current

I_e = excitation current

I_m, X_m = magnetizing current and reactance

I_c, R_c = current and resistance representing core loss

R_1 = resistance of the primary winding

R_2 = resistance of the secondary winding

X_1 = primary leakage reactance

X_2 = secondary leakage reactance

In the transformer equivalent circuit of Fig. 4.12, the ideal transformer can be moved out to the right or to the left of the equivalent circuit by referring all quantities to the primary or secondary, respectively, as shown in Fig. 4.14. This is almost always done because of the great simplicity it introduces in transformer performance analysis.

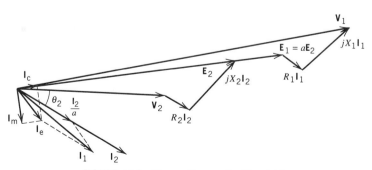

FIGURE 4.13 Phasor diagram for Fig. 4.12.

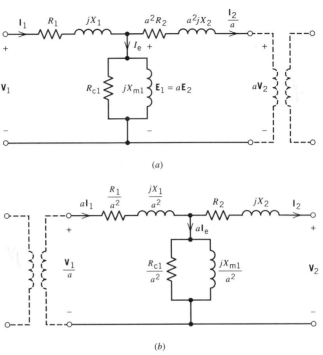

FIGURE 4.14 Referred transformer equivalent circuit: (*a*) referred to primary; (*b*) referred to secondary.

EXAMPLE 4.6

A 25-kVA, 440/220-V, 60-Hz transformer has the following parameters:

$$R_1 = 0.16 \ \Omega \qquad R_2 = 0.04 \ \Omega \qquad R_{c1} = 270 \ \Omega$$
$$X_1 = 0.32 \ \Omega \qquad X_2 = 0.08 \ \Omega \qquad X_{mI} = 100 \ \Omega$$

The transformer delivers 20 kW at 0.8 power factor lagging to a load on the low-voltage side with 220 V across the load. Find the primary terminal voltage.

Solution The voltage across the load is taken as reference phasor; thus,

$$\mathbf{V}_2 = 220 \underline{/0°} \ \text{V}$$

For a load $P_2 = 20{,}000$ W at 0.8 power factor lagging, the secondary current is computed as follows:

$$\mathbf{I}_2 = \frac{20{,}000}{(220)(0.8)} \underline{/-\cos^{-1} 0.8} = 113.64 \underline{/-36.9°} \ \text{A}$$

The transformer turns ratio is $a = 440/220 = 2$. Thus, the secondary voltage and current and the winding resistance and reactance are referred to the primary side as follows:

$$aV_2 = 2(220 \underline{/0^\circ}) = 440 \underline{/0^\circ} \text{ V}$$

$$I_2/a = (113.64 \underline{/-36.9^\circ})/2 = 56.82 \underline{/-36.9^\circ} \text{ A}$$

$$a^2R_2 = (2)^2(0.04) = 0.16 \ \Omega$$

$$a^2X_2 = (2)^2(0.08) = 0.32 \ \Omega$$

Referring to the phasor diagram of Fig. 4.13, the primary induced voltage is calculated as follows:

$$\begin{aligned} \mathbf{E}_1 &= a\mathbf{V}_2 + (\mathbf{I}_2/a)(a^2R_2 + ja^2X_2) \\ &= 440 \underline{/0^\circ} + (56.82 \underline{/-36.9^\circ})(0.16 + j0.32) \\ &= 458.2 + j9.07 = 458.3 \underline{/1^\circ} \text{ V} \end{aligned}$$

The shunt branch currents are

$$\begin{aligned} \mathbf{I}_c &= \mathbf{E}_1/R_{c1} = (458.2 + j9.07)/270 = 1.7 + j0.03 \text{ A} \\ \mathbf{I}_m &= \mathbf{E}_1/jX_{m1} = (458.2 + j9.07)/j100 = 0.09 - j4.58 \text{ A} \\ \mathbf{I}_e &= \mathbf{I}_c + \mathbf{I}_m = 1.79 - j4.55 \text{ A} \end{aligned}$$

Thus, the primary current is

$$\begin{aligned} \mathbf{I}_1 &= \mathbf{I}_e + \mathbf{I}_2/a \\ &= (1.79 - j4.55) + (56.82 \underline{/-36.9^\circ}) = 61.04 \underline{/-39.3^\circ} \text{ A} \end{aligned}$$

Therefore, the primary voltage is found from

$$\begin{aligned} \mathbf{V}_1 &= \mathbf{E}_1 + \mathbf{I}_1(R_1 + jX_1) \\ &= (458.2 + j9.07) + (61.04 \underline{/-39.3^\circ})(0.16 + j0.32) \\ &= 478.1 + j18 = 478.4 \underline{/2.2^\circ} \text{ V} \end{aligned}$$

4.5.3 Approximate Equivalent Circuits

The derivation of approximate equivalent circuits begins with the diagrams shown in Fig. 4.14. All quantities have been referred to the same side of the transformer, and the ideal transformer may be omitted from the equivalent circuit.

The first step in the simplification process is to move the shunt magnetization branch from the middle of the T circuit to either the primary or secondary terminal, as shown in Fig. 4.15a and b. This step neglects the voltage drop across the primary or secondary winding caused by the exciting current. The voltage drop caused by the load component of the current is still included, of course. The error introduced in this step is generally very small in most problems involving power transformers.

The primary and secondary winding resistances are combined to give either the equivalent resistance referred to the primary side $R_{e1} = R_1 + a^2R_2$ or the equivalent resistance referred to the secondary side $R_{e2} = R_1/a^2 + R_2$. Similarly, the primary and secondary winding reactances are combined to obtain either the equivalent reactance referred to the primary side $X_{e1} = X_1 + a^2X_2$ or the equivalent reactance referred to the secondary side $X_{e2} = X_1/a^2 + X_2$.

The next step in deriving the approximate equivalent circuit is the deletion of the shunt magnetizing branch completely. Thus, the transformer equivalent circuit reduces to a simple equivalent series impedance referred to either primary or secondary, as shown in Fig. 4.15c and d.

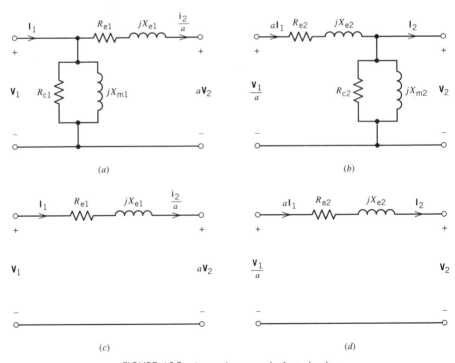

FIGURE 4.15 Approximate equivalent circuits.

4.5.4 Voltage Regulation

Distribution and power transformers are often used to supply loads that are designed to operate at essentially constant voltage. The amount of secondary (load) current drawn depends on the load connected to the transformer secondary terminals. As this current changes with changing load, with the same applied primary voltage, the load voltage likewise changes. This change is due to the voltage drop across the internal impedance of the transformer. A measure of how much the voltage will change as load is varied is called *voltage regulation*.

The voltage regulation of a transformer is defined as the change in the magnitude of the secondary voltage as the current changes from full load to no load with the primary voltage held fixed. This is expressed mathematically as

$$\text{Voltage regulation} = \frac{|\mathbf{V}_{2,\text{nl}}| - |\mathbf{V}_{2,\text{fl}}|}{|\mathbf{V}_{2,\text{fl}}|} 100\% \qquad (4.50a)$$

$$= \frac{|\mathbf{V}_1| - |a\mathbf{V}_2|}{|a\mathbf{V}_2|} 100\% \qquad (4.50b)$$

$$= \frac{|\mathbf{V}_1/a| - |\mathbf{V}_2|}{|\mathbf{V}_2|} 100\% \qquad (4.50c)$$

The use of Eq. 4.50 in calculating the voltage regulation implies that the primary voltage V_1 is adjusted to a value that provides rated secondary voltage V_2 across the load when rated secondary current is supplied to the load. Then as current is reduced to zero or no load, the increase in secondary voltage is determined.

4.5.5 Efficiency

The percent *efficiency* of the transformer is defined as the ratio of the power output P_{output} to the power input P_{input}, both expressed in watts, multiplied by 100%. Thus,

$$\eta = \frac{P_{\text{output}}}{P_{\text{input}}} 100\% \qquad (4.51)$$

This can also be expressed as

$$\eta = \frac{P_{\text{output}}}{P_{\text{output}} + \Sigma(\text{losses})} 100\% \qquad (4.52)$$

or

$$\eta = \frac{P_{\text{input}} - \Sigma(\text{losses})}{P_{\text{input}}} \ 100\% \tag{4.53}$$

where

$$\Sigma(\text{losses}) = \text{core loss} \ + \ \text{copper loss} \tag{4.54a}$$
$$\Sigma(\text{losses}) = P_{\text{core}} + (I_1^2 R_1 + I_2^2 R_2) \tag{4.54b}$$
$$\Sigma(\text{losses}) = P_{\text{core}} + I_1^2 R_{\text{e1}} \tag{4.54c}$$
$$\Sigma(\text{losses}) = P_{\text{core}} + I_2^2 R_{\text{e2}} \tag{4.54d}$$

EXAMPLE 4.7

A 150-kVA, 2400/240-V transformer has the following parameters referred to the primary side: $R_{\text{e1}} = 0.5 \ \Omega$ and $X_{\text{e1}} = 1.5 \ \Omega$. The shunt magnetizing impedance is very large and can be neglected. At full load, the transformer delivers rated kVA at 0.85 lagging power factor and the secondary voltage is 240 V. Calculate (a) the voltage regulation and (b) the efficiency assuming core losses amount to 600 W.

Solution

a. The transformer turns ratio is

$$a = 2400/240 = 10$$

Take the secondary voltage $\mathbf{V}_2$ as the reference phasor.

$$\mathbf{V}_2 = 240 \underline{/0°} \ \text{V}$$
$$a\mathbf{V}_2 = 2400 \underline{/0°} \ \text{V}$$

At rated load and 0.85 PF lagging:

$$\mathbf{I}_2 = (150{,}000/240) \underline{/-\cos^{-1} 0.85°} = 625 \underline{/-31.8°} \ \text{A}$$
$$\mathbf{I}_1 = \mathbf{I}_2/a = (625/10) \underline{/-31.8°} \ \text{A} = 62.5 \underline{/-31.8°} \ \text{A}$$

The primary voltage is calculated as follows:

$$\mathbf{V}_1 = a\mathbf{V}_2 + (\mathbf{I}_2/a)(R_{\text{e1}} + jX_{\text{e1}})$$
$$= 2400 \underline{/0°} + (62.5 \underline{/-31.8°})(0.5 + j1.5) = 2476.8 \underline{/1.5°} \ \text{V}$$

Thus, the percent voltage regulation is found by using Eq. 4.50b as follows:

$$\text{Voltage regulation} = \frac{V_1 - aV_2}{aV_2}\, 100\%$$

$$= \frac{2476.8 - 2400}{2400}\, 100\% = 3.2\%$$

b. At rated output,

$$P_{\text{output}} = (150{,}000)(0.85) = 127{,}500 \text{ W}$$
$$P_{\text{cu}} = I_1^2 R_{e1} = (62.5)^2(0.5) = 1950 \text{ W}$$
$$P_{\text{core}} = 600 \text{ W}$$
$$P_{\text{input}} = P_{\text{output}} + \Sigma(\text{losses}) = 130{,}050 \text{ W}$$

Therefore, the efficiency is found by using Eq. 4.51 as follows:

$$\eta = \frac{\text{power output}}{\text{power input}}\, 100\%$$

$$= \frac{127{,}500}{130{,}050}\, 100\% = 98\%$$

4.5.6   Determination of Equivalent Circuit Parameters

Two simple tests are used to determine the values for the parameters of the transformer equivalent circuit of Fig. 4.15a and b. If it is desired to find the parameters of the exact equivalent circuit of Fig. 4.12, it is customary to assume $R_1 = a^2 R_2$ and $X_1 = a^2 X_2$. This assumption allows the decomposition of the equivalent resistance and reactance into the primary and secondary components.

The two tests are the short-circuit and open-circuit tests. Let the primary (winding 1) be the high-voltage side and the secondary (winding 2) the low-voltage side for the transformer of Fig. 4.10.

Open-Circuit Test   In the *open-circuit test*, the transformer rated voltage is applied to the low-voltage side of the transformer with the high-voltage side left open. Measurements of power, current, and voltage are made on the low-voltage side as shown in Fig. 4.16.

Since the high-voltage side is open, the input current I_{oc} is equal to the exciting current through the shunt excitation branch as shown in the equivalent circuit of Fig. 4.17. Because this current is very small, about 5% of rated value, the voltage drop across the low-voltage winding and the winding copper losses are neglected.

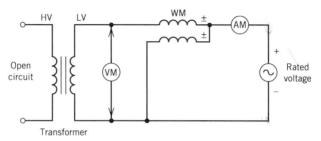

FIGURE 4.16 Connections for open-circuit test.

The magnitude of the admittance of the shunt excitation branch of the equivalent circuit referred to the low-voltage side is calculated as follows:

$$|Y_{o2}| = \frac{I_{oc}}{V_{oc}} \tag{4.55}$$

The phase angle of the admittance is found as

$$-\theta_{o2} = -\cos^{-1}\left(\frac{P_{oc}}{V_{oc}I_{oc}}\right) \tag{4.56}$$

Thus, the complex admittance may be expressed as

$$Y_{o2} = |Y_{o2}|\underline{/-\theta_{o2}} = G_{c2} - jB_{m2} \tag{4.57}$$

The corresponding resistance and reactance parameters of Fig. 4.15b are derived from the conductance and susceptance, respectively:

$$R_{c2} = \frac{1}{G_{c2}} \tag{4.58}$$

$$jX_{m2} = \frac{1}{-jB_{m2}} \tag{4.59}$$

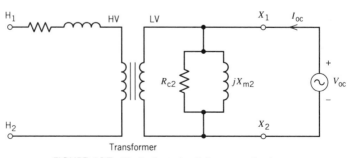

FIGURE 4.17 Equivalent circuit for open-circuit test.

These parameters may be referred to the high-voltage side to give the parameters of the equivalent circuit of Fig. 4.15a.

$$R_{c1} = a^2 R_{c2} \qquad (4.60)$$

$$X_{m1} = a^2 X_{m2} \qquad (4.61)$$

Note that when rated voltage at rated frequency is applied during the open-circuit test, the power input P_{oc} is practically equal to the rated core loss. In most applications, this value of core loss is typically assumed to remain constant for different load levels.

Short-Circuit Test In the *short-circuit test*, the low-voltage side is short-circuited and the high-voltage side is connected to a variable, low-voltage source. Measurements of power, current, and voltage are made on the high-voltage side as shown in Fig. 4.18.

The applied voltage is adjusted until rated short-circuit current flows in the windings. This voltage is generally much smaller than the rated voltage, in the range of 0.05 to 0.10 per unit. Therefore, the current through the magnetizing branch is negligible, and the applied voltage may be assumed to occur wholly as a voltage drop across the transformer series impedance as shown in the equivalent circuit of Fig. 4.19.

The magnitude of the series impedance referred to the high-voltage (primary) side may be calculated as follows:

$$|Z_{e1}| = \frac{V_{sc}}{I_{sc}} \qquad (4.62)$$

The equivalent series resistance referred to the high-voltage side is

$$R_{e1} = \frac{P_{sc}}{I_{sc}^2} = R_1 + a^2 R_2 \qquad (4.63)$$

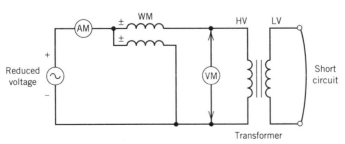

FIGURE 4.18 Connections for short-circuit test.

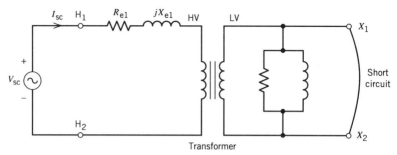

FIGURE 4.19 Equivalent circuit for short-circuit test.

Correspondingly, the equivalent series reactance referred to the high-voltage side is

$$X_{e1} = \sqrt{|Z_{e1}|^2 - R_{e1}^2} = X_1 + a^2 X_2 \qquad (4.64)$$

The values of these parameters derived from the short-circuit test are used in conjunction with Fig. 4.15a. These parameters may be referred to the low-voltage (secondary) side to derive the corresponding values for use with Fig. 4.15b as follows:

$$R_{e2} = \left(\frac{1}{a^2}\right) R_{e1} \qquad (4.65)$$

$$X_{e2} = \left(\frac{1}{a^2}\right) X_{e1} \qquad (4.66)$$

Note that when rated current flows through the windings during the short-circuit test, the power input P_{sc} is equal to the rated copper loss.

Tests on Three-Phase Transformers   When a three-phase transformer undergoes open-circuit and short-circuit tests, it must be remembered that the power being measured is total three-phase power, the measured voltage is line-to-line voltage, and the measured current is line current. The impedance parameters of interest are to be calculated on a per-phase basis. Therefore, before using the formulas derived above to calculate the resistances and reactances, the measured values must also be converted to per-phase values. Three-phase transformers are discussed later in Section 4.7.

EXAMPLE 4.8

A 50-kVA, 2400/240-V, 60-Hz single-phase transformer has a short-circuit test performed on its high-voltage side. An open-circuit test is performed on the low-voltage side. The following test results were obtained:

	Voltage (V)	Current (A)	Power (W)
Short-circuit (LV shorted)	48	20.8	620
Open-circuit (HV open)	240	5.4	186

a. Draw the transformer's equivalent circuit.

b. Determine its voltage regulation, and efficiency at rated load, 0.8 power factor lagging, and rated voltage at the secondary terminals.

Solution   The ratings of the transformer, with the high-voltage side as primary, are as follows:

Rated primary voltage $V_1 = 2400$ V
Rated secondary voltage $V_2 = 240$ V
Rated primary current $I_1 = 50,000/2400 = 20.83$ A
Rated secondary current $I_2 = 50,000/240 = 208.3$ A
Turns ratio $a = 2400/240 = 10$

a. The equivalent circuit for the short-circuit test is shown in Fig. 4.19. The series impedance parameters are calculated as follows:

$$Z_{e1} = V_{sc}/I_{sc} = 48/20.8 = 2.30 \ \Omega$$
$$R_{e1} = P_{sc}/I_{sc}^2 = 620/(20.8)^2 = 1.43 \ \Omega$$
$$X_{e1} = \sqrt{Z_{e1}^2 - R_{e1}^2} = \sqrt{(2.30)^2 - (1.43)^2} = 1.80 \ \Omega$$

The equivalent circuit for the open-circuit test is shown in Fig. 4.17. The parameters of the shunt magnetizing branch are computed as follows:

$$Y_{o2} = \frac{I_{oc}}{V_{oc}} = \frac{5.4}{240} = 0.0225 \ \text{S}$$
$$-\theta_2 = \cos^{-1}\left(\frac{P_{oc}}{V_{oc}I_{oc}}\right) = -\cos^{-1}\left[\frac{186}{(240)(5.4)}\right] = -81.8°$$
$$Y_{o2} = 0.0225 \ \underline{/-81.8°} = (3.23 - j22.3) \times 10^{-3} \ \text{S} = G_{c2} - jB_{m2}$$
$$R_{c2} = \frac{1}{G_{c2}} = \frac{1}{3.23 \times 10^{-3}} = 309.6 \ \Omega$$
$$X_{m2} = \frac{1}{B_{m2}} = \frac{1}{22.3 \times 10^{-3}} = 44.8 \ \Omega$$

The shunt magnetizing branch may be referred to the high-voltage side as

$$R_{c1} = a^2 R_{c2} = (10)^2(309.6) = 30.96 \ \text{k}\Omega$$
$$X_{m1} = a^2 X_{m2} = (10)^2(44.8) = 4.48 \ \text{k}\Omega$$

Finally, the transformer equivalent circuit is drawn as shown in Fig. 4.20.

b. At rated secondary conditions, and 0.8 power factor lagging:

$$aV_2 = (10)(240\,\underline{/0^\circ}) = 2400\,\underline{/0^\circ}\ \text{V}$$

$$\frac{I_2}{a} = \frac{208.3\,\underline{/-36.9^\circ}}{10} = 20.83\,\underline{/-36.9^\circ}\ \text{A}$$

The primary voltage is computed as follows:

$$V_1 = aV_2 + (I_2/a)(R_{e1} + jX_{e1})$$
$$= 2400\,\underline{/0^\circ} + (20.83\,\underline{/-36.9^\circ})(1.43 + j1.80)$$
$$= 2446.3 + j12.1 = 2446.6\,\underline{/0.3^\circ}\ \text{V}$$

The percent voltage regulation is found by using Eq. 4.50b as follows:

$$\text{Voltage regulation} = \frac{V_1 - aV_2}{aV_2}\,100\%$$
$$= \frac{2446.6 - 2400}{2400}\,100\% = 1.9\%$$

At rated load and 0.8 power factor, the power input is computed as the sum of the power output and losses as follows:

$$P_{\text{output}} = (50{,}000)(0.08) = 40{,}000\ \text{W}$$
$$\Sigma(\text{losses}) = \text{core loss} + \text{copper loss} = P_{\text{oc}} + I_1^2 R_{e1}$$
$$= 186 + (20.83)^2(1.43) = 806\ \text{W}$$
$$P_{\text{input}} = P_{\text{output}} + \Sigma(\text{losses}) = 40{,}000 + 806 = 40{,}806\ \text{W}$$

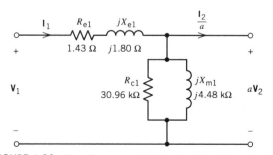

FIGURE 4.20   Transformer equivalent circuit for Example 4.7.

Therefore, the efficiency of the transformer is found by using Eq. 4.51 as follows:

$$\eta = \frac{\text{power output}}{\text{power input}} 100\%$$

$$= \frac{40,000}{40,806} 100\% = 98\%$$

4.5.7 Polarity

Transformer windings are marked to identify terminals with the same polarity. Polarity marks may either be dots or + marks. Alternatively, polarity marks may be shown by assigning the same subscripts to corresponding primary and secondary labels, H and X, respectively. Thus, when the primary terminals are named H_1 and H_2, the corresponding secondary terminals are identified as X_1 and X_2, respectively.

Primary and secondary terminals have the same polarity when, at a given instant of time, current enters the primary terminal and leaves the secondary terminal as shown in Fig. 4.21.

The polarity marks also indicate that at the instant the primary dotted terminal H_1 is positive with respect to the undotted end H_2, so is the secondary dotted terminal X_1 positive with respect to its undotted end X_2.

The terms additive polarity and subtractive polarity merely have reference to the relative position of the locations of the H and X terminals. Figure 4.22 illustrates both conditions.

If the top terminals of the transformer of Fig. 4.22 are connected together and one winding is excited by a sinusoidal voltage source, the voltage measured across the bottom terminals will be either

a. The difference between the induced voltage across the H and X windings as in Fig. 4.22*a* or

b. The sum of the induced voltages as in Fig. 4.22*b*.

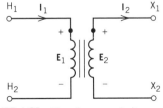

FIGURE 4.21   Transformer polarity markings.

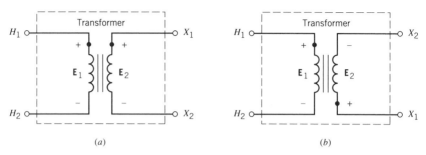

(a)   (b)

FIGURE 4.22 Transformer polarity: (a) subtractive polarity; (b) additive polarity.

DRILL PROBLEMS

D4.9 A 240/120-V 60-Hz ideal transformer is rated at 5 kVA.

  a. Calculate the turns ratio.

  b. Calculate the rated primary and secondary currents.

  c. Calculate the primary and secondary currents when the transformer delivers 3.2 kW at rated secondary voltage and at 0.8 lagging power factor.

D4.10 A 50-kVA, 2400/240-V, 60-Hz distribution transformer has the following resistances and leakage reactances in ohms referred to its own side.

$$R_1 = 3.5 \qquad X_1 = 5.7$$
$$R_2 = 0.035 \qquad X_2 = 0.057$$

The subscript 1 denotes the primary or high-voltage winding and subscript 2 denotes the secondary or low-voltage winding.

  a. Find the equivalent impedance Z_{e1} referred to the primary.

  b. Find the equivalent impedance Z_{e2} referred to the secondary.

  c. The transformer secondary is connected to an electrical load, and it delivers its rated current at rated voltage and 0.85 power factor lagging. Find the primary voltage.

D4.11 A single-phase, 10-kVA, 2200/220-V transformer has the following parameters:

$$R_1 = 4.0 \ \Omega \qquad R_2 = 0.04 \ \Omega$$
$$X_1 = 5.0 \ \Omega \qquad X_2 = 0.05 \ \Omega$$
$$R_{c1} = 35 \ k\Omega \qquad X_{m1} = 4.0 \ k\Omega$$

The transformer is supplying its rated current to a load at 220 V and 0.8 PF lagging.

a. Draw the equivalent circuit of this transformer, showing values of the elements referred to the primary side.

b. Determine the input voltage of the transformer.

c. Calculate the transformer voltage regulation.

d. Find the efficiency of the transformer.

D4.12 Solve Problem D4.11 by using the approximate equivalent circuit of a transformer given in Fig. 4.15c or Fig. 4.15d.

D4.13 A single-phase, 25-kVA, 2300/230-V, 60-Hz distribution transformer has the following characteristics:

$$\text{Core loss at full voltage} = 250 \text{ W}$$
$$\text{Copper loss at half load} = 300 \text{ W}$$

a. Determine the efficiency of the transformer when it delivers rated load at 0.866 power factor lagging.

b. The transformer has the following (24-h) load cycle:

1/4 full load for 4 h at 0.8 PF
1/2 full load for 10 h at 0.8 PF
3/4 full load for 6 h at 0.8 PF
Full load for 4 h at 0.9 PF

Find the all-day (or energy) efficiency, that is, the ratio of the energy output to the energy input over a 24-h period.

D4.14 A 5-kVA, 440/220-V, single-phase transformer was subjected to the short-circuit and open-circuit tests, and the following test results were obtained:

	Voltage (V)	Current (A)	Power (W)
Short-circuit test	28.5	11.4	65
Open-circuit test	220	1.25	50

a. Find the circuit parameters of the transformer.

b. Draw the equivalent circuit of the transformer.

4.6   AUTOTRANSFORMER

The *autotransformer* serves a function similar to that of the ordinary transformer: to raise or lower voltage. It consists of a single continuous winding with a tap brought out at some intermediate point as shown in Fig. 4.23. Because the primary and secondary windings of the autotransformer are

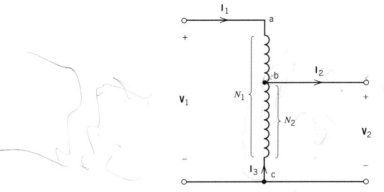

FIGURE 4.23 Autotransformer.

physically connected, the supply and output voltages are not insulated from each other.

When a voltage V_1 is applied to the primary of the autotransformer, the induced voltages are related by

$$\frac{E_1}{E_2} = \frac{E_{ac}}{E_{bc}} = \frac{N_1}{N_2} = a \qquad (4.67)$$

Neglecting voltage drops in the windings,

$$\frac{V_1}{V_2} = a \qquad (4.68)$$

When a load is connected to the secondary of the autotransformer, a current I_2 flows in the direction shown in Fig. 4.23. By Kirchhoff's current law,

$$I_2 = I_1 + I_3 \qquad (4.69)$$

As in the ordinary transformer, the primary and secondary ampere-turns balance each other, except for the small current required for core magnetization:

$$N_1 I_1 = N_2 I_2 \qquad (4.70)$$

Equation 4.70 may also be written as

$$\frac{I_2}{I_1} = \frac{N_1}{N_2} = a \qquad (4.71)$$

Substituting Eq. 4.71 into Eq. 4.69, the ratio of the winding currents is found as

$$\frac{I_3}{I_1} = a - 1 \tag{4.72}$$

In an autotransformer, the total power transmitted from the primary to the secondary does not actually pass through the whole winding. This means that a greater amount of power can be transferred without exceeding the current ratings of the windings of the transformer.

The input apparent power is given by

$$S_1 = V_1 I_1 \tag{4.73}$$

Similarly, the output apparent power is given by

$$S_2 = V_2 I_2 \tag{4.74}$$

However, the apparent power in the transformer windings is

$$S_w = V_2 I_3 = (V_1 - V_2)I_2 = S_{ind} \tag{4.75}$$

This power is the component of the power transferred by transformer action or by electromagnetic induction.

The difference $(S_2 - S_w)$ between the output apparent power and the apparent power in the windings is the component of the output transferred by electrical conduction. This is equal to

$$S_{cond} = V_2 I_2 - V_2 I_3 = V_2 I_1 \tag{4.76}$$

EXAMPLE 4.9   440/550

A single-phase, 10-kVA, 440/110-V, two-winding transformer is connected as an autotransformer to supply a load at 550 V from a 440 V supply as shown below. Calculate the following:

a. kVA rating as an autotransformer
b. Apparent power transferred by conduction
c. Apparent power transferred by electromagnetic induction

Solution

The single-phase, two-winding transformer is reconnected as an autotransformer as shown in Fig. 4.24. The current ratings of the windings are given by

$$I_{ab} = 10,000/110 = 90.9 \text{ A}$$
$$I_{bc} = 10,000/440 = 22.7 \text{ A}$$

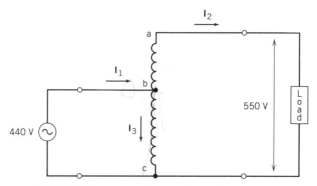

FIGURE 4.24 Autotransformer circuit of Example 4.9.

At full or rated load, the primary and secondary terminal currents are

$$I_2 = 90.9 \text{ A}$$
$$I_1 = I_2 + I_3 = 90.9 + 22.7 = 113.6 \text{ A}$$

Therefore, the kVA rating of the autotransformer is

$$\text{kVA}_1 = (440)(113.6)/1000 = 50 \text{ kVA}$$
$$\text{kVA}_2 = (550)(90.9)/1000 = 50 \text{ kVA}$$

Note that this transformer, whose rating as an ordinary two-winding transformer is only 10 kVA, is capable of handling 50 kVA as an autotransformer. However, not all of the 50 kVA is transformed by electromagnetic induction. A large part is merely transferred electrically by conduction.

The apparent power transformed by induction is

$$S_{\text{ind}} = V_1 I_3 = (440)(22.7) \text{ VA} = 10 \text{ kVA}$$

The apparent power transferred by conduction is

$$S_{\text{cond}} = V_1 I_2 = (440)(90.9) \text{ VA} = 40 \text{ kVA}$$

4.7 THREE-PHASE TRANSFORMERS

Three-phase transformers are used quite extensively in power systems to transform a balanced set of three-phase voltages at a particular voltage level into a balanced set of voltages at another level. Transformers used between generators and transmission systems, between transmission and subtransmission systems, and between subtransmission and distribution systems are all three-phase transformers. Most commercial and industrial loads require three-phase transformers to transform the three-phase distribution voltage to the ultimate utilization level.

Three-phase transformers are formed in either of two ways. The first method is to connect three single-phase transformers to form a three-phase bank. The second method is to manufacture a three-phase transformer bank with all three phases located on a common multilegged core. As far as analysis is concerned, there is no difference between the two methods.

The primary windings and secondary windings of the three-phase transformer may be independently connected in either a wye (Y) or delta (Δ) connection. As a result, four types of three-phase transformers are in common use:

1. Wye-wye (Y-Y)
2. Wye-delta (Y-Δ)
3. Delta-wye (Δ-Y)
4. Delta-delta (Δ-Δ)

The four possible connections are shown in Fig. 4.25. It should be noted that to form a wye connection, the undotted ends of the three windings (three primaries or three secondaries) are joined together and form the neutral point and the dotted ends become the three line terminals. In forming a delta connection, the three windings belonging to the same side are connected in series in such a way that the sum of the phase voltages in the closed delta is equal to zero; then the line terminals are taken off the junctions of the windings.

In Fig. 4.25, the primary and secondary windings that are drawn parallel to each other belong to the same phase. Also shown in the figure are the various voltages and currents, where V and I are secondary line-to-line voltage and line current, respectively, and a is the turns ratio of the single-phase transformer.

The Y-Δ connection is commonly used in stepping down from a high voltage to a medium or low voltage level, as in distribution transformers. Conversely, the Δ-Y connection is used for stepping up to a high voltage, as in generation station transformers.

The Y-Y connection is seldom used because of possible voltage unbalances and problems with third harmonic voltages. The Δ-Δ connection is used because of its advantage that one of the three single-phase transformers can be removed for repair or maintenance. The remaining two transformers continue to function as a three-phase bank, although the kVA rating of the bank is reduced to 58% of the original three-phase bank rating. This mode of operation is known as open-delta connection, or V-V connection.

The open-delta connection is also used when the load is presently small but is expected to grow in the future. Thus, instead of installing a three-phase bank of three single-phase transformers right away, only two single-phase transformers are used for three-phase voltage transformation. The third single-phase transformer serves as a spare and is connected at a later date when the load has grown.

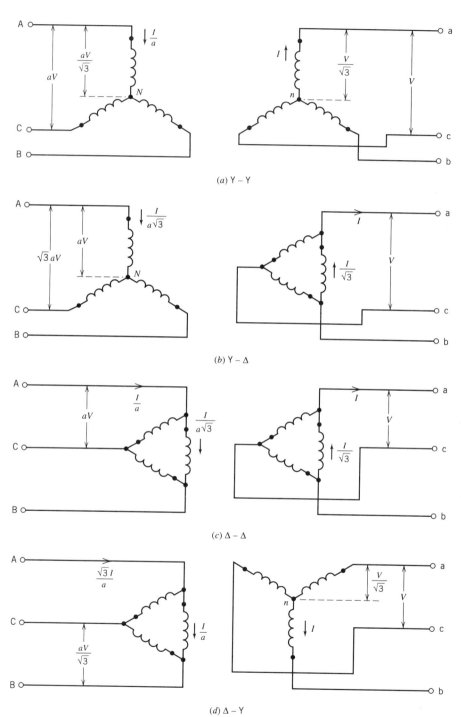

(a) Y – Y

(b) Y – Δ

(c) Δ – Δ

(d) Δ – Y

FIGURE 4.25 Three-phase transformer connections.

EXAMPLE 4.10

Three identical single-phase transformers are each rated 30 MVA, 200/40 kV, 60 Hz. They are connected to form a three-phase Y-Y transformer bank as shown in Fig. 4.26.

The bank is energized by a 345-kV three-phase source. A 60-MVA three-phase load, 0.9 PF lagging, is connected to the secondary of the transformer bank. Neglect exciting currents and voltage drops across the transformer. Choose V_{AB} at the primary as reference phasor.

a. Determine primary and secondary voltages and currents for this configuration.

b. Repeat part (a) when the secondary is Δ connected as shown in Fig. 4.27.

Solution

a. Assume an abc phase sequence. The primary line-to-line and phase voltages expressed in kV are as follows:

$$V_{AB} = 345\,\underline{/0°} \qquad V_{AN} = 200\,\underline{/-30°}$$
$$V_{BC} = 345\,\underline{/-120°} \qquad V_{BN} = 200\,\underline{/-150°}$$
$$V_{CA} = 345\,\underline{/120°} \qquad V_{CN} = 200\,\underline{/90°}$$

Since the turns ratio is $a = 200/40 = 5$ for each single-phase transformer, the secondary voltages in kV of the bank are given by

$$V_{an} = 40\,\underline{/-30°} \qquad V_{ab} = 69\,\underline{/0°}$$
$$V_{bn} = 40\,\underline{/-150°} \qquad V_{bc} = 69\,\underline{/-120°}$$
$$V_{cn} = 40\,\underline{/90°} \qquad V_{ca} = 69\,\underline{/120°}$$

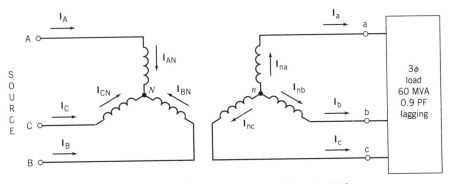

FIGURE 4.26 Y-Y transformer bank of Example 4.10.

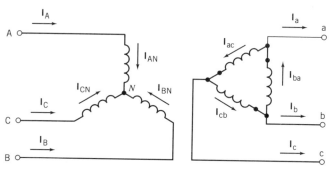

FIGURE 4.27 Y-Δ transformer bank.

The line currents are equal to the phase currents in the wye-connected transformer. At the secondary, the currents expressed in amperes are given by the following:

$$\mathbf{I}_a = \mathbf{I}_{na} = \frac{60,000}{\sqrt{3}\ 69} \underline{/-30° - \cos^{-1}0.9} = 500\ \underline{/-55.8°}$$

$$\mathbf{I}_b = \mathbf{I}_{nb} = 500\ \underline{/-175.8°}$$

$$\mathbf{I}_c = \mathbf{I}_{nc} = 500\ \underline{/64.2°}$$

The primary currents expressed in amperes are

$$\mathbf{I}_A = \mathbf{I}_{AN} = 100\ \underline{/-55.8°}$$

$$\mathbf{I}_B = \mathbf{I}_{BN} = 100\ \underline{/-175.8°}$$

$$\mathbf{I}_C = \mathbf{I}_{CN} = 100\ \underline{/64.2°}$$

b. For the Y-Δ connected transformer bank, the primary line-to-line and phase voltages are the same as in part (a). The secondary phase voltages are identical to the line-to-line voltages, and they are expressed in kV as follows.

$$\mathbf{V}_{ab} = 40\ \underline{/-30°}$$

$$\mathbf{V}_{bc} = 40\ \underline{/-150°}$$

$$\mathbf{V}_{ca} = 40\ \underline{/90°}$$

The secondary line currents in amperes are

$$\mathbf{I}_a = \frac{60,000}{\sqrt{3}\ 40} \underline{/-30° - 30° - 25.8°} = 866\ \underline{/-85.8°}\ A$$

$$I_b = 866 \underline{/-205.8°} = 866 \underline{/154.2°}$$

$$I_c = 866 \underline{/34.2°}$$

and the secondary phase currents in amperes are

$$I_{ba} = 500 \underline{/-55.8°}$$

$$I_{cb} = 500 \underline{/-175.8°}$$

$$I_{ac} = 500 \underline{/64.2°}$$

The primary line currents in amperes are equal to the phase currents. Thus,

$$I_A = I_{AN} = 100 \underline{/-55.8°}$$

$$I_B = I_{BN} = 100 \underline{/-175.8°}$$

$$I_C = I_{CN} = 100 \underline{/64.2°}$$

In either Y-Y or Δ-Δ connections, corresponding phase voltages are in phase. Similarly, corresponding line-to-line voltages in the primary and secondary are in phase. In other words, V_{AN} is in phase with V_{an}, and V_{AB} is in phase with V_{ab}. On the other hand, for both Y-Δ and Δ-Y connections, it is customary in the United States to have the primary phase or line-to-line voltage lead by 30°; thus, V_{AN} leads V_{an} by 30°, and V_{AB} leads V_{ab} by the same amount of phase shift.

Circuit analysis involving three-phase transformers under balanced conditions can be performed on a per-phase basis. This follows from the relationship that the per-phase real power and reactive power are one-third of the total real power and reactive power, respectively, of the three-phase transformer bank. It is convenient to carry out computations in a per-phase wye line-to-neutral basis. When Δ-Y, or Y-Δ connections are present, the parameters are referred to the Y side. In dealing with Δ-Δ connections, the Δ-connected impedances are converted to equivalent Y-connected impedances. The Δ-Y impedance conversion formula was given in Chapter 3 and is repeated here as Eq. 4.77.

$$Z_Y = \frac{1}{3} Z_\Delta \tag{4.77}$$

EXAMPLE 4.11

Three single-phase 100-kVA, 2400/240-V, 60-Hz transformers are connected to form a three-phase, 4160/240-V transformer bank. The equivalent impedance of each transformer referred to its low-voltage side is $0.045 + j0.16\ \Omega$. The transformer bank is connected to a three-phase source through a three-phase feeder with an impedance of $0.5 + j1.5\ \Omega$ /phase. The transformer delivers 250 kW at 240 V and 0.866 lagging power factor.

a. Determine the transformer winding currents.

b. Determine the sending end voltage at the source.

Solution

a. The three-phase power system is connected as shown in Fig. 4.28.

The high-voltage windings are connected in wye so that the primary can be connected to the 4160-V source. The low-voltage windings are connected in delta to supply 240 V to the load.

The line current delivered to the load is

$$I_s = \frac{250,000}{\sqrt{3}(240)(0.866)} = 694.5\ A$$

The transformer secondary winding current is

$$I_2 = I_s/\sqrt{3} = 694.5/\sqrt{3} = 400\ A$$

The transformer winding ratio is

$$a = 2400/240 = 10$$

Therefore, the primary current is found as

$$I_p = I_1 = I_2/a = 400/10 = 40\ A$$

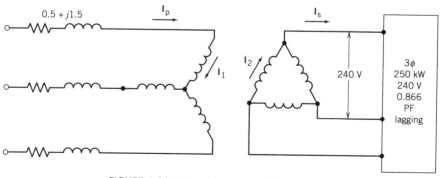

FIGURE 4.28 Three-phase connection diagram.

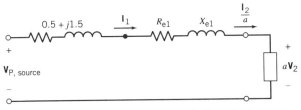

FIGURE 4.29   Single-phase equivalent circuit.

b. The equivalent impedance of the transformer referred to the high-voltage side is

$$Z_{e1} = a^2 Z_{e2} = (10)^2(0.045 + j0.16) = 4.5 + j16 \; \Omega/\text{phase}$$

The single-phase equivalent circuit is shown in Fig. 4.29.
The sending end voltage is found as follows:

$$V_{P,\text{source}} = 2400\underline{/0^\circ} + (40\underline{/-30^\circ})(0.5 + j1.5 + 4.5 + j16)$$
$$= 2923.2 + j506.2 = 2966.7\underline{/9.8^\circ} \; \text{V (line-to-neutral)}$$

The line-to-line voltage is given by

$$V_{L,\text{source}} = \sqrt{3} \; V_{P,\text{source}} = \sqrt{3} \; 2966.7\underline{/39.8^\circ}$$
$$= 5138.5\underline{/39.8^\circ} \; \text{V (line-to-line)}$$

DRILL PROBLEMS

D4.15   Three single-phase transformers are connected to make a three-phase Δ-Y transformer bank. The rated line-to-line voltages of the bank are 2400 V and 208 V on the primary and secondary, respectively. The transformer supplies a wye-connected load that takes 18 kW at 208 V and 0.85 power factor lagging. Determine the primary and secondary line and phase currents.

D4.16   A three-phase transformer bank is formed by interconnecting three single-phase transformers. The high-voltage terminals are connected to a three-phase 69-kV feeder, and the low-voltage terminals are connected to a three-phase load rated at 1000 kVA and 4.16 kV. Specify the voltage, current, and kVA ratings of each transformer, both high-voltage and low-voltage windings, for the following connections:

a. Y-Y   **b.** Y-Δ

c. Δ-Y   **d.** Δ-Δ

REFERENCES

1. Bergseth, F. R., and S. S. Venkata. *Introduction to Electric Energy Devices.* Prentice Hall, Englewood Cliffs, N.J., 1987.

2. Brown, David, and E. P. Hamilton III. *Electromechanical Energy Conversion.* Macmillan, New York, 1984.

3. Chapman, Stephen J. *Electric Machinery Fundamentals.* 2nd ed. McGraw-Hill, New York, 1991.

4. Del Toro, Vincent. *Electric Machines and Power Systems.* Prentice Hall, Englewood Cliffs, N.J., 1985.

5. El-Hawary, Mohamed E. *Principles of Electric Machines with Power Electronics Applications.* Prentice Hall, Englewood Cliffs, N.J., 1986.

6. Fitzgerald, A. E., Charles Kingsley, and Stephen Umans. *Electric Machinery.* 5th ed. McGraw-Hill, New York, 1990.

7. Kosow, Irving L. *Electric Machinery and Transformers.* Prentice Hall, Englewood Cliffs, N.J., 1991.

8. Kraus, John D. *Electromagnetics.* 3rd ed. McGraw-Hill, New York, 1984.

9. Macpherson, George, and Robert D. Laramore. *An Introduction to Electrical Machines and Transformers.* 2nd ed. Wiley, New York, 1990.

10. Matsch, Leander W., and J. Derald Morgan. *Electromagnetic and Electromechanical Machines.* 3rd ed. Harper & Row, New York, 1986.

11. Nasar, Syed A. *Electric Energy Conversion and Transmission.* Macmillan, New York, 1985.

12. Ramshaw, Raymond, and R. G. van Heeswijk. *Energy Conversion Electric Motors and Generators.* Saunders College Publishing, Philadelphia, 1990.

13. Sen, P. C. *Principles of Electrical Machines and Power Electronics.* Wiley, New York, 1989.

14. Wildi, Theodore. *Electrical Machines, Drives, and Power Systems.* 2nd ed. Prentice Hall, Englewood Cliffs, N.J., 1991.

PROBLEMS

4.1 A long solenoid coil has its length much greater than its diameter as shown in Fig. 4.30. The magnetic field inside the coil may, therefore, be considered uniform.

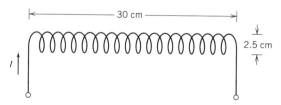

FIGURE 4.30 Solenoid coil of Problem 4.1.

The coil has 150 turns, its length is 30 cm, and its diameter is 2.5 cm. A current $I = 25$ A is supplied to the coil. Neglect the magnetic field outside the coil.

    a. Determine the magnetic field intensity H and the magnetic flux density B inside the solenoid.

    b. Determine the inductance of the coil.

4.2 A toroid has a circular cross section as shown in Fig. 4.31. It is made from cast steel with a relative permeability of 2500. Its outer diameter is 24 cm, and its inner diameter is 16 cm. The magnetic flux density in the core is 1.25 tesla measured at the mean diameter of the toroid.

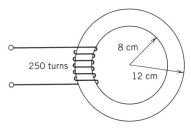

FIGURE 4.31   Magnetic circuit of Problem 4.2.

    a. Find the current that must be supplied to the coil, which consists of 250 turns.

    b. Determine the magnetic flux in the core.

    c. A 10-mm air gap is cut across the toroid. Determine the current that must be supplied to the coil to produce the same value of magnetic flux density as in part (a).

4.3 A ferromagnetic circuit has a magnetic core with infinitely high relative permeability. It has three legs, and air gaps of 2 mm and 1 mm are cut from sections A and C as shown in Fig. 4.32. A coil is wound on the center leg B, and it has 200 turns and a resistance of 2.5 Ω. The magnetic core has a 5 × 5 cm uniform cross-sectional area. A DC voltage is applied to the coil.

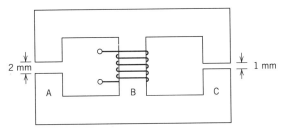

FIGURE 4.32   Magnetic circuit of Problem 4.3.

    a. Determine the voltage that will produce a flux density of 0.75 T in the right leg C, which contains the 1-mm air gap.

    b. Find the magnetic flux in the other two legs of the core.

4.4 Repeat Problem 4.3 if the coil is removed from the center leg B and placed on the right leg C, which contains the 1-mm air gap.

4.5 The electromagnetic system shown in Fig. 4.33 has a magnetic core of infinite relative permeability. Neglect fringing and leakage flux. Determine the expressions for

a. The magnetic flux in each of the three legs

b. The flux linkage and self-inductance of each coil

c. The mutual inductance

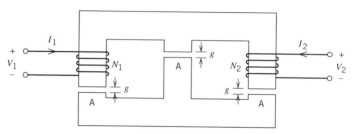

FIGURE 4.33 Magnetic circuit of Problem 4.5.

4.6 For the electromagnetic system of Fig. 4.33, the polarity of the voltage source V_2 is reversed so that the current I_2 flows in the opposite direction. The magnetic core has infinite relative permeability. Determine the expressions for

a. The magnetic flux in each of the three legs

b. The flux linkage of each coil

c. The self-inductance of each coil

d. The mutual inductance

4.7 The magnetic circuit shown in Fig. 4.34 has an infinitely permeable magnetic core. The following are given:

$$g_1 = 5 \text{ mm} \qquad A_1 = 5 \text{ cm}^2 \qquad N_1 = 80 \text{ turns}$$
$$g_2 = 5 \text{ mm} \qquad A_2 = 5 \text{ cm}^2 \qquad N_2 = 100 \text{ turns}$$
$$g_3 = 10 \text{ mm} \qquad A_3 = 10 \text{ cm}^2 \qquad N_3 = 125 \text{ turns}$$

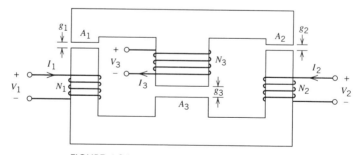

FIGURE 4.34 Magnetic circuit of Problem 4.7.

A current of 12 A flows in the first coil, N_1. The second and third coils, N_2 and N_3, respectively, are unexcited. Determine the flux densities in each of the air gaps g_1, g_2, and g_3.

4.8 In the electromagnetic circuit of Fig. 4.34, the three coils are excited simultaneously such that $I_1 = 12$ A, $I_2 = 10$ A, and $I_3 = 8$ A, with the directions of currents as shown. Determine the magnetic flux densities in the three air gaps.

4.9 In the electromagnetic circuit of Fig. 4.34, assume that there is no air gap in the center leg containing coil N_3, and only the first coil, N_1, carries a current of 12 A.

   a. Determine the flux densities in the air gaps g_1 and g_2.

   b. Suppose that the length of air gap g_2 is also reduced to zero; that is, there is also no air gap in the leg containing coil N_2. Determine the flux density in the first air gap g_1.

4.10 Refer to the electromagnetic system described in Problem 4.7 and shown in Fig. 4.34. Determine the self- and mutual inductances of the three coils.

4.11 The magnetic circuit shown in Fig. 4.35 has an iron core with infinite permeability. The core dimensions are:

$$A_c = 16 \text{ cm}^2 \qquad g = 2 \text{ mm} \qquad l_c = 80 \text{ cm}$$

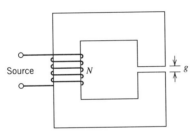

FIGURE 4.35 Magnetic circuit of Problem 4.11.

The coil has 500 turns and draws a current $I = 4$ A from the source. Neglect magnetic leakage and fringing. Calculate

   a. The total magnetic flux

   b. The flux linkages of the coil

   c. The coil inductance

4.12 Repeat Problem 4.11 assuming that the core has a relative permeability of $\mu = 2000$.

4.13 A toroidal magnetic circuit consists of a coil of $N = 200$ turns each of circular cross section of radius $\rho = 0.25$ m, as shown in Fig. 4.36. The radius of the toroid is $R = 5$ m, as measured to the center of each circular turn. Assume that the magnetic field intensity is zero outside the toroid and the magnetic field intensity inside the toroid is given by

$$H = \frac{NI}{2\pi r}$$

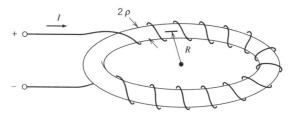

FIGURE 4.36   Magnetic circuit of Problem 4.13.

  a. Calculate the coil inductance L.

  b. A voltage source is connected to the coil, and the current is adjusted to produce a magnetic flux density of 1.0 T. Calculate the total stored magnetic energy.

4.14   A 4800/240-V, single-phase transformer is rated at 10 kVA, and it has an equivalent impedance referred to the primary side of $Z_{e1} = 120 + j300\ \Omega$.

  a. Find the equivalent impedance referred to the secondary side Z_{e2}.

  b. Calculate the voltage at the primary terminals if the secondary supplies rated secondary current at 230 V and unity power factor.

4.15   A 100-kVA, 2300/230-V, single-phase transformer has the following parameters:

$$R_1 = 0.30\ \text{ohm} \qquad R_2 = 0.0030\ \text{ohm} \qquad R_c = 4.5\ \text{k}\Omega$$
$$X_1 = 0.65\ \text{ohm} \qquad X_2 = 0.0065\ \text{ohm} \qquad X_m = 1.0\ \text{k}\Omega$$

The transformer delivers 75 kW at 230 V and 0.85 power factor lagging. Determine

  a. The input current

  b. The input voltage

  c. The input power and power factor

4.16   A 15-kVA, 2400/240-V, single-phase transformer has an equivalent series impedance $Z_{e1} = 6 + j8.5\ \Omega$. The shunt magnetizing branches are given as $R_{c1} = 50\ \text{k}\Omega$ and $X_{m1} = 15\ \text{k}\Omega$. The transformer is delivering rated current to a load at 240 V and 0.8 lagging power factor. Find (a) the primary current and (b) the applied voltage.

4.17   A 25-kVA, 2300/230-V, single-phase transformer has a high-voltage winding with a resistance of 1.5 Ω and a leakage reactance of 2.4 Ω. The low-voltage winding has a resistance of 0.015 Ω and a leakage reactance of 0.024 Ω. Find the following:

  a. Equivalent impedance referred to the high-voltage winding

  b. Equivalent impedance referred to the low-voltage winding

  c. Voltage regulation of the transformer for full load at 230 V and 0.866 power factor lagging

  d. The efficiency of the transformer under the conditions of part (c)

4.18 The transformer of Problem 4.15 has a secondary voltage of 230 V and the load on the transformer is 100 kVA. By using the approximate equivalent circuit of a transformer, determine the voltage that must be applied to the primary terminals if

a. The power factor of the load is 0.8 lagging

b. The power factor of the load is 0.8 leading

4.19 A 25-kVA, 2400/240-V, single-phase transformer has an equivalent resistance and reactance, both referred to the primary side, of $R_{el} = 3.45$ Ω and $X_{el} = 5.75$ Ω, respectively. The core loss is 120 W. The transformer delivers rated kVA to a load at rated secondary voltage and 0.85 power factor lagging.

a. Determine the voltage applied to the primary side.

b. Determine the percent voltage regulation.

c. Find the efficiency of the transformer.

4.20 A 25-kVA, 2200/220-V, single-phase transformer has an equivalent series impedance of $Z_{el} = 3.5 + j4.0$ Ω referred to the primary side. The transformer is connected to a load whose power factor varies.

a. Calculate the voltage regulation at full load, 0.8 PF lagging.

b. Determine the highest value of voltage regulation for full-load output at rated secondary terminal voltage.

c. Determine the efficiency when the transformer delivers full-load output at rated secondary voltage and 0.8 power factor lagging.

4.21 A 10-kVA transformer has an iron loss of 150 W and a full-load copper loss of 250 W. Calculate the transformer efficiency for the following load conditions:

a. Full load at 0.8 power factor lagging

b. 75% of full load at unity power factor

c. 50% of full load at 0.6 power factor lagging

4.22 The transformer of Problem 4.21 operates on full load at 80% power factor for 4 h, on 75% of full load at unity power factor for 8 h, and on 50% of full load at 60% power factor for 12 h during 1 day. Determine the all-day efficiency. Refer to Drill Problem 4.13.

4.23 A 25-kVA, 2400/240-V, 60-Hz, distribution transformer was tested at 60 Hz, and the following data were obtained.

	Voltage (V)	Current (A)	Power (W)
Open-circuit test	240	3.2	165
Short-circuit test	55	10.4	375

a. Compute the efficiency for full-load output at rated voltage and 0.8 power factor lagging.

b. The power factor of the load is varied while the magnitudes of the current and the secondary voltage are held constant. Determine the largest value of voltage regulation and the power factor at which it occurs. Draw a phasor diagram depicting this condition.

4.24 A 50-kVA, 2400/240-V, single-phase transformer was tested, and the following test data were obtained.

	Voltage (V)	Current (A)	Power (W)
Short-circuit test	55	20.8	600
Open-circuit test	240	5.0	450

a. Calculate the voltage regulation and efficiency when the transformer is connected to a load that takes 156 A at 220 V and 0.8 power factor lagging.

b. Calculate the voltage regulation and efficiency at rated load conditions and 0.8 power factor lagging.

4.25 Two single-phase transformers are each rated 2400/120 V. Draw a circuit diagram showing the interconnections and polarity markings of these transformers.

a. When they are used for 4800/240-V operation

b. When they are used for 2400/120-V operation

4.26 A single-phase load is supplied through a 34.5-kV feeder and a 34.5/2.4-kV transformer. The feeder has an impedance of $50 + j180$ ohms, and the transformer has an equivalent impedance of $24 + j120$ ohms referred to its high-voltage side. The load takes 260 kW at 2.3 kV and 0.866 lagging power factor.

a. Find the voltage at the primary side of the transformer.

b. Determine the voltage at the sending end of the feeder.

c. Calculate the real and reactive power input at the sending end of the feeder.

4.27 A 10-kVA, 4160/240-V, single-phase transformer has a per-unit resistance of 0.01 and a per-unit reactance of 0.05. It has a core loss at rated voltage of 150 W. The transformer supplies 7.5 kVA at 240 V and 0.6 power factor lagging to a load connected to its secondary terminals.

a. Determine the equivalent impedance in ohms referred to the primary side.

b. Find the input voltage.

c. Determine the efficiency of the transformer.

4.28 A 15-kVA, 2200/220-V, single-phase transformer is connected to act as a booster from 2200 to 2420 V. Without exceeding the rated current of any winding and assuming an ideal transformer, determine (a) the kVA input and output, (b) the kVA transformed, and (c) the kVA conducted.

4.29 A 5-kVA, 480/120-V, single-phase transformer is to be used as an autotransformer to transform a 600-V source to a 480-V supply. As a single-phase transformer delivering rated load at 0.80 power factor lagging, its efficiency is 0.95%.

a. Show a connection diagram as an autotransformer.

b. Determine the kVA rating as an autotransformer.

c. Find the efficiency as an autotransformer delivering rated capacity at 480 V and 0.80 power factor lagging.

4.30 A 5-kVA, 220/220-V, 60-Hz, two-winding transformer has a full-load efficiency of 95% at unity power factor. The iron loss at 60 Hz is 100 W. This transformer is connected as an autotransformer and supplied by a 440-V source. The autotransformer delivers the maximum possible kVA without overloading the windings. Calculate (a) the primary current and (b) the efficiency of the autotransformer.

4.31 Three single-phase, 20/2.4-kV, ideal transformers are connected to form a three-phase, 10-MVA, 34.5/2.4-kV transformer bank. The transformer bank supplies a load of 6 MW at 2.4 kV and 0.85 power factor lagging.

a. Determine the line and phase currents at the primary and secondary sides of the transformer.

b. Determine the line-to-line and line-to-neutral voltages at the primary and secondary sides of the transformer.

4.32 Repeat Problem 4.31 if the three single-phase, 20/2.4-kV, ideal transformers are connected to form a three-phase, 10 MVA, 20/2.4-kV transformer bank and the bank supplies 6 MW at 2.4 kV and 0.85 power factor lagging.

4.33 Three single-phase, 10-kVA, 2400/120-V, 60-Hz transformers are connected to form a three-phase, 4160/208-V transformer bank. The equivalent impedance of each single-phase transformer referred to the primary side is $10 + j25 \ \Omega$. The transformer bank delivers 27 kW at 208 V and 0.9 power factor leading.

a. Draw the three-phase schematic diagram showing the transformer connection. Draw a per-phase equivalent circuit.

b. Determine the primary current and power factor.

c. Determine the primary voltage.

d. Determine the voltage regulation.

4.34 Repeat Problem 4.33 if the three single-phase 2400/120-V transformers are connected to form a three-phase 2400/208-V transformer bank and the bank supplies 27 kW at 208 V and 0.9 power factor leading.

4.35 A three-phase, 300-kVA, 2300/230-V, wye-wye transformer bank has an iron loss of 2200 W and a full-load copper loss of 3800 W. Determine the efficiency of the transformer for 70% full load at 230 V and 0.85 power factor.

4.36 A three-phase transformer bank is formed by interconnecting three single-phase transformers. The three-phase bank is designed to be rated at 300 MVA and 230/34.5 kV. Find the voltage, current, and kVA ratings of each single-phase transformer, both high-voltage and low-voltage windings, if the transformer bank is connected:

a. Δ-Δ

b. Y-Δ

c. Y-Y

d. Δ-Y

Five

Fundamentals of Rotating Machines

5.1 INTRODUCTION

A transformer may be described as an energy transfer device; that is, energy is transferred from the primary to the secondary without changing its form. Both sides of the transformer, primary and secondary, carry energy in electrical form. On the other hand, rotating machines such as synchronous machines, induction machines, and DC machines are energy converters. They convert either mechanical energy to electrical energy, in the case of generators, or electrical energy to mechanical energy in the case of motors.

In converting energy from mechanical to electrical or electrical to mechanical, part of the energy is lost. If losses are neglected, however, the machine becomes an ideal energy converter and may be represented as shown in Fig. 5.1.

From the law of conservation of energy, the mechanical input is equal to the electrical output, since losses are neglected; that is,

$$\omega_m T = vi \tag{5.1}$$

where

$\omega_m T$ = input mechanical power

vi = output electrical power

T = mechanical input torque, expressed in N-m

ω_m = rotor speed, in rad/s

v = voltage (V)

i = current (A)

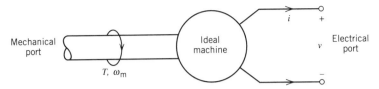

FIGURE 5.1   Ideal energy converter.

The energy losses consist of mechanical losses, which include windage and friction, and electrical losses, which include winding copper losses and magnetic core losses. These losses can be modeled externally. Therefore, in this chapter, the discussion will concentrate on the ideal energy converter. Another assumption made is infinite permeability for magnetic cores, both rotors and stators, used for the energy converters.

5.2   BASIC CONCEPTS OF ENERGY CONVERTERS

The source of a magnetic field may be a permanent magnet or an electric current. Hence, a magnetic field is produced near a conducting loop carrying a current i. Both magnitude and direction of the magnetic field will vary with position for a given current and loop dimensions. A current-carrying loop is shown in Fig. 5.2.

The magnetic field $\mathbf{H}$ at point P for the current loop of Fig. 5.2 is given by

$$\mathbf{H} = \oint \frac{i \, \mathbf{dl} \times \mathbf{u}_r}{4\pi r^2} \qquad (5.2)$$

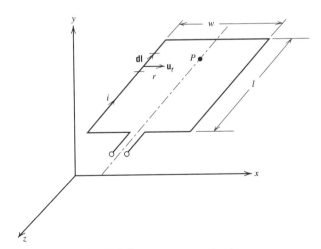

FIGURE 5.2   A current-carrying loop.

The integral in Eq. 5.2 is quite complicated. However, if length l is very long compared to the width w, then $\mathbf{H}$ along the centerline can be found using Ampère's law.

The magnetic field intensity $\mathbf{H}$ is directed downward, in the negative y direction, and has a magnitude given by

$$H = \frac{i}{2\pi w} \tag{5.3}$$

Hence, the magnetic flux density $\mathbf{B}$ may be expressed as

$$\mathbf{B} = \mu\mathbf{H} = -\frac{\mu i}{2\pi w}\mathbf{u}_y \tag{5.4}$$

where
μ = permeability
$\mathbf{u}_y$ = unit vector in the y direction

The foregoing discussion describes how a magnetic flux density is produced by an electric current in a conductor. Next, consider what will happen if the conducting loop shown in Fig. 5.3 is rotated about its axis (shown by the broken line) at a speed of ω_m rad/s under a magnetic field.

It can be seen from Fig. 5.3 that the magnetic flux ϕ linking the conducting loop varies from a minimum value to zero, to a maximum value, and back to the minimum, as the loop rotates. Therefore in accordance with Faraday's law, a voltage is induced between terminals a and a′ and is given by

$$e_{aa'} = \frac{d\lambda}{dt} = N\frac{d\phi}{dt} = \frac{d\phi}{dt} \tag{5.5}$$

since the loop contains $N = 1$ turn.

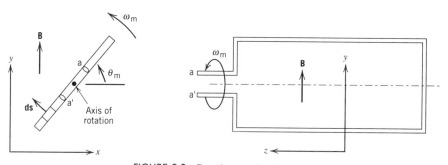

FIGURE 5.3 Rotating conducting loop.

From the given magnetic flux density **B**, the magnetic flux ϕ is found as

$$\phi = \int_S \mathbf{B} \cdot \mathbf{dS} \tag{5.6}$$

where **dS** is a differential area of the loop, and directed perpendicular to the loop and outward. Thus, **dS** makes an angle of θ_m with **B**, and Eq. 5.6 reduces to

$$\phi = \int B \cos \theta_m \, dS = Bwl \cos \theta_m \tag{5.7}$$

Since θ_m is equal to $\int_0^t \omega_m \, dt + \theta_m(0)$, and assuming θ_m is zero at $t = 0$, Eq. 5.7 may also be written as

$$\phi = Bwl \cos \omega_m t \tag{5.8}$$

Substituting Eq. 5.8 into Eq. 5.5 yields the expression for the induced emf.

$$e_{aa'} = -Bwl\omega_m \sin \omega_m t = Bwl\omega_m \cos(\omega_m t + 90°) \tag{5.9}$$

If a resistor R is connected across terminals a–a', a current ($i = e_{aa'}/R$) will flow through both the resistor and the conducting loop. Hence, electrical power ($ie_{aa'} = i^2R$) is delivered to the resistor. This electrical power originates from the mechanical power required to keep the loop rotating at a speed of ω_m. The mechanical power is given by

$$P_m = \omega_m T_m = i e_{aa'} \tag{5.10}$$

The direction of the current i is such that the magnetic field produced by the loop current will oppose the change in the flux linking this loop due to the external magnetic flux density **B**. For example, when the loop is horizontal (z–x plane), current i will be clockwise to oppose the decrease in external flux linked by the loop.

Since the loop carries current and is situated in a magnetic flux density **B**, there will be an electromagnetic torque T_e acting on the loop. This torque tends to line up the magnetic axis of the loop with the external magnetic flux density **B**. When the loop is horizontal, T_e is clockwise. To maintain the speed ω_m, the applied mechanical torque T_m should be equal in magnitude and opposite in direction to T_e.

The preceding discussion describes the basic principle of operation of a generator. For motor action, on the other hand, a current i is supplied by an electric source. This current sets up a magnetic field. Interaction of this field

122 CHAPTER 5 FUNDAMENTALS OF ROTATING MACHINES

and the external magnetic flux density **B** results in a torque T_e. This torque can be used to rotate a mechanical load that requires a torque T_m at some speed ω_m.

The basic concepts and principles of operation described in this section apply to both AC and DC machines. Although these concepts were discussed and derived with respect to the simple energy converter, the same concepts and principles remain valid for the more complicated and practical rotating machines.

EXAMPLE 5.1

A coil is formed by connecting 10 conducting loops, or turns, in series. Each turn has a length $l = 2$ m and width $w = 10$ cm. The 10-turn coil is rotated at a constant speed of 30 revolutions per second in a magnetic flux density $B = 2$ T directed upward.

a. Find an expression for the induced emf across the coil.

b. A resistor $R = 500 \ \Omega$ is connected between the terminals of the coil. Determine the average power delivered to this resistor.

c. Calculate the average mechanical torque needed to turn the coil and generate power for the resistor.

Solution

a. The induced emf across a single loop, or turn, is found by using Eq. 5.9 as follows:

$$e_{\text{turn}} = (2)(0.10)(2)[(2\pi)(30)] \cos(60\pi t + \pi/2)$$
$$= 75.4 \cos(188.5t + \pi/2) \text{ V}$$

Thus, the induced voltage across the coil of 10 turns is

$$e_{\text{coil}} = 10e_{\text{turn}} = 754 \cos(188.5t + \pi/2) \text{ V}$$

b. The rms value of the induced voltage across the coil is

$$E_{\text{coil}} = 754/\sqrt{2} = 533 \text{ V}$$

The rms current flowing through the resistor and coil is

$$I = E_{\text{coil}}/R = 533/500 = 1.066 \text{ A}$$

Thus, the average power delivered to the resistor is given by

$$P = I^2R = (1.066)^2(500) = 568 \text{ W}$$

c. The average mechanical torque required to rotate the coil of part (b) is obtained by using Eq. 5.10 as follows:

$$T_{\mathrm{m}} = P/\omega_{\mathrm{m}} = 568/188.5 = 3.0 \text{ N-m}$$

DRILL PROBLEMS

D5.1 The machine of Example 5.1 can be used as a motor. The terminals of the coil are connected to a voltage source whose rms voltage is 600 V. The motor runs at 1800 revolutions per minute (rpm) and draws a current of 1.0 A. Find the torque supplied to the mechanical load.

D5.2 The coil of Fig. 5.3 has 100 turns and is rotated at a constant speed of 300 rpm. The axis of rotation is perpendicular to a uniform magnetic flux density of 0.1 T in the vertical direction. The coil has width $w = 10$ cm and length $l = 20$ cm. Calculate

a. The maximum flux passing through the coil

b. The flux linkage as a function of time

c. The maximum instantaneous voltage induced in the coil

d. The time-average value of the induced voltage

e. The induced voltage when the plane of the coil is 30° from the vertical

5.3   ROTATING MACHINES

In the last section, a description of an elementary electromechanical energy converter was presented. However, practical machines are quite different in construction from the simple energy converter. In these machines, voltages are generated in coils that each consist of several turns of conductors. These coils can be rotated mechanically through a magnetic field, or a magnetic field can be rotated mechanically past the coils. This relative motion between the coils and the magnetic field results in a time-varying voltage generated across the coils. A group of such coils interconnected so that their generated voltages add up to the desired value is called an *armature winding*. The armature of a DC machine is the rotating member, or *rotor*. The armature of an AC machine (e.g., a synchronous generator) is the stationary member, or *stator*.

Another group of coils is also present in the rotating machine. These are the excitation coils, or *field windings*. These field windings act as the primary source of magnetic flux for voltage generation in the armature. Also, the magnetic field produced by the field windings interacts with the field produced by the armature windings for torque production. In a DC machine the field windings are located in the stationary member, whereas in an AC machine the field winding is in the rotor.

These machines are described briefly in this section. Later chapters give more detailed discussions of each machine.

5.3.1   Induction Machines

Induction machines are used mostly as motors. They are seldom used as generators because their performance characteristics as generators are unsatisfactory for most applications. As motors, on the other hand, they are called the workhorse of the industry.

An *induction machine* consists of a stator and a rotor. The stator consists of a laminated core, as in transformers, with conductors embedded in slots. These conductors, when energized by an AC source of power, provide a magnetic flux density that is varying. The rotor is mounted on bearings and is separated from the stator by a small air gap. The rotor is cylindrical and carries either

a. Windings whose terminals are brought out to slip rings for external connections, as in wound-rotor machines, or

b. Conductor bars that are short-circuited at both ends for squirrel-cage motors.

In either case, voltage is induced in the rotor conductors. This voltage then causes current to flow. The current, in turn, creates a magnetic field. Interaction of the stator and rotor magnetic fields produces an electromagnetic torque. This torque, in turn, rotates a mechanical load.

Induction motors are sometimes called rotating transformers. This is because the rotor emfs are induced by transformer action. Squirrel-cage motors are cheaper than comparable wound-rotor motors and are highly reliable. These factors contribute to their immense popularity and widespread use. Induction motors are either three-phase or single-phase.

5.3.2   Synchronous Machines

The bulk of electric power is produced by three-phase synchronous generators. Synchronous generators with ratings of 1000 MVA are fairly common in the power industry. *Synchronous machines* may be used as generators or motors. From an analytical point of view, there is no difference between a synchronous generator and a motor.

A simple three-phase, two-pole synchronous generator is shown in Fig. 5.4. The field winding is excited by direct current conducted to it by means of stationary carbon brushes. The DC power required for excitation is 1% to 2% of the machine's rating.

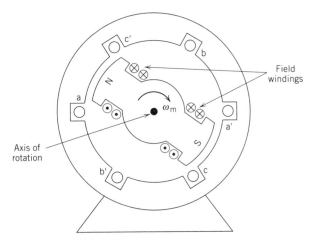

FIGURE 5.4 Elementary synchronous generator.

The armature winding consists of three coils of N turns each, shown by terminals aa', bb', and cc'. The conducting loops that form coils of each phase are parallel to the shaft and are connected in series by end connections (end connections are not shown in the figure). The rotor is rotated at a constant speed by the source of mechanical torque, the prime mover, connected to its shaft.

Consider phase a. As the rotor rotates, the flux linking coil aa' changes, and therefore a voltage is induced between terminals a and a'. For every complete cycle of rotor motion, the induced voltage goes through one complete cycle. Therefore, the angular frequency ω of the induced voltage $e_{aa'}$ is the same as the mechanical speed ω_m (in rad/s). Thus, the frequency in Hz is the same as the speed of the rotor in revolutions per second; that is, the frequency is synchronized with the mechanical speed. This is why the adjective "synchronous" is used with synchronous machines. In order to produce a 60-Hz voltage, the rotor has to rotate at 3600 rpm.

Most synchronous machines have more than two poles. Figure 5.5 shows a four-pole machine. For every complete rotation of the prime mover, the flux linking coil aa' goes through two complete cycles. Therefore, the angular frequency ω (in rad/s) of $e_{aa'}$ will be twice the speed of rotation ω_m of the prime mover. Thus, in general,

$$\omega = \left(\frac{p}{2}\right)\omega_m \qquad (5.11)$$

where

ω = electrical angular frequency

ω_m = mechanical angular speed

p = number of magnetic poles set up by the field winding

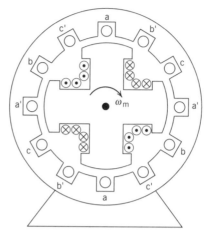

FIGURE 5.5   A four-pole synchronous machine.

In the United States, electric utilities supply their customers at 60 Hz (377 rad/s). From Eq. 5.11, it may be seen that to generate a 60-Hz voltage, the required speed of the prime mover is dependent on the number of poles. Since poles come in pairs, by law of nature, doubling the number of poles reduces the required speed by half. Therefore, to generate a 60-Hz voltage, a four-pole machine runs at 1800 rpm and an eight-pole machine runs at 900 rpm. Similarly, a six-pole machine runs at 1200 rpm and a 12-pole machine runs at 600 rpm.

The rotors shown in Figs. 5.4 and 5.5 have *salient*, or *projecting*, poles with concentrated windings. Another type of rotor configuration is called *nonsalient*, *round*, or *cylindrical*. The field winding is distributed over the surface of the rotor. A round-rotor synchronous machine is shown in Fig. 5.6.

Salient-pole construction is a characteristic of synchronous machines that have a large number of poles and operate at low speeds to produce the desired frequency of 60 Hz. Hydroelectric generators are salient-pole synchronous machines. On the other hand, the generators of steam turbines, such as those in coal and nuclear generating stations, and the generators of gas turbines are nonsalient synchronous machines. They operate best at high speeds, and they have few poles—generally two or four.

Almost all synchronous machines are three-phase machines. To generate a set of three voltages that are phase displaced by 120 electrical degrees, three coils phase displaced in space by 120 electrical degrees must be used. In Fig. 5.4, coils aa', bb', and cc' are 120 mechanical degrees apart for the two-pole machine. The flux linking coil bb' reaches its maximum 120 mechanical degrees after the flux linking coil aa' reaches its maximum, and the flux linking cc' reaches its maximum 120 mechanical degrees after the flux linking bb' reaches its maximum. The induced coil voltages $e_{aa'}$, $e_{bb'}$, and $e_{cc'}$ are the

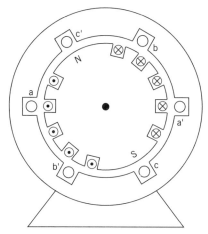

FIGURE 5.6   A two-pole, round-rotor synchronous machine.

derivatives of their respective flux linkages. Therefore, these voltages will also be 120 mechanical degrees apart. In this case, electrical degrees and mechanical degrees are the same.

In the four-pole machine shown in Fig. 5.5, the 360 mechanical degrees are equivalent to 720 electrical degrees. Hence, the coils are placed in the stator separated by 60 mechanical degrees, which are equivalent to 120 electrical degrees. Therefore, the induced coil voltages $e_{aa'}$, $e_{bb'}$, and $e_{cc'}$ are 120 electrical degrees apart just the same.

When a synchronous generator supplies electric power to an electrical load, the armature (stator) current creates a magnetic flux. This flux reacts with the magnetic flux produced by the field (rotor) current, and electromagnetic torque is produced. This torque tends to align the resultant fluxes of the rotor and stator currents. If an external mechanical torque is applied to counteract the electromagnetic torque, the rotor can keep rotating at constant speed. As a result, mechanical work is done on the rotor and electrical energy is delivered to the electrical load, for example, a resistor. Therefore, electromechanical energy conversion has occurred.

Synchronous motors are the counterparts of synchronous generators. A three-phase AC voltage source supplies three-phase currents to the armature windings, which are located on the stator. A DC voltage source provides DC current to the field winding on the rotor. Both stator and rotor windings produce magnetic fields and fluxes. The interaction of the stator and rotor magnetic fields (fluxes) results in an electromagnetic torque, which is used to rotate a mechanical load requiring a torque equal to the developed torque. The speed of the motor is determined by Eq. 5.11, where the electrical frequency ω is the frequency of the voltages applied to the stator windings, mostly 377 rad/s (60 Hz) in the United States.

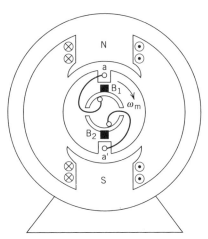

FIGURE 5.7 Elementary DC machine.

5.3.3 Direct-Current Machines

A DC machine can be either a generator or a motor. The armature of this machine is on the rotor, and the field winding is on the stator. This arrangement is opposite to that of an AC synchronous machine, or induction machine, where the armature is on the stator and the field is on the rotor. A simple two-pole DC machine is shown in Fig. 5.7.

For *DC machines*, the field and armature windings both have DC voltages at their terminals. Although the voltage induced in the armature winding is AC, the commutator segments are used to rectify this AC voltage. Hence, the voltage between brushes B_1 and B_2 is a DC voltage. This need for rectification is why the armature winding is located on the rotor.

The energy conversion process in DC machines is similar to that in AC machines. That is, the interaction of stator and rotor fields produces electromagnetic torque. In a DC generator, the rotor is rotated at constant speed by a mechanical energy source sufficient to overcome the electromagnetic torque to produce electricity, and thus mechanical energy is converted to electrical energy. On the other hand, in a DC motor, a DC voltage is applied to both armature and field windings and the electromagnetic torque produced is used to turn a mechanical load; thus, electrical energy is converted to mechanical energy.

DRILL PROBLEMS

D5.3 A synchronous generator has a rotor with six poles and operates at 50 Hz.

  a. Determine the speed of the prime mover of the generator.

b. Repeat part (a) if the generator rotor has 12 poles.

c. Repeat part (a) if the generator rotor has two poles.

D5.4 A three-phase AC motor is connected to a 60-Hz voltage supply. It is used to drive a draft fan. At no load, the speed is 1188 rpm; at full load, the speed drops to 1128 rpm.

a. Determine the number of poles of this AC motor.

b. State whether this motor is an induction motor or a synchronous motor.

5.4   ARMATURE MMF AND MAGNETIC FIELD

Most armatures have distributed windings, that is, windings that are spread over a number of slots around the periphery of the machine. This is illustrated in Fig. 5.8.

To derive the magnetic field of a distributed winding, the single N-turn coil is considered first. Such a coil spanning 180 mechanical degrees is shown in Fig. 5.9. The dot and cross indicate current toward and away from the reader, respectively. The rotor is of cylindrical type.

By applying Ampère's law to the semicircular path shown by broken lines in Fig. 5.9, the magnetic field is related to the mmf F as follows:

$$\oint \mathbf{H} \cdot \mathbf{dl} = F \tag{5.12}$$

Assuming that the relative permeability of the iron core is infinitely high, Eq. 5.12 reduces to

$$H(2g) = Ni \tag{5.13}$$

where

g = length of air gap

N = number of turns in the coil

i = current through coil aa$'$

It is apparent that the mmf associated with the closed path takes the form of an mmf drop across the total air-gap length. This is because the mmf drop inside the iron, on both the stator and the rotor, is negligible, since the core is assumed to have infinitely high permeability. Half of the mmf appears as an mmf drop across the top half of the air gap, and the other half appears across the lower half of the air gap. As a result, the air-gap mmf looks like that of Fig. 5.9b.

The rectangular waveform of the mmf shown in Fig. 5.9b can be resolved into a Fourier series composed of a fundamental component and a series of

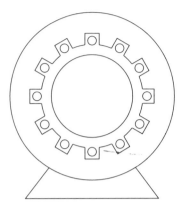

FIGURE 5.8   Distributed armature windings.

odd harmonics. For ease of calculations, the mmf is approximated as

$$F = F_{a1} = \frac{4}{\pi} \frac{Ni}{2} \cos \theta_m = F_m \cos \theta_m \qquad (5.14)$$

where F_{a1} is the fundamental component of F and $F_m = (4/\pi)(Ni/2)$.
This sinusoidal approximation of the armature winding mmf is an excellent
choice for AC machines. If the winding is distributed among a number of slots,
the resultant mmf will have a peak value F_p that is smaller than F_m.
The mmf waveform of the armature of a DC machine approximates a sawtooth waveform. This sawtooth waveform can also be approximated by a sinusoidal waveform.

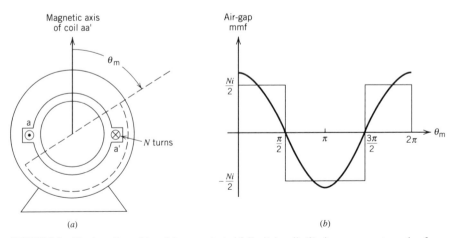

(a)   (b)

FIGURE 5.9   Single coil machine: (a) concentrated full-pitch coil; (b) air-gap magnetomotive force.

Therefore, in the analysis of AC and DC machines, the mmf of the armature windings will be assumed to have a sinusoidal space distribution. Also, for modeling of AC and DC machines, a two-pole machine will be considered.

From Eq. 5.13, it may be seen that the magnetic field intensity $\mathbf{H}$ is equal to the mmf drop $(Ni/2)$ across the air gap divided by the air-gap length g. Therefore, the fundamental component of $\mathbf{H}$ may be expressed as

$$H = H_{a1} = \frac{F_{a1}}{g} = \frac{F_m}{g} \cos \theta_m = H_m \cos \theta_m \tag{5.15}$$

where $H_m = F_m/g = (4/\pi)(Ni/2g)$.

For a distributed winding such as that shown in Fig. 5.10, the air-gap magnetic field intensity will have a peak value H_p, which is less than H_m.

5.5   ROTATING MMF IN AC MACHINES

To understand the theory of polyphase AC machines, it is necessary to study the nature of the mmf produced by a polyphase winding. Before analyzing the three-phase situation, the single-phase situation is considered.

5.5.1   Single-Phase Winding mmf

In the last section, the mmf of a single-phase winding was derived. It is rewritten here as

$$F_{a1} = F_m \cos \theta_m = \frac{4}{\pi} \frac{Ni}{2} \cos \theta_m \tag{5.16}$$

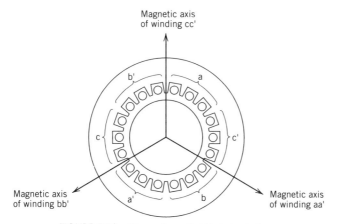

FIGURE 5.10   Three-phase distributed winding.

When the winding is excited by a sinusoidal current, $i = I_a \cos \omega t$, the expression for the mmf becomes

$$F_{a1} = \frac{2NI_a}{\pi} \cos \theta_m \cos \omega t = F_a \cos \theta_m \cos \omega t \qquad (5.17)$$

where $F_a = 2NI_a/\pi$.

Applying trigonometric identities for the product of cosines to Eq. 5.17 yields

$$F_{a1} = \tfrac{1}{2}F_a \cos(\theta_m - \omega t) + \tfrac{1}{2}F_a \cos(\theta_m + \omega t) = F^+ + F^- \qquad (5.18)$$

Equation 5.18 contains two variables: θ_m (space variable) and t (time variable). The first term, F^+, is a traveling wave with amplitude $\tfrac{1}{2}F_a$, traveling in the direction of increasing θ_m. This phenomenon can be seen by plotting the first term as a function of θ_m for two particular values of t as shown in Fig. 5.11.

Similarly, the second term of Eq. 5.18 represents a traveling wave in the negative θ_m direction. Hence, the single-phase winding mmf can be represented by two vectors traveling in opposite directions as shown in Fig. 5.12.

As seen from Figs. 5.11 and 5.12, the sum of the forward- and backward-traveling waves, F^+ and F^-, respectively, yields a pulsating wave F_{a1}. At any set of values of t and θ_m, the net mmf F_{a1} is found by adding the components of F^+ and F^- along the line that makes an angle θ_m with respect to the magnetic axis of winding aa'. For example at $t = t_1$ and $\theta_m = \pi/3$, the value of F_{a1} is

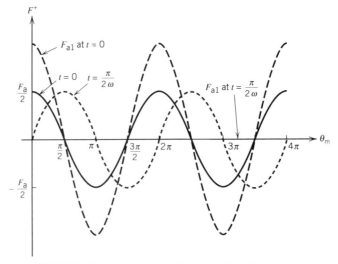

FIGURE 5.11   Traveling wave in the positive θ_m direction.

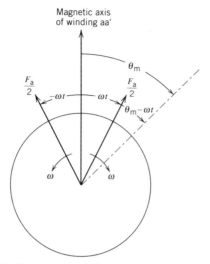

FIGURE 5.12 Double rotating waveforms representation of single-phase mmf.

given by

$$F_{a1} = \frac{F_a}{2}\cos(\frac{\pi}{3} - \omega t_1) + \frac{F_a}{2}\cos(\frac{\pi}{3} + \omega t_1)$$

$$= F_a \cos\frac{\pi}{3}\cos\omega t_1 \tag{5.19}$$

5.5.2   Three-Phase Winding mmf

Armature windings of synchronous machines and induction machines are typically three-phase windings. Such an armature is illustrated in Fig. 5.13.

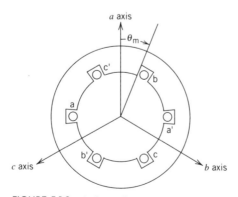

FIGURE 5.13 A three-phase armature winding.

The three phases have equal numbers of coils. At θ_m, the resultant mmf F due to all three phases is given by the sum of the individual phase mmfs. Thus,

$$
\begin{aligned}
F &= F_{a1} + F_{b1} + F_{c1} \\
&= \frac{2}{\pi} N i_a \cos \theta_m + \frac{2}{\pi} N i_b \cos(\theta_m - 120°) \\
&\quad + \frac{2}{\pi} N i_c \cos(\theta_m - 240°)
\end{aligned}
\tag{5.20}
$$

Under balanced three-phase conditions, the instantaneous currents are expressed as follows:

$$
i_a = I_p \cos \omega t \tag{5.21}
$$

$$
i_b = I_p \cos(\omega t - 120°) \tag{5.22}
$$

$$
i_c = I_p \cos(\omega t - 240°) \tag{5.23}
$$

where I_p is the maximum current.

Substituting these current expressions into Eq. 5.20 yields

$$
\begin{aligned}
F(\theta_m, t) = \frac{2}{\pi} N I_p \big[&\cos \theta_m \cos \omega t \\
&+ \cos(\theta_m - 120°) \cos(\omega t - 120°) \\
&+ \cos(\theta_m - 240°) \cos(\omega t - 240°) \big]
\end{aligned}
\tag{5.24}
$$

By using trigonometric identities to simplify the expression inside the brackets in Eq. 5.24, the resultant mmf may be expressed as

$$
F(\theta_m, t) = \frac{3}{2} F_p \cos(\theta_m - \omega t) \tag{5.25}
$$

where $F_p = 2 N I_p / \pi$.

Equation 5.25 represents a traveling wave moving in the direction of positive θ_m. It is shown graphically in Fig. 5.14.

At any time t and position θ_m, the resultant mmf due to all three windings is found by taking the projection of the vector $\frac{3}{2} F_p$ on the line θ_m. For example, at time $t = \pi/(4\omega)$ seconds, the mmf along the a axis is the component of the vector $\frac{3}{2} F_p$ along the line $\theta_m = 0$. Thus,

$$
F(\theta_m, t) = F\left(0, \frac{\pi}{4\omega}\right) = \frac{3}{2} F_p \cos\left(0 - \frac{\pi}{4}\right) = \frac{3}{2} F_p \cos \frac{\pi}{4} \tag{5.26}
$$

In this derivation, it has been assumed that at $t = 0$, the mmf due to coil aa' is maximum along the line $\theta_m = 0$.

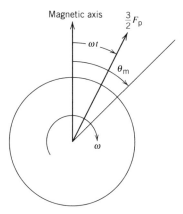

FIGURE 5.14 Graphical representation of three-phase resultant mmf.

DRILL PROBLEMS

D5.5 The windings of a three-phase machine are supplied with currents i_a, i_b, and i_c. The mmfs produced by these currents are given as follows:

$$F_a = Ni_a \cos \theta_m$$
$$F_b = Ni_b \cos(\theta_m - 120°)$$
$$F_c = Ni_c \cos(\theta_m - 240°)$$

a. Assume that the three-phase windings are connected in series and supplied by one voltage source, that is, $i_a = i_b = i_c$. Find the resultant mmf due to all three windings as a function of θ_m.

b. The three-phase windings are connected to a balanced three-phase voltage supply, and they take the following currents:

$$i_a = I_p \cos \omega t$$
$$i_b = I_p \cos(\omega t - 120°)$$
$$i_c = I_p \cos(\omega t - 240°)$$

Find the resultant mmf.

c. Let $\theta_m = \omega_m t + \alpha$. Determine the relationship between ω_m and ω that results in maximum mmf.

D5.6 Show that

$$F \cos \alpha \cos \beta + F \cos(\alpha - 120°) \cos(\beta - 120°)$$
$$+ F \cos(\alpha - 240°) \cos(\beta - 240°) = \tfrac{3}{2}F \cos(\alpha - \beta)$$

D5.7 A synchronous generator has a three-phase winding with 10 turns per phase. The three phase currents are given by

$$i_a = 100 \cos 377t \ \text{A}$$
$$i_b = 100 \cos(377t - 120°)$$
$$i_c = 100 \cos(377t - 240°)$$

Determine (a) the fundamental component of the mmf of each winding and (b) the resultant mmf.

5.6   GENERATED VOLTAGE IN ROTATING MACHINES

In Section 5.2, the calculation of the induced voltage (emf) in a coil was described. By Faraday's law, an emf is produced when the flux linking a coil changes. In this section, the generated voltage for a three-phase rotating machine will be derived. An elementary three-phase synchronous machine is shown in Fig. 5.15.

The rotor coil receives DC current through the slip rings. This current I_f produces an mmf that is maximum along the rotor field's magnetic axis. Its fundamental component is distributed sinusoidally around the rotor's periphery.

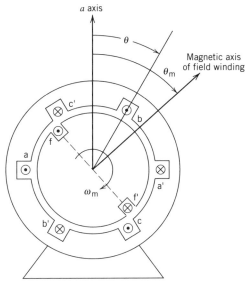

FIGURE 5.15   A three-phase AC synchronous machine.

Therefore, the mmf in the air gap along the line θ may be expressed as

$$F_{r1} = \frac{4}{\pi} \frac{N_f I_f}{2} \cos(\theta_m - \theta) \qquad (5.27)$$

where N_f is the number of turns of the field winding.

The magnetic field and flux density are then obtained as follows:

$$H = \frac{F_{r1}}{g} = \frac{2N_f I_f}{\pi g} \cos(\theta_m - \theta) \qquad (5.28)$$

$$B = \mu_0 H = \frac{2\mu_0 N_f I_f}{\pi g} \cos(\theta_m - \theta) \qquad (5.29)$$

where g is the length of the air gap. Thus, the flux linking the stator windings aa', bb', and cc' is computed as

$$\phi_{aa'} = \int_S \mathbf{B} \cdot d\mathbf{S} = \int_{-\pi/2}^{\pi/2} \frac{2\mu_0 N_f I_f}{\pi g} [\cos(\theta_m - \theta)] l r \, d\theta$$

$$= \frac{4\mu_0 N_f I_f}{\pi g} l r \, \cos \theta_m = \Phi_p \cos \theta_m \qquad (5.30)$$

where

$\Phi_p = 4\mu_0 N_f I_f l r / (\pi g)$

$\quad$ = maximum flux linking one loop of winding aa' due to I_f

l = axial length of the machine

r = rotor radius

Denoting by N_a the number of turns in each armature phase winding, the total flux linkages are obtained as follows:

$$\lambda_{aa'} = N_a \Phi_p \cos \theta_m \qquad (5.31)$$

$$\lambda_{bb'} = N_a \Phi_p \cos(\theta_m - 120°) \qquad (5.32)$$

$$\lambda_{cc'} = N_a \Phi_p \cos(\theta_m - 240°) \qquad (5.33)$$

The angle θ_m is the relative position of the rotor magnetic axis with respect to the magnetic axis of the winding aa'. Since the rotor turns at a constant speed of ω rad/s, the angle θ_m may be expressed as

$$\theta_m = \int_0^t \omega \, dt + \theta_m(0) = \omega t + \theta_m(0) \qquad (5.34)$$

Thus, the total flux linkages may be written as

$$\lambda_{aa'} = N_a\Phi_p \cos[\omega t + \theta_m(0)] \tag{5.35}$$

$$\lambda_{bb'} = N_a\Phi_p \cos[\omega t + \theta_m(0) - 120°] \tag{5.36}$$

$$\lambda_{cc'} = N_a\Phi_p \cos[\omega t + \theta_m(0) - 240°] \tag{5.37}$$

From these equations, and since the flux linkages are time varying, the induced voltages are derived as follows:

$$e_{aa'} = \frac{d\lambda_{aa'}}{dt} = -\omega N_a\Phi_p \sin[\omega t + \theta_m(0)]$$

$$= \omega N_a\Phi_p \cos[\omega t + \theta_m(0) + 90°] \tag{5.38}$$

$$e_{bb'} = \frac{d\lambda_{bb'}}{dt} = \omega N_a\Phi_p \cos[\omega t + \theta_m(0) - 30°] \tag{5.39}$$

$$e_{cc'} = \frac{d\lambda_{cc'}}{dt} = \omega N_a\Phi_p \cos[\omega t + \theta_m(0) - 150°] \tag{5.40}$$

Equations 5.38–5.40 represent a set of balanced induced voltages. The maximum value of the voltage is given by

$$E_{max} = \omega N_a\Phi_p = 2\pi f N_a\Phi_p \tag{5.41}$$

and its rms value is

$$E_{rms} = \frac{2\pi}{\sqrt{2}} f N_a\Phi_p = 4.44 f N_a\Phi_p \tag{5.42}$$

where f is the frequency in hertz.

It has been assumed that the three-phase windings were full-pitch, concentrated windings. In full-pitch windings, the two sides of each coil are placed in slots that are 180 electrical degrees apart. Concentrated windings have the conductors of each phase winding placed in just one pair of slots.

In actual machines, the coils of each armature phase winding are distributed among a number of slots with corresponding coil sides in slots possibly less than 180 electrical degrees apart. This results in a reduction in the generated voltages for each phase because the induced voltages in the coils are no longer in time phase. Hence, the phasor sum of the coil voltages is less than it would be if all the coils in one phase were concentrated in one pair of full-pitch slots. Therefore, for distributed and fractional-pitch windings, a reduction factor K_W is introduced. K_W is called the machine *winding factor* and has a value from 0.85 to 0.95. Thus, Eq. 5.42 is modified as follows:

$$E_{rms} = 4.44 K_W f N_a\Phi_p \tag{5.43}$$

When the terminals of the machine are connected to a balanced electrical load, a set of balanced currents will flow. These balanced currents produce a magnetic field that can be represented as a rotating mmf at an angular velocity ω, the frequency of the generated voltage, which is also the angular velocity of the rotor in rad/s. If there are more than two poles, the rotor speed and the radian frequency of the voltage are related by Eq. 5.11.

The voltage induced in DC machines can also be analyzed by the same principles as in AC machines. In DC machines, however, the field winding is on the stator and the armature winding is on the rotor. The rotor coils are distributed over the periphery of the rotor. The voltage induced in each turn of the rotor winding is a sinusoidal function of time. Mechanical rectification is provided by a commutator and brush combination. The average, or DC, value of the voltage between brushes is given by

$$E_a = \frac{1}{\pi} \int_0^\pi \omega N \Phi_p \sin \omega t \, d(\omega t) = \frac{2}{\pi} \omega N \Phi_p \qquad (5.44)$$

Equation 5.44 is more conveniently expressed in terms of the mechanical speed ω_m in rad/s, or n in rev/min (rpm). Thus, E_a may be written as

$$E_a = \frac{2}{\pi} \left(\frac{p}{2} \omega_m \right) N \Phi_p = \frac{p \omega_m N \Phi_p}{\pi} \qquad (5.45)$$

Since $\omega_m = 2\pi n/60$, Eq. 5.45 may also be written as

$$E_a = \frac{2}{\pi} \left(\frac{p}{2} \frac{2\pi n}{60} \right) N \Phi_p = \frac{p n N \Phi_p}{30} \qquad (5.46)$$

where

p = number of poles

N = total number of turns in series between armature terminals

In terms of the total number of armature conductors Z and the number of parallel paths a between armature terminals, the number of series turns is given by

$$N = \frac{Z}{2a} \qquad (5.47)$$

Substituting Eq. 5.47 into Eq. 5.45 gives

$$E_a = \frac{pZ}{2\pi a} \Phi_p \omega_m \qquad (5.48)$$

The commutator and brush combination converts the AC to a DC voltage in the case of a DC generator. In the DC motor, each conducting loop on the rotor is in contact with the applied DC voltage through the commutator and brush. Hence, the current in each loop will be a pulse that can be approximated to a sinusoidal function, that is, the fundamental component in a Fourier series representation.

DRILL PROBLEMS

D5.8 A two-pole, three-phase, wye-connected, round-rotor synchronous generator has the following data:

$$N_a = 24 \text{ turns per phase} \qquad N_f = 500 \text{ turns}$$
$$l = 4 \text{ m} \qquad r = 50 \text{ cm} \qquad g = 20 \text{ mm}$$

The field current, or rotor current, is adjusted to $I_f = 8$ A. Find (a) the peak fundamental mmf due to the field current and (b) the induced voltage in each phase and their rms values.

D5.9 The two-pole machine shown in Fig. 5.16 has a field winding located on the stator and an armature winding on the rotor, with $N_f = 800$ turns and $N_a = 50$ turns, respectively. It has a uniform air gap of 0.5 mm. The armature has a diameter of 0.5 m, and it is 1.5 m long. The field winding carries a

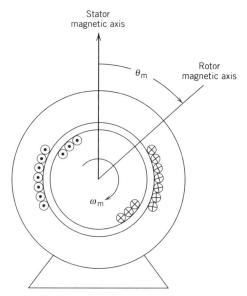

Stator
magnetic axis

θ_m

Rotor
magnetic axis

ω_m

FIGURE 5.16 Rotating machine with uniform air gap.

current of 2 A. The rotor is driven at 3600 rpm. Determine (a) the frequency of the rotor induced voltage and (b) the instantaneous voltage and rms voltage induced in the rotor coil.

5.7   TORQUE IN ROUND-ROTOR MACHINES

The behavior of an electromechanical energy conversion device can be described in terms of its equivalent circuit, which is governed by Kirchhoff's voltage law, and its torque equation. The previous section derived expressions for generated voltage that are sufficient to model equivalent circuits. In this section, the torque equations are derived.

Two points of view are presented. The first considers the machine as a set of coupled coils. The second point of view considers the machine as two groups of windings producing magnetic fields in the air gap—one group on the rotor and the other group on the stator. In this manner, the torque is expressed as the tendency for two magnetic fields to line up in the same manner as permanent magnets tend to align themselves. The generated voltage is expressed as the result of relative motion between a winding and a magnetic field, or mmf.

Before the two approaches are described in greater detail, the expression for electromagnetic torque for an electromechanical energy conversion device will be derived. This expression can be obtained from the energy conservation law.

The model of a lossless electromechanical energy conversion device is shown in Fig. 5.17. This schematic represents a motor.

For the energy conversion system shown, the energy conservation law may be expressed as

$$W_e = W_f + W_m \qquad (5.49)$$

where

W_e = electrical energy input from the electrical source

W_f = energy stored in the magnetic field of the two coils associated with the electrical inputs

W_m = mechanical energy output

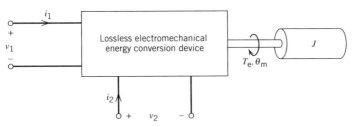

FIGURE 5.17   Model of electromechanical energy converter.

The electrical energy and mechanical energy are given by the following expressions:

$$W_e = \int_0^t v_1 i_1 \, d\tau + \int_0^t v_2 i_2 \, d\tau = \int_0^t p_1 \, d\tau + \int_0^t p_2 \, d\tau \qquad (5.50)$$

$$W_m = \int_{\theta_m(0)}^{\theta_m} T_m \, d\phi_m = \int_{\theta_m(0)}^{\theta_m} T_e \, d\phi_m \qquad (5.51)$$

Equations 5.50 and 5.51 can be written in their differential forms as Eqs. 5.52 and 5.53, respectively.

$$dW_e = v_1 i_1 \, dt + v_2 i_2 \, dt \qquad (5.52)$$

$$dW_m = T_m \, d\theta_m = T_e \, d\theta_m \qquad (5.53)$$

Since the windings are assumed to have negligible resistances, the terminal voltages are equal to the induced voltages in the coils:

$$v_1 = e_1 = \frac{d\lambda_1}{dt} \qquad (5.54)$$

$$v_2 = e_2 = \frac{d\lambda_2}{dt} \qquad (5.55)$$

Substituting Eqs. 5.54 and 5.55 into Eq. 5.52 yields

$$dW_e = i_1 \, d\lambda_1 + i_2 \, d\lambda_2 \qquad (5.56)$$

Substituting Eqs. 5.53 and 5.56 into the differential form of Eq. 5.49, and solving for dW_f gives

$$dW_f = i_1 \, d\lambda_1 + i_2 \, d\lambda_2 - T_e \, d\theta_m \qquad (5.57)$$

In Fig. 5.17, one independent variable can be specified for each of the terminal pairs; therefore, two electrical variables and one mechanical (or spatial) variable are specified. In Eq. 5.57, the variables λ_1, λ_2, and θ_m are selected as independent variables; thus, the magnetic *field energy* is expressed as a function of these variables: $W_f(\lambda_1, \lambda_2, \theta_m)$. However, the derivation of an expression for the energy function is not an easy task inasmuch as the flux linkages are not physically measurable.

At this point, a new function called *coenergy* is defined. It is designated as W_f', and it is associated with the field energy function W_f as follows:

$$W_f' + W_f = \lambda_1 i_1 + \lambda_2 i_2 \tag{5.58}$$

Equation 5.58 can be written in differential form, and the expression for the differential of the coenergy function dW_f' is obtained as follows:

$$
\begin{aligned}
dW_f' &= d(\lambda_1 i_1) + d(\lambda_2 i_2) - dW_f \\
&= \lambda_1 \, di_1 + i_1 \, d\lambda_1 + \lambda_2 \, di_2 + i_2 \, d\lambda_2 - dW_f
\end{aligned} \tag{5.59}
$$

Substituting Eq. 5.57 into Eq. 5.59 and simplifying yield

$$dW_f' = \lambda_1 \, di_1 + \lambda_2 \, di_2 + T_e \, d\theta_m \tag{5.60}$$

The coenergy function is seen to be dependent on i_1, i_2, and θ_m, and it is expressed as a function of these variables. Thus,

$$W_f' = W_f'(i_1, i_2, \theta_m) \tag{5.61}$$

The total differential of the coenergy function may be written as

$$dW_f' = \frac{\partial W_f'}{\partial i_1} \, di_1 + \frac{\partial W_f'}{\partial i_2} \, di_2 + \frac{\partial W_f'}{\partial \theta_m} \, d\theta_m \tag{5.62}$$

Upon term-by-term comparison of Eqs. 5.60 and 5.62, the expression for electromagnetic torque is obtained.

$$T_e = \frac{\partial W_f'(i_1, i_2, \theta_m)}{\partial \theta_m} \tag{5.63}$$

The torque expression of Eq. 5.63 can be generalized for the case of n windings, or n currents, as follows:

$$T_e = \frac{\partial W_f'(i_1, i_2, \dots, i_n, \theta_m)}{\partial \theta_m} \tag{5.64}$$

When the magnetic cores on the rotor and stator are assumed to have linear characteristics, the energy function W_f may be expressed as

$$W_f = \tfrac{1}{2}\lambda_1 i_1 + \tfrac{1}{2}\lambda_2 i_2 \tag{5.65}$$

Substituting this expression for W_f into Eq. 5.58 yields

$$W_f' = \tfrac{1}{2}\lambda_1 i_1 + \tfrac{1}{2}\lambda_2 i_2 \tag{5.66}$$

Thus, it is seen that the field energy function and the coenergy function are the same in a linear magnetic circuit.

In *linear electromagnetic systems*, the relationships between flux linkages and currents are given by

$$\lambda_1 = L_{11}i_1 + L_{12}i_2 \tag{5.67}$$

$$\lambda_1 = L_{21}i_1 + L_{22}i_2 \tag{5.68}$$

where

L_{11} = self-inductance of winding 1

L_{22} = self-inductance of winding 2

$L_{12} = L_{21}$ = mutual inductance between windings 1 and 2

These inductances are functions of only the mechanical variable, or spatial variable, θ_m. Substituting Eqs. 5.67 and 5.68 into Eq. 5.66 yields the expression for the coenergy function.

$$W'_f = \tfrac{1}{2}L_{11}i_1^2 + L_{12}i_1i_2 + \tfrac{1}{2}L_{22}i_2^2 \tag{5.69}$$

Thus, the expression for the electromagnetic torque may be written as follows:

$$
T_e = \left(\frac{\partial W'_f(i_1, i_2, \theta_m)}{\partial \theta_m} \right)\Bigg|_{i_1, i_2 = \text{ fixed}}
$$

$$
= \left(\frac{1}{2}\frac{\partial \left(L_{11}i_1^2\right)}{\partial \theta_m} + \frac{\partial (L_{12}i_1i_2)}{\partial \theta_m} + \frac{1}{2}\frac{\partial \left(L_{22}i_2^2\right)}{\partial \theta_m} \right)\Bigg|_{i_1, i_2 = \text{ fixed}} \tag{5.70}
$$

Since the inductances are functions of only the spatial variable θ_m, Eq. 5.70 reduces to

$$
T_e = \frac{1}{2}i_1^2 \frac{dL_{11}(\theta_m)}{d\theta_m} + i_1 i_2 \frac{dL_{12}(\theta_m)}{d\theta_m} + \frac{1}{2}i_2^2 \frac{dL_{22}(\theta_m)}{d\theta_m} \tag{5.71}
$$

Note that no electromagnetic torque is developed if none of the inductances is a function of the spatial variable θ_m.

EXAMPLE 5.2

The two-pole machine shown in Fig. 5.16 has two mutually coupled coils. The first coil, located on the stator, is held fixed, and the second coil, on the rotor, is free to rotate. The inductances of the coils are given by

$$L_{11} = 10 \qquad L_{22} = 8 \qquad L_{12} = L_{21} = 6 \cos \theta_m$$

Let the current in the first coil be denoted by i_1 and the current in the second coil by i_2. The magnetic axes of the two coils are initially displaced by an angle $\theta_m(0) = 30°$. Find the electrical torque for the following conditions:

a. $i_1 = 2$ A and $i_2 = 0$
b. $i_1 = i_2 = 2$ A
c. $i_1 = 2$ A and $i_2 = \sqrt{2} \sin \omega t$ A
d. $i_1 = i_2 = \sqrt{2} \sin \omega t$ A
e. $i_1 = 2$ A and coil 2 short-circuited

Solution Using Eq. 5.69, the expression for the coenergy function is given by

$$W_f' = \tfrac{1}{2}(10)i_1^2 + 6 \cos \theta_m i_1 i_2 + \tfrac{1}{2}(8)i_2^2$$

The general expression for the electromagnetic torque can be found by using Eq. 5.71 as follows:

$$T_e = \frac{1}{2}i_1^2 \frac{d}{d\theta_m}(10) + i_1 i_2 \frac{d}{d\theta_m}(6 \cos \theta_m) + \frac{1}{2}i_2^2 \frac{d}{d\theta_m}(8) = -6i_1 i_2 \sin \theta_m$$

a. $T_e = -6i_1 i_2 \sin \theta_m = -(6)(2)(0) \sin 30° = 0$
b. $T_e = -6i_1 i_2 \sin \theta_m = -(6)(2)(2) \sin 30° = -12$ N-m
c. $T_e = -6i_1 i_2 \sin \theta_m = -(6)(2)(\sqrt{2} \sin \omega t) \sin 30°$
$\qquad = -6\sqrt{2} \sin \omega t$ N-m
d. $T_e = -6i_1 i_2 \sin \theta_m = -(6)(\sqrt{2} \sin \omega t)(\sqrt{2} \sin \omega t) \sin 30°$
$\qquad = -6 \sin^2 \omega t$
e. Since the rotor is short-circuited, the flux linkage of the rotor may be set equal to zero; thus,

$$0 = L_{21}i_1 + L_{22}i_2$$

Therefore, the rotor current can be expressed as

$$i_2 = (L_{21}/L_{22})i_1 = [(6 \cos \theta_m)/8](2) = 1.5 \cos \theta_m \text{ A}$$

Substituting in the general expression for torque yields

$$T_e = -6i_1 i_2 \sin \theta_m = -(6)(2)(1.5 \cos 30°) \sin 30° = -7.8 \text{ N-m}$$

EXAMPLE 5.3

A linear electromechanical energy conversion system has two electrical inputs and one mechanical output. The self- and mutual inductances of the coils are

$$L_{11} = 5 + \cos 2\theta_m \text{ mH}$$
$$L_{22} = 50 + 10 \cos 2\theta_m \text{ H}$$
$$L_{12} = L_{21} = 100 \cos \theta_m \text{ mH}$$

The first coil is supplied with a current $i_1 = 1$ A, and the second draws a current $i_2 = 10$ mA. Determine (a) the general expression for electromagnetic torque T_e and (b) the maximum torque.

Solution

a. For this linear system, the coenergy function is found by using Eq. 5.69. Substituting the expressions for the inductances yields

$$W_f' = W_f = \tfrac{1}{2}L_{11}i_1^2 + L_{12}i_1i_2 + \tfrac{1}{2}L_{22}i_2^2$$
$$= \tfrac{1}{2}(5 + \cos 2\theta_m) \times 10^{-3}i_1^2 + 0.10 \cos \theta_m i_1 i_2$$
$$+ \tfrac{1}{2}(50 + 10 \cos 2\theta_m)i_2^2$$

The expression for the electromagnetic torque is obtained as follows:

$$T_e = \left(\frac{\partial W_f'(i_1, i_2, \theta_m)}{\partial \theta_m}\right)\bigg|_{i_1 = 1.0, i_2 = 0.01}$$
$$= \left(\frac{1}{2}\frac{\partial(i_1^2 L_{11})}{\partial \theta_m} + \frac{\partial(i_1 i_2 L_{12})}{\partial \theta_m} + \frac{1}{2}\frac{\partial(i_2^2 L_{22})}{\partial \theta_m}\right)\bigg|_{i_1 = 1.0, i_2 = 0.01}$$
$$= \tfrac{1}{2}(1.0)^2(-2 \sin 2\theta_m) \times 10^{-3} + (1.0)(0.01)(-0.1 \sin \theta_m)$$
$$+ \tfrac{1}{2}(0.01)^2(-20 \sin 2\theta_m)$$
$$= -(2 \sin 2\theta_m + \sin \theta_m) \times 10^{-3}$$

b. At maximum torque

$$\frac{dT_e}{d\theta_m} = 0$$

Differentiating T_e from part (a),

$$4 \cos 2\theta_m + \cos \theta_m = 0$$

Solving for θ_m by the quadratic formula,

$$\theta_m = 49.7°, 140.6° \quad \text{(extraneous)}$$

Substituting the value of $\theta_m = 49.7°$ into the expression for the electromagnetic torque yields

$$T_{e,max} = -[2 \sin 2(49.7°) + \sin 49.7°] \times 10^{-3} = -2.74 \times 10^{-3} \text{ N-m}$$

DRILL PROBLEMS

D5.10 An electromagnetic system has two windings and a nonuniform air gap. The self- and mutual inductances are given by

$$L_{11} = 6 + 1.5 \cos 2\theta \qquad L_{22} = 4 + \cos 2\theta \qquad M = 5 \cos \theta$$

The windings are connected to a DC voltage source. The first winding takes a current $I_1 = 10$ A, and the second draws a current $I_2 = 2$ A.

a. Determine the developed torque as a function of θ.

b. Find the stored energy as a function of θ.

c. For $\theta = 30°$, find the magnitude and direction of the torque on the rotor.

D5.11 An electromagnetic system consists of two mutually coupled coils as shown in Fig. 5.16. The inductances of the coils are

$$L_{ss} = 5 \text{ H}$$
$$L_{rr} = 3 \text{ H}$$
$$L_{sr} = 2 \cos \theta \text{ H}$$

Find the electromagnetic torque for the following cases.

a. $I_s = 10$ A and $I_r = 0$

b. $I_s = I_r = 10$ A

c. $I_s = 10$ A and $i_r = 10 \sin \omega t$

d. $i_s = i_r = 10 \sin \omega t$

5.7.1   The Coupled-Coils Approach

An elementary machine is shown in Fig. 5.18. It has one winding on the stator and one winding on the rotor. The slot openings for the stator and rotor coils

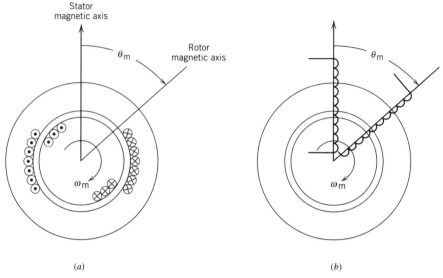

(a)   (b)

FIGURE 5.18   A round-rotor machine: (a) winding distribution; (b) schematic representation.

are not shown in the figure. Machines that have more than one coil on the stator may also be analyzed by representing the stator coils with one equivalent winding.

For this machine containing two coils, the electromagnetic torque T_e is given by

$$T_e = \frac{\partial W'(i_s, i_r, \theta_m)}{\partial \theta_m} \qquad (5.72)$$

Since linearity has been assumed for both stator and rotor magnetic cores, the coenergy function W'_f may be expressed as

$$W'_f = \tfrac{1}{2}L_{ss}i_s^2 + \tfrac{1}{2}L_{rr}i_r^2 + L_{sr}i_s i_r \qquad (5.73)$$

where
L_{ss} = self-inductance of the stator coil
L_{rr} = self-inductance of the rotor coil
L_{sr} = mutual inductance between stator and rotor coils
i_s = stator current
i_r = rotor current

Assuming infinite permeability for the magnetic cores, the inductance parameters are given as

$$L_{ss} = \frac{4\mu_0 N_s^2 l r}{\pi g} \qquad (5.74)$$

$$L_{rr} = \frac{4\mu_0 N_r^2 l r}{\pi g} \tag{5.75}$$

$$L_{sr} = \frac{4\mu_0 N_s N_r l r}{\pi g} \cos \theta_m = M \cos \theta_m \tag{5.76}$$

where

μ_0 = permeability of free space

M = $(4\mu_0 N_s N_r l r)/(\pi g)$ = maximum mutual inductance that occurs when stator and rotor magnetic axes are aligned

N_s = number of turns on the stator

N_r = number of turns on the rotor

l = axial length

r = radius of the rotor

g = length of the air gap

Substituting Eqs. 5.74–5.76 into Eq. 5.73 and differentiating the result with respect to θ_m give the expression for the electromagnetic torque.

$$\begin{aligned} T_e &= -M i_s i_r \sin \theta_m \\ &= M i_s i_r \cos(\theta_m + 90°) \end{aligned} \tag{5.77}$$

It may be noted from Eq. 5.77 that the electromagnetic torque is directly proportional to the product of the stator current and the rotor current. In steady state operations, this torque is balanced by its counterpart mechanical torque T_m. For generator action, T_m is an input quantity; while for motor action, T_m is the output quantity.

EXAMPLE 5.4

For the two-pole machine of Fig. 5.18, the currents flowing in the stator and rotor windings are given by $i_s = I_s \cos \omega_s t$ and $i_r = I_r$ = constant, respectively, where ω_s is the angular frequency of the stator current.

a. Find the expression for the electromagnetic torque.

b. Find the speed at which average torque is nonzero.

c. Find the average torque $T_{e,ave}$ if $I_r = 4$ A, $I_s = 50$ A, $\theta_m(0) = 10°$, and $M = 2$ H in part (b).

Solution

a. The angle θ_m is found by using Eq. 5.34. Thus,

$$\theta_m = \int_0^t \omega_m \, dt + \theta_m(0) = \omega_m t + \theta_m(0)$$

Assume ω_m remains constant. Then the torque may be expressed as

$$T_e = -MI_s \cos \omega_s t \, I_r \sin \theta_m = -MI_s I_r \cos \omega_s t \, \sin[\omega t + \theta_m(0)]$$

b. The expression for T_e may also be written as

$$T_e = -MI_s I_r \{ \tfrac{1}{2} \sin[(\omega_m + \omega_s)t + \theta_m(0)] + \tfrac{1}{2} \sin[(\omega_m - \omega_s)t + \theta_m(0)] \}$$

From this expression for torque, it is seen that a nonzero average will be present if and only if $\omega_m = \omega_s$. For a 60-Hz stator current frequency, the speed is

$$n = 120f/p = (120)(60)/2 = 3600 \text{ rpm}$$

c. For a nonzero average torque, $\omega_m = \omega_s$; thus,

$$T_e = -\tfrac{1}{2} MI_s I_r \{ \sin[2\omega_s t + \theta_m(0)] + \sin \theta_m(0) \}$$

Since the average of a sinusoidal function of time is zero, the first term on the right-hand side of the preceeding expression reduces to zero; therefore,

$$T_{e,\text{ave}} = -\tfrac{1}{2} MI_s I_r \sin \theta_m(0) = -\tfrac{1}{2}(2)(50)(4) \sin 10° = -34.7 \text{ N-m}$$

DRILL PROBLEMS

D5.12 Derive Eqs. 5.74, 5.75, and 5.76. Assume that the permeabilities of rotor and stator ferromagnetic cores are infinitely high.

D5.13 Two mutually coupled coils, one mounted on a stator and the other on a rotor, are shown in Fig. 5.16. The self- and mutual inductances are given as follows:

$$L_{ss} = 0.50 \text{ H}$$
$$L_{rr} = 0.25 \text{ H}$$
$$L_{sr} = 1.0 \cos \theta \text{ H}$$

The coils are connected in series, and the combination is excited by a sinusoidal voltage source.

a. Determine the instantaneous torque if the coils draw a current $i(t) = 10\sqrt{2} \sin 377t$ A.

b. Determine the time-averaged torque T_{ave} as a function of the angular displacement θ.

c. Find the value of T_{ave} for $\theta = 60°$.

d. The series-connected coils are supplied from a DC source, and they take a current $I = 10$ A. Compute the torque at $\theta = 60°$, and compare with the result of part (c).

5.7.2 The Magnetic Field Approach

In this section, the machine model is based on the interaction of magnetic fields of the stator and rotor windings in the air gap. Since the core is assumed to have infinite permeability, the core magnetic field, or mmf drop, is approximately zero.

A two-pole synchronous machine is shown in Fig. 5.19. The currents in the stator and rotor windings create magnetic flux in the air gap. The stator and rotor mmfs can each be represented by a rotating mmf as shown in Fig. 5.19b. For the synchronous machine, the expressions for the mmfs of the stator and rotor are given by $F_s = \frac{3}{2}(2N_sI_s/\pi)$ and $F_r = 2N_rI_r/\pi$, respectively.

The interaction of the stator and rotor mmfs, or magnetic fields, creates an electromagnetic torque T_e. This torque tends to align the stator and rotor mmfs. As long as the angle between the two mmfs, δ_{sr}, is nonzero, an electromagnetic torque will exist.

At steady state, the angle δ_{sr} is a constant. In other words, ω and ω_m are equal. Note that ω and ω_m represent the angular speed of F_s and F_r, respectively, from the same reference line. In the derivation to follow, all losses and leakages are neglected and the permeability of the stator and rotor cores is assumed to be infinite.

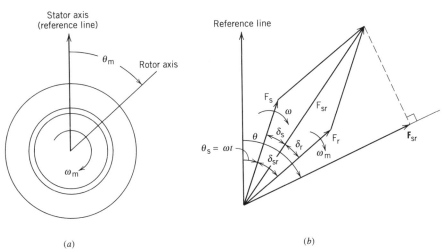

(a) (b)

FIGURE 5.19 A two-pole machine: (a) elementary model; (b) vector diagram of mmf waves.

From Ampère's law, the magnetic field intensity **H** in the air gap is given by

$$\mathbf{H} = \frac{\mathbf{F}_{sr}}{g} \tag{5.78}$$

where

$\mathbf{F}_{sr}$ = resultant mmf at the air gap along the line θ

g = length of the air gap

It is seen from Fig. 5.19b that $\mathbf{F}_{sr}$ is the component of F_{sr} along the line θ. The expression for F_{sr} is found from Eq. 5.79.

$$F_{sr}^2 = F_s^2 + F_r^2 + 2F_s F_r \cos \delta_{sr} \tag{5.79}$$

Hence, the resultant mmf $\mathbf{F}_{sr}$ is given by

$$\mathbf{F}_{sr} = F_{sr} \cos(\theta - \omega t - \delta_s) \tag{5.80}$$

Substituting Eq. 5.80 into Eq. 5.78 yields

$$\mathbf{H} = \frac{F_{sr}}{g} \cos(\theta - \omega t - \delta_s) \tag{5.81}$$

Because of the assumption of magnetic core linearity, the coenergy function is equal to the energy function. The energy density W in the air gap is given by

$$W = \tfrac{1}{2}\mathbf{B} \cdot \mathbf{H} = \tfrac{1}{2}\mu_0 \mathbf{H} \cdot \mathbf{H} = \tfrac{1}{2}\mu_0 H^2 \tag{5.82}$$

Substituting the expression for **H** from Eq. 5.81 into 5.82 yields

$$W = \frac{\mu_0 F_{sr}^2}{2g^2} \cos^2(\theta - \omega t - \delta_s) \tag{5.83}$$

The average energy density (or coenergy density) is found from Eq. 5.83 as

$$W_{ave} = W'_{ave} = \frac{\mu_0 F_{sr}^2}{4g^2} \tag{5.84}$$

Finally, the total coenergy W' may be expressed as

$$W' = \frac{\mu_0 F_{sr}^2}{4g^2}(2\pi r l g) \tag{5.85}$$

where

r = radius of the rotor

l = axial length of the machine

g = air-gap clearance

Substituting the expression for F_{sr} from Eq. 5.79 into Eq. 5.85 yields

$$W' = \frac{\pi\mu_0 r l}{2g}(F_s^2 + F_r^2 + 2F_sF_r \cos \delta_{sr}) \tag{5.86}$$

The electromagnetic torque T_e is the rate of change of the coenergy W' with respect to rotor position θ_m, that is,

$$T_e = \frac{\partial W'}{\partial \theta_m} \tag{5.87}$$

Since θ_m can be expressed as the sum of θ_s and δ_{sr},

$$\theta_m = \theta_s + \delta_{sr} \tag{5.88}$$

Therefore, the partial derivative of W' with respect to δ_{sr} also gives the electromagnetic torque. Thus,

$$T_e = -\frac{\pi\mu_0 r l}{g}F_sF_r \sin \delta_{sr} \tag{5.89}$$

Equation 5.89 states that T_e is proportional to the peak value of the stator and rotor mmfs and the sine of δ_{sr}, which is the angle between the two rotating mmfs. The negative sign means that the fields tend to align themselves.

Equal and opposite torques are exerted on the stator and rotor. The torque on the stator is transmitted through the machine frame to the foundation. The torque on the rotor is balanced by the mechanical torque.

Equations 5.89 and 5.77 are similar in form. They represent alternative ways of calculating the torque. If F_s and F_r are expressed in terms of i_s and i_r, Eq. 5.89 reduces to Eq. 5.77.

EXAMPLE 5.5

A two-pole, three-phase, 60-Hz synchronous generator has a rotor radius of 10 cm, an air-gap length of 0.25 mm, and a rotor length of 40 cm. The rotor field winding has 300 turns, and the stator windings have 80 turns/phase. When connected to a three-phase electrical load, a balanced set of stator currents flow. At t = 0, the current in phase a is 15 A, its maximum value. At this same instant, the rotor axis makes an angle of 20 degrees with the reference line (refer to Fig. 5.19). The rotor current is constant at 2 A.

a. Calculate electromagnetic torque using Eq. 5.77.

b. Calculate electromagnetic torque using Eq. 5.89.

Solution

a. The mutual inductance is found by using Eq. 5.76 as follows:

$$M = \frac{4\mu_0 N_s N_r l r}{\pi g}$$

$$= \frac{4(4\pi \times 10^{-7})(80)(300)(0.40)(0.10)}{\pi(0.00025)} = 6.14 \text{ H}$$

In deriving Eq. 5.77, it was assumed that there is only one winding on the stator. In this example, there are three windings on the stator. Hence, the one winding equivalent for the stator should carry a constant current of $\frac{3}{2}I_a$, where I_a is the current in phase a. Therefore, the electromagnetic torque T_e is given by

$$T_e = -M\left[\tfrac{3}{2}I_a\right]I_r \sin \theta_m(0)$$
$$= -(6.14)\left[\tfrac{3}{2}(15)\right](2) \sin 20° = -94.5 \text{ N-m}$$

b. From Eqs. 5.24 and 5.25, the stator mmf is given as

$$F_s = \frac{3}{2}F_p = \frac{3}{2}\left(\frac{2N_s I_a}{\pi}\right) = \frac{3}{2}\left[\frac{(2)(80)(15)}{\pi}\right] = 1146 \text{ A-t}$$

The rotor mmf is

$$F_r = \frac{2N_f I_f}{\pi} = \frac{(2)(300)(2)}{\pi} = 382 \text{ A-t}$$

Since $\omega_m = \omega_s$, and referring to Fig. 5.19,

$$\delta_{sr} = \theta_m - \theta_s = \omega_m t + \theta_m(0) - \omega_s t = \theta_m(0) = 20°$$

Therefore, according to Eq. 5.89,

$$T_e = -\frac{\pi(4\pi \times 10^{-7})(0.10)(0.40)}{0.00025}(1146)(382) \sin 20° = -94.5 \text{ N-m}$$

Thus, it is seen that Eqs. 5.77 and 5.89 yield the same value for electromagnetic torque.

DRILL PROBLEMS

D5.14   In Drill Problem D5.7, the mmf due to the generator rotor field winding may be expressed in terms of the rotor position θ_m and the peak value of rotor mmf F_r as follows:

$$F_{r1} = F_r \cos(\theta - \theta_m)$$

Let $\theta_m = (377t + \pi/2)$ radians and $F_r = 3000$ A-t. Find an expression for mmf due to both armature and field windings.

D5.15   Show that Eq. 5.89 is equivalent to Eq. 5.77.

REFERENCES

1. Bergseth, F. R., and S. S. Venkata. *Introduction to Electric Energy Devices*. Prentice Hall, Englewood Cliffs, N.J., 1987.
2. Del Toro, Vincent. *Electric Machines and Power Systems*. Prentice-Hall, Englewood Cliffs, N.J., 1985.
3. Fitzgerald, A. E., Charles Kingsley, and Stephen Umans. *Electric Machinery*. 4th ed. McGraw-Hill, New York, 1983.
4. Krause, Paul C., and Oleg Wasynczuk. *Electromechanical Motion Devices*. McGraw-Hill, New York, 1989.
5. Nasar, Syed A. *Electric Energy Conversion and Transmission*. Macmillan, New York, 1985.
6. Ramshaw, Raymond, and R. G. van Heeswijk. *Energy Conversion Electric Motors and Generators*. Saunders College Publishing, Philadelphia, 1990.
7. Sen, P. C. *Principles of Electrical Machines and Power Electronics*. Wiley, New York, 1989.
8. Shultz, Richard D., and Richard A. Smith. *Introduction to Electric Power Engineering*. Wiley, New York, 1988.

PROBLEMS

5.1   An electromagnet with two air gaps is shown in Fig. 5.20. The relative permeability of the iron is assumed to be infinite, and flux fringing in the air gaps can be neglected. The coil has 500 turns, and it is supplied with a current of 2.5 A. Find

a. The magnetic flux in each air gap
b. The total flux linkage of the coil
c. The magnetic energy stored in this system

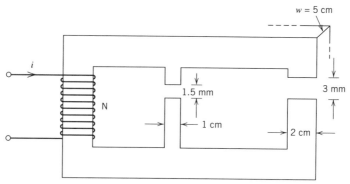

FIGURE 5.20 Electromagnet of Problem 5.1.

5.2 A cylindrical electromagnet is shown in Fig. 5.21. The plunger is free to move in the vertical direction. The air gap between the core shell and the plunger is uniform at 0.5 mm. It may be assumed that magnetic flux leakage and flux fringing in the air gaps are negligible and the relative permeability of the core is infinitely high. The coil contains 600 turns and is supplied with a DC current of 8 A.

   a. Calculate the magnetic flux density in the air gap between the surfaces of the center core and plunger when the air-gap length $g = 1.0$ mm.

   b. Find the inductance of the coil.

   c. Determine the stored magnetic energy.

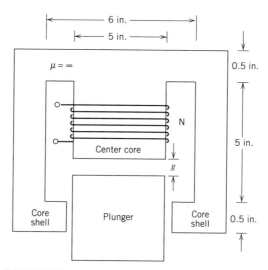

FIGURE 5.21 Cylindrical electromagnet of Problem 5.2.

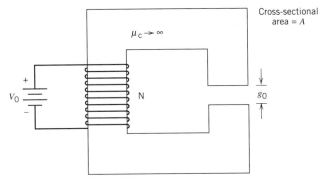

FIGURE 5.22   Inductor circuit of Problem 5.3.

5.3   An inductor is made from a magnetic core with an air gap as shown in Fig. 5.22. The winding has resistance R and is connected to a source with voltage V_0.

   a. Show that the coil inductance is given by the expression

$$L = \frac{\mu_0 N^2 A}{g_0}$$

   b. Calculate the magnetic stored energy in the inductor.
   c. With the voltage held constant at V_0, the air-gap length is varied from g_0 to g_1. Calculate the change in the stored magnetic energy.

5.4   The magnetic circuit of Fig. 5.23 is connected to a source whose voltage varies sinusoidally with time, that is,

$$v(t) = V_0 \cos \omega t$$

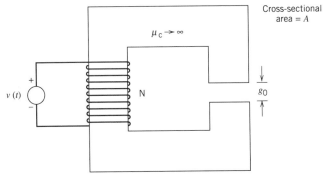

FIGURE 5.23   Magnetic circuit of Problem 5.4.

Assume that the air gap is fixed at g_0 and that the resistance of the coil is R. Calculate

a. The time-averaged magnetic energy stored in the inductor

b. The instantaneous power output of the voltage source as a function of R

c. The time-averaged power output of the voltage source as a function of R

d. The time-averaged power dissipated in the resistor

5.5 Twenty-five conducting loops, or turns, are connected in series to form a coil. Each turn has a length $l = 2.5$ m and width $w = 20$ cm. The coil is rotated at a constant speed of 1200 rpm in a magnetic flux density **B** directed upward. The induced voltage across the coil has an rms value of 1000 V. Determine the required value of flux density.

5.6 A coil of 50 turns is in the shape of a square with 15 cm on each side. It is driven at a constant speed of 600 rpm. The coil is placed in such a way that its axis of revolution is perpendicular to a uniform magnetic flux density **B** of 0.15 T in the vertical direction. The coil is in a position of maximum flux linkage at time $t = 0$. Determine (a) the time variation of the flux linkage and (b) the instantaneous voltage induced in the coil.

5.7 Two mutually coupled coils are shown in Fig. 5.24. The first coil is held fixed, and the second coil is free to rotate. The inductances of the coils are given by

$$L_{11} = A \qquad L_{22} = B \qquad L_{12} = L_{21} = M \cos \theta$$

Let the current in the first coil be denoted by i_1 and the current in the second coil by i_2. Find the electrical torque for the following conditions:

a. $i_1 = I_0$ and $i_2 = 0$

b. $i_1 = i_2 = I_0$

c. $i_1 = I_0$ and $i_2 = I_m \sin \omega t$

d. $i_1 = i_2 = I_m \sin \omega t$

e. $i_1 = I_0$ and coil 2 short-circuited

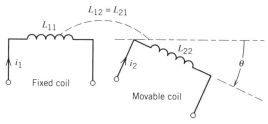

FIGURE 5.24 Mutually coupled coils of Problem 5.7.

5.8 Two mutually coupled coils are shown in Fig. 5.25. The first coil is held stationary, while the second is free to rotate. The self-inductances of the two coils are L_{11} and L_{22}, and the mutual inductance is L_{12}. The values of these inductances in henrys are given in the following table for two angular positions θ of the rotor, where θ is measured from the axis of the stationary coil used as reference. The inductances may be assumed to vary linearly with θ over the range $45° < \theta < 75°$.

θ	L_{11}	L_{22}	L_{12}
45°	0.6	1.1	0.3
75°	1.0	2.0	1.0

For each of the following cases, compute the electromagnetic torque when the rotor is at angular position $\theta = 60°$. State whether the torque tends to turn the rotor clockwise or counterclockwise.

a. $I_1 = 10$ A and $I_2 = 0$

b. $I_1 = 0$ and $I_2 = 10$ A

c. $I_1 = 10$ A and $I_2 = 10$ A along the arrow directions

d. $I_1 = 10$ A in the arrow direction and $I_2 = 10$ A in the reverse direction

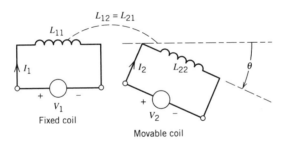

FIGURE 5.25   Magnetic system of Problem 5.8.

5.9   An electromagnetic system consists of two windings, one mounted on a stator and the other on a rotor. The self- and mutual inductances are given as

$$L_{11} = 1.0 \text{ H} \qquad L_{22} = 0.5 \text{ H} \qquad L_{12} = 1.414 \cos \theta \text{ H}$$

where θ is the angle between the axes of the windings. The resistances of the windings are assumed to be negligible. Winding 2 is short-circuited, and the current supplied to winding 1 as a function of time is $i_1 = 14.14 \sin \omega t$.

a. Derive an expression for the instantaneous torque on the rotor in terms of the angle θ.

b. Compute the time-averaged torque when $\theta = 45°$.

c. If the rotor is allowed to move, will it rotate continuously or will it tend to come to rest? If it comes to rest, at what angular position θ?

5.10 The two mutually coupled coils described in Problem 5.9 are connected in parallel across a 208-V, 60-Hz source. The coil resistances are negligible.

    a. Derive an expression for the instantaneous torque on the rotor in terms of the angle θ.

    b. Compute the average torque when $\theta = 30°$.

5.11 The doubly excited rotating system shown in Fig. 5.16 has a coil on the stator and a coil on the rotor. The self- and mutual inductances are given by

$$L_{ss} = L_{rr} = 2 \text{ H}$$
$$L_{sr} = \cos \theta \text{ H}$$

The resistances of the windings may be neglected. The rotor winding is short-circuited, and the stator winding is connected to a current source $i_s(t) = 14.14 \sin \omega t$ A. The rotor is held stationary at an angular position of $\theta = 15°$. Determine

    a. The instantaneous electromagnetic torque

    b. The time-averaged electromagnetic torque

    c. The equilibrium positions of the rotor coil

5.12 Two coils, one mounted on a stator and the other on a rotor, have self- and mutual inductances of

$$L_{11} = 0.40 \text{ mH} \qquad L_{22} = 0.20 \text{ mH} \qquad L_{12} = 0.5 \cos \theta \text{ mH}$$

where θ is the angle between the axes of the coils. The coils are connected in series and carry a current

$$i(t) = \sqrt{2} I \sin \omega t$$

    a. Determine the instantaneous torque T on the rotor as a function of the angular position θ.

    b. Find the time-averaged torque T_{ave} as a function of θ.

    c. Compute the value of T_{ave} for $I = 5$ A, and $\theta = 90°$.

5.13 Two windings are mounted in a machine with a uniform air gap such as that shown in Fig. 5.16. The winding resistances are negligible. The self- and mutual inductances are given as follows:

$$L_{ss} = 2.5 \text{ H}$$
$$L_{rr} = 1.0 \text{ H}$$
$$M_{sr} = \sqrt{2} \cos \theta \text{ H}$$

The first winding is connected to a voltage source, and the second winding is short-circuited. The current in winding 1 is known to be $i_s(t) = 10\sqrt{2} \sin \omega t$, where ω is the frequency of the source. The rotor is held stationary.

a. Derive an expression for the instantaneous torque in terms of the angular position θ.

b. Compute the time-averaged torque at $\theta = 45°$.

c. If the rotor is released, will it rotate continuously or will it tend to come to rest? If it comes to rest, at what angle θ?

5.14 A machine has self- and mutual inductances that can be described by

$$M_{12} = 1 - \sin \theta \text{ H}$$
$$L_{11} = 1 + \sin \theta \text{ H}$$
$$L_{22} = 2(1 + \sin \theta) \text{ H}$$

The two coils are connected to separate sources. Coil 1 is supplied with a constant current of $I_1 = 15$ A, and coil 2 is supplied with a constant current of $I_2 = -4$ A. Assume that the value of $\theta = 45°$.

a. Find the value and direction of the developed torque.

b. Compute the amount of energy supplied by each source.

c. The current of coil 1 of part (a) is changed to a sinusoidal current of 10 A rms, and coil 2 is short-circuited. Find the rms value of the current in coil 2. The source has a voltage of 120 V at 60 Hz.

d. Determine the instantaneous torque produced in part (c).

e. Find the average value of the torque in part (c).

5.15 If the number of turns on the stator of Example 5.5 is doubled, calculate the electromagnetic torque using Eqs. 5.77 and 5.89.

Six

DC Machines

6.1   INTRODUCTION

The DC machine is a versatile electromechanical energy conversion device characterized by superior torque characteristics and a wide range of speed. Its efficiency is very good over its speed range. DC currents are required for both its field winding, located on the stator, and its armature winding on the rotor.

The DC machine is more costly than comparable AC machines, and its maintenance costs are also higher. Because of their costs, DC machines are not widely used in industry. Their use is limited to tough jobs, such as in steel mills and paper mills. They are also used as motors for control purposes.

Basic torque production and induced voltage were discussed in the previous chapter. In this chapter, the operational characteristics of various types of DC machines are discussed.

6.2   BASIC PRINCIPLES OF OPERATION

DC machines, like other electromechanical energy conversion devices, have two sets of electrical windings: field and armature windings. The field winding is on the stator, and the armature winding is on the rotor. A two-pole DC machine is shown in Fig. 6.1.

The magnetic field of the field winding is approximately sinusoidal. Thus, AC voltage is induced in the armature winding as the rotor turns under the magnetic field of the stator. This induced or generated voltage is also approximately sinusoidal. Since the armature windings are distributed over the armature

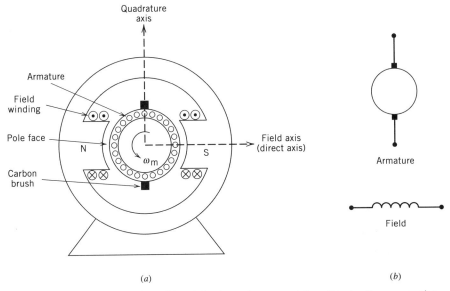

FIGURE 6.1 A two-pole DC machine: (*a*) schematic representation; (*b*) circuit representation.

periphery, the generated voltages of the armature turns reach their maxima at different times.

The commutator and brush combination converts the AC generated voltages to DC. The commutator is located on the same shaft as the armature and rotates together with the armature windings. The brushes are stationary and are located so that commutation occurs when the coil sides are in the neutral zone; that is, the potentials of the conductor loop that leaves the brush and the loop that comes in contact with the brush are the same.

The axis of the armature mmf is 90° from the axis of the field winding. Denoting the stator (field) winding and the rotor (armature) winding mmfs by F_s and F_r, respectively, these mmfs are shown in Fig. 6.2.

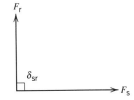

FIGURE 6.2 Stator and rotor mmf vector representation.

The electromagnetic torque T_e is produced by the interaction of the stator and rotor mmfs. Since they are in quadrature with each other, T_e may be expressed as follows:

$$T_e = -\frac{\pi\mu_0 r l}{g} F_s F_r \tag{6.1}$$

The rotor mmf F_r is a linear function of the armature current I_a, and the stator mmf F_s is similarly a linear function of the field current I_f. Hence, Eq. 6.1 may be written as follows:

$$T_e = K_T I_a I_f \tag{6.2}$$

where K_T = torque constant.

The DC induced voltage E_a appearing between the brushes is a function of the field current I_f and the speed of rotation ω_m of the machine. This generated voltage is given by

$$E_a = K'_a I_f \omega_m \tag{6.3}$$

where K'_a = voltage constant.

If the losses of the DC machine are neglected, from the energy conservation principle, the electrical power is equal to the mechanical power:

$$E_a I_a = \omega_m T_m \tag{6.4}$$

where

$E_a I_a$ = electrical power

$\omega_m T_m$ = mechanical power

At steady state, the mechanical torque T_m is equal to the electromagnetic torque T_e.

6.3   GENERATION OF UNIDIRECTIONAL VOLTAGE

A DC generator with two poles in the stator and a single conducting loop on the rotor is shown in Fig. 6.3.

As the rotor is rotated at an angular velocity ω_m, the armature flux linkages change and a voltage $e_{aa'}$ is induced between terminals a and a'. The expression for the voltage induced is given by Faraday's law as

$$e_{aa'} = -\frac{d\lambda}{dt} \tag{6.5}$$

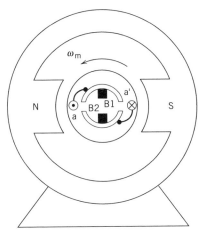

FIGURE 6.3   A two-pole DC generator.

This induced voltage is plotted against time in Fig. 6.4b, where at time $t = 0$ the conductors a and a$'$ are as shown in Fig. 6.3. The plot of the flux $\phi_{aa'}$ and the plot of the rectified voltage (across brushes B1 and B2) are given in Figs. 6.4a and c, respectively.

It is seen from Fig. 6.4 that although the flux and the coil voltage are both sinusoidal functions of time, the voltage across the brushes is a unidirectional voltage.

Suppose that a second winding bb$'$ is placed on the armature displaced from the aa$'$ winding by 90°. Two new commutator segments are also added as illustrated in Fig. 6.5.

The induced voltages $e_{aa'}$ and $e_{bb'}$ across terminals aa$'$ and bb$'$, respectively, and the voltage across the brushes, e_{B1B2}, are plotted in Fig. 6.6.

It may be seen from Fig. 6.6c that the armature voltage is closer to a DC voltage for the generator of Fig. 6.4. Therefore, it may be concluded that by increasing the number of conducting loops on the armature and correspondingly increasing the number of commutator segments, the quality of the armature terminal voltage is greatly improved. In the limit, a pure DC voltage between brushes is obtained as shown in Fig. 6.7.

The generated voltage of a DC machine having p poles and Z conductors on the armature with a parallel paths between brushes is given in Eq. 5.48 and is repeated here as Eq. 6.6.

$$E_a = \frac{pZ}{2\pi a}\Phi_p\omega_m = K_a\Phi_p\omega_m \qquad (6.6)$$

where $K_a = pZ/(2\pi a)$ = machine constant.

By substituting Eq. 6.6 into Eq. 6.4, the mechanical torque, which is also equal to the electromagnetic torque, is found as follows:

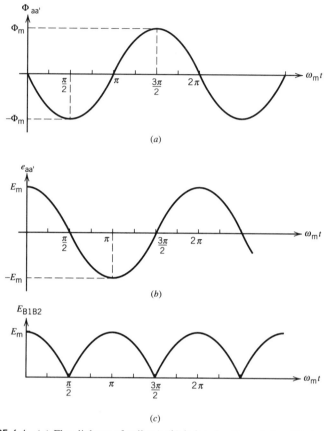

FIGURE 6.4 (a) Flux linkage of coil aa'; (b) induced voltage; (c) rectified voltage.

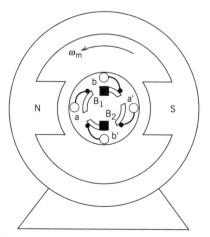

FIGURE 6.5 A two-pole, two-coil DC generator.

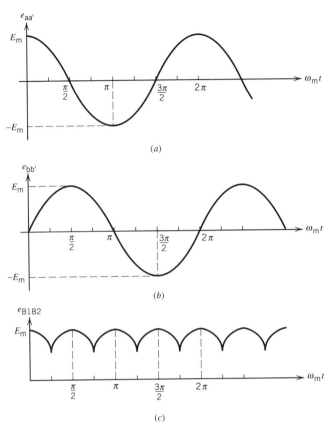

FIGURE 6.6 (a) Voltage of coil aa'; (b) voltage of coil bb';
(c) voltage across armature terminals (between brushes).

$$T_e = T_m = \frac{E_a I_a}{\omega_m} = K_a \Phi_p I_a \qquad (6.7)$$

In the case of a generator, T_m is the input mechanical torque, which is converted to electrical power. For a motor, T_e is the developed electromagnetic torque, which is used to drive the mechanical load.

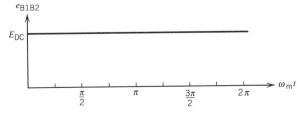

FIGURE 6.7 DC terminal voltage.

EXAMPLE 6.1

A six-pole DC machine has a flux per pole of 30 mWb. The armature has 536 conductors connected as a lap winding in which the number of parallel paths a is equal to the number of poles p. The DC machine runs at 1050 revolutions per minute (rpm), and it delivers a rated armature current of 225 A to a load connected to its terminals. Calculate the following:

    a. Machine constant K_a
    b. Generated voltage E_a
    c. Conductor current I_c
    d. Electromagnetic torque T_e
    e. Power developed, P_{dev}, by the armature

Solution

    a. The machine constant is given by

$$K_a = pZ/(2\pi a) = (6)(536)/[(2\pi)(6)] = 85.31$$

    b. The angular speed of the machine is

$$\omega_m = 2\pi n/60 = 2\pi(1050)/60 = 109.96 \text{ rad/s}$$

Thus, the generated voltage is found by using Eq. 6.6 as follows:

$$E_a = K_a\Phi_p\omega_m = (85.31)(0.030)(109.96) = 281.4 \text{ V}$$

    c. For a lap winding, there are $a = 6$ parallel paths; therefore, the conductor current is

$$I_c = I_{coil} = I_a/a = 225/6 = 37.5 \text{ A}$$

    d. The electromagnetic torque is found by using Eq. 6.7 as follows:

$$T_e = K_a\Phi_pI_a = (85.31)(0.030)(225) = 575.84 \text{ N-m}$$

    e. The power developed by the armature is

$$P_{dev} = T_e\omega_m = (575.84)(109.96) = 63.32 \text{ kW}$$
$$= E_aI_a = (281.4)(225) = 63.32 \text{ kW}$$

EXAMPLE 6.2

Repeat Example 6.1 if the armature is reconnected as a wave winding such that the rated conductor current remains the same. For a wave winding, there are two parallel paths; that is, $a = 2$.

Solution

a. The machine constant is

$$K_a = pZ/(2\pi a) = (6)(536)/[(2\pi)(2)] = 255.92$$

b. At the same angular speed $\omega_m = 109.96$ rad/s, the generated voltage is

$$E_a = K_a\Phi_p\omega_m = (255.92)(0.030)(109.96) = 844.2 \text{ V}$$

c. The rated conductor current is $I_c = 37.5$ A; thus, since a wave winding has $a = 2$ parallel paths, the armature current is

$$I_a = aI_c = 2(37.5) = 75 \text{ A}$$

d. The electromagnetic torque is

$$T_e = K_a\Phi_pI_a = (255.92)(0.030)(75) = 575.82 \text{ N-m}$$

e. The power developed is

$$P_{dev} = T_e\omega_m = (575.82)(109.96) = 63.32 \text{ kW}$$
$$= E_aI_a = (844.2)(75) = 63.32 \text{ kW}$$

DRILL PROBLEMS

D6.1 A four-pole DC machine has a flux per pole of 15 mWb. The armature has 75 coils with four turns per coil. The armature is connected as a wave winding that has $a = 2$ parallel paths. Find the generated voltage at a speed of 1050 rpm.

D6.2 A DC generator has six poles and is running at 1150 rpm. The armature has 120 slots with eight conductors per slot and is connected as a lap winding; that is, parallel paths = number of poles. The generated voltage is 230 V, and the armature current is 25 A.

a. Determine the required flux per pole.

b. Determine the electromagnetic torque developed.

6.4 TYPES OF DC MACHINES

DC machines are classified according to the electrical connections of the armature winding and the field windings. The operating characteristics of the specific DC machine being considered depend on the particular interconnection of the armature and field windings. There are generally four means of interconnection, giving rise to the following types of DC machines:

1. Shunt machine
2. Separately excited machine
3. Series machine
4. Compound machine

6.4.1 Shunt DC Machine

The armature and field windings are connected in parallel. The shunt field winding consists of several turns of small-diameter conductors, since the field current is normally a low current. The armature conductors are considerably larger because they are designed to carry rated current. The armature voltage and the field voltage are the same.

The interconnection for the DC shunt machine is illustrated in Fig. 6.8.

6.4.2 Separately Excited DC Machine

The armature and field windings are electrically separate from one another. Thus, the field winding is excited by a separate DC source.

The schematic representation of the separately excited DC machine is shown in Fig. 6.9.

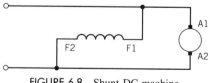

FIGURE 6.8   Shunt DC machine.

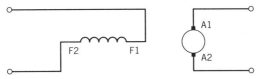

FIGURE 6.9   Separately excited DC machine.

6.4.3   Series DC Machine

The field winding and the armature are electrically connected in series. The series field winding consists of a few turns of large diameter conductors since it carries the same current as the armature.

The interconnection for the DC series machine is illustrated in Fig. 6.10.

6.4.4   Compound DC Machine

The compound DC machine has two field windings: one is connected in series with the armature, and the other is connected in parallel with the armature. The first is called the series field, and the second is called the shunt field. If the magnetic fluxes produced by both series field and shunt field windings are in the same direction, that is, additive, the machine is cumulative compound. If the magnetic fluxes are in opposition, the machine is differential compound. The cumulative and differential compound machines are illustrated in Fig. 6.11a and b, respectively.

A cumulative compound or differential compound machine may be connected either long-shunt compound or short-shunt compound. In a long-shunt compound machine, the series field is connected in series with the armature and the combination is in parallel with the shunt field. In the short-shunt compound machine, the shunt field is in parallel with the armature and the combination is connected in series with the series field. The long-shunt and short-shunt connections are shown in Fig. 6.12.

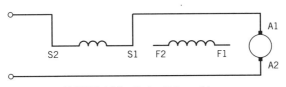

FIGURE 6.10   Series DC machine.

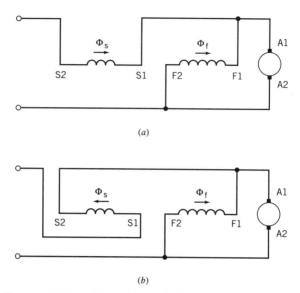

FIGURE 6.11 Compound DC machines: (a) cumulative compound; (b) differential compound.

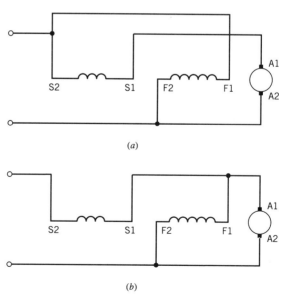

FIGURE 6.12 Compound DC machines: (a) long-shunt compound; (b) short-shunt compound.

6.5   DC MACHINE ANALYSIS

The armature and field windings of DC machines are modeled in steady state as shown in Fig. 6.13. The direction of the armature current I_a shown in the figure is for the case of a DC motor, and the torque T_e is the electromagnetic torque developed, which is available for driving a mechanical load. The voltage V_t is the applied terminal voltage.

In the case of a generator, the torque is the mechanical torque applied to the machine shaft, and the direction of the armature current is reversed. The voltage V_t is the output terminal voltage.

In the equivalent circuit of the DC machine shown, the mechanical losses, such as friction and windage losses, are not included. However, the electrical copper losses are modeled.

For shunt and compound machines, the voltage V_f applied to the field circuit is the same as the terminal voltage V_t. However, in the case of a separately excited generator, a separate voltage source is required to provide the excitation current I_f to the shunt field.

The generated or excitation voltage E_a is the emf induced in the armature of the machine. The expression for E_a derived in Section 6.3 may also be written as

$$E_a = K_a \Phi_p \omega_m = K_a' I_f \omega_m \tag{6.8}$$

Similarly, the electromagnetic torque may also be expressed as

$$T_e = K_a \Phi_p I_a = K_a' I_f I_a \tag{6.9}$$

In Equations 6.8 and 6.9 a linear relationship between Φ_p and I_f is assumed; that is, the saturation effect is neglected. Depending on whether a DC generator or DC motor is under consideration, the steady-state issues analyzed are different, as shown in Table 6.1.

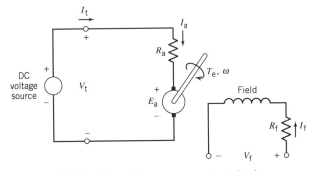

FIGURE 6.13   DC machine equivalent circuit.

Table 6.1 Scope of DC Machine Problems

Generator Problems	Motor Problems
• Speed fixed by prime mover	• Applied voltage fixed by source
• Issues:	• Issues:
Determine terminal voltage for a given load and fixed excitation current.	Determine speed for a given mechanical load and fixed excitation current.
Determine generated voltage for a given load and fixed terminal voltage.	Determine excitation voltage for a given mechanical load and a required speed.

EXAMPLE 6.3

A 12-kW, 240-V, 1200-rpm, separately excited DC generator has armature and field winding resistances of 0.20 Ω and 200 Ω, respectively. At no load, the terminal voltage is 240 V, the field current is 1.2 A, and the machine runs at 1200 rpm. When the generator delivers rated current to a load at 240 V, calculate

a. The generated voltage E_a
b. The field circuit voltage V_f
c. The developed torque T_e

Solution

a. At no load, the armature current $I_a = 0$; therefore, the generated voltage is equal to the terminal voltage.

$$E_a = V_t + I_a R_a = 240 + (0)(0.20) = 240 \text{ V}$$

The machine speed is

$$\omega_m = 2\pi n/60 = 2\pi(1200)/60 = 40\pi$$

The generated voltage may also be expressed in terms of the field current and the angular speed as given in Eq. 6.8. Thus,

$$E_a = K_a' I_f \omega_m = K_a'(1.2)(40\pi) = 240 \text{ V}$$

Solving for K_a' yields

$$K_a' = 1.592$$

At rated load, the armature current is

$$I_a = 12,000/240 = 50 \text{ A}$$

Therefore, the generated voltage is given by

$$E_a = V_t + I_a R_a = 240 + (50)(0.20) = 250 \text{ V}$$

b. Expressing the generated voltage in terms of the field current and the angular speed,

$$250 = (1.592)I_f(40\pi)$$

Solving for the field current yields

$$I_f = 1.25 \text{ A}$$

Thus, the field circuit voltage is found as follows:

$$V_f = I_f R_f = (1.25)(200) = 250 \text{ V}$$

c. The electromagnetic torque developed is obtained as follows:

$$T_e = E_a I_a / \omega_m = (250)(50)/(40\pi) = 99.5 \text{ N-m}$$

DRILL PROBLEMS

D6.3 A separately excited generator has a no-load voltage of 125 V at a field current of 2 A and a speed of 1780 rpm. Assume that the generator is operating on the straight-line portion of its saturation curve.

a. Calculate the generated voltage when the field current is increased to 2.5 A.

b. Calculate the generated voltage when the speed is reduced to 1650 rpm and the field current is increased to 2.75 A.

D6.4 A DC shunt generator has armature and field resistances of 0.2 Ω and 150 Ω, respectively. The generator supplies 10 kW to a load connected to its terminals at 230 V. Assuming that the total brush-contact voltage drop is 2 V, determine the induced voltage.

D6.5 A 100-kW, 600-V, 1200-rpm, long-shunt compound generator has a total brush voltage drop of 4 V, a series field winding resistance of 0.05 Ω, a shunt field winding resistance of 200 Ω, and an armature resistance of 0.1 Ω. The generator supplies rated output power to a load at rated voltage and rated speed. Calculate (a) the armature current and (b) the armature induced voltage.

6.6   DC GENERATOR PERFORMANCE

In this section, the performance characteristics of DC generators are described. Among these are the no-load or open-circuit characteristic, the terminal or load characteristic, voltage regulation, and efficiency.

6.6.1   No-Load Characteristics

Equations 6.8 and 6.9 assume a linear relationship between the flux Φ_p and the field current I_f. If the assumption of linearity is removed, these expressions have to be modified as follows:

$$E_a = K_a\Phi_p(I_f)\omega_m = K_a''\Phi_p(I_f)n \tag{6.10}$$

$$T_e = K_a\Phi_p(I_f)I_a \tag{6.11}$$

Equation 6.10 is referred to as the magnetization equation, and the corresponding curve is called the *saturation curve,* or *magnetization curve,* which is illustrated in Fig. 6.14. These curves are also referred to as the *no-load characteristics* of a separately excited generator.

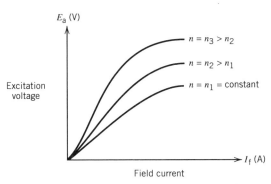

FIGURE 6.14   No-load characteristics of a separately excited generator.

6.6.2   Voltage Buildup in a Shunt Generator

In a shunt generator, the field winding is connected across the armature terminals. Thus, the generator itself provides its own field excitation. The equivalent circuit of the self-excited shunt generator is shown in Fig. 6.15.

When the generator is rotated by its prime mover, a small *residual voltage* E_{res} is generated because of the presence of residual flux in the magnetic field poles. This voltage is given by

$$E_{res} = K_a \Phi_{res} \omega_m \qquad (6.12)$$

The induced voltage E_{res} is essentially applied to the field circuit, and it causes a current I_f to flow in the field coils. The resultant mmf in the field coils produces more flux Φ_p in the poles, causing an increase in the generated voltage E_a, which increases the terminal voltage V_t. The higher V_t causes an increased I_f, further increasing the flux Φ_p, which increases E_a, and so forth. The final operating voltage is determined by the intersection of the field resistance line and the saturation curve. This voltage buildup process is depicted in Fig. 6.16. If the resistance of the field circuit is decreased, the resistance line is rotated clockwise, which results in a higher operating voltage, and vice versa.

Three factors affect the proper buildup of a shunt generator:

1. *Residual magnetism.* If there is no residual flux in the poles, there is no residual voltage; that is, if $\Phi_{res} = 0$, then $E_{res} = 0$, and the voltage will never build up.

2. *Critical resistance.* Normally, the shunt generator builds up to a voltage determined by the intersection of the field resistance line and the saturation curve. If the field resistance is greater than the *critical resistance*, the generator fails to build up and the voltage remains at the residual level. To solve this problem, the field resistance is reduced to a value less than the critical resistance.

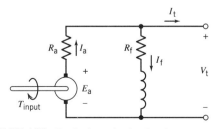

FIGURE 6.15   Equivalent circuit of a shunt generator.

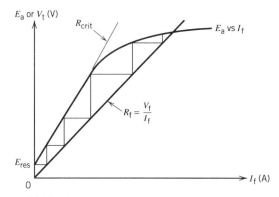

FIGURE 6.16 Buildup of voltage in a shunt generator.

3. *Relative polarity* of the field winding and the terminal voltage. For the shunt generator to build up properly, the current supplied to the field winding should produce flux that aids the residual flux. Otherwise, the flux produced by the field current will tend to neutralize the residual flux Φ_{res}, and the generator will not build up. When this happens, reversal of either the field connection or the direction of rotation will enable the generator to build up. Reversing both field connection and direction of rotation will cause the generator not to build up.

6.6.3 Terminal or Load Characteristics

The *terminal characteristic* of a device is a plot of the output, or terminal, quantities with respect to one another. For DC generators, the terminal characteristic is in the form of a plot of terminal voltage V_t versus load current I_L. Thus, it is also called the *load characteristic*.

For a shunt generator, the main causes of the decrease in terminal voltage as load is increased are

1. *Armature resistance drop.* This is an $I_a R_a$ drop due to the armature resistance R_a.

2. *Brush contact drop.* The brushes are pressed on the commutator by mechanical springs. The nonideal contact, therefore, offers an electrical resistance and will cause a voltage drop when current flows through the brush. A constant value of 2 V is usually assumed for brush contact voltage drop.

3. *Armature reaction voltage drop.* When a load is connected to the terminals of the generator, a current will flow in the armature windings. This resulting armature mmf will produce its own magnetic field, which will affect the original magnetic field produced by the field poles. This interaction of the two fields is called *armature reaction*. Armature reaction is manifested in two ways.

First, the uniform flux distribution under the pole faces is altered. In some areas under the pole faces, the armature magnetic field subtracts from the pole flux; in other areas, the armature field adds to the pole flux. This gives rise to commutation problems, as evidenced by arcing and sparking at the brushes.

The second problem caused by armature reaction is flux weakening, resulting in a reduced generated voltage. In the areas of the pole faces where the armature magnetic field adds to the pole flux, because of saturation of the magnetic pole, there is only a small increase in flux. In the areas where the armature field subtracts from the field flux, there is a larger decrease in flux. Thus, there is a net decrease in the average flux under the entire pole face. This has the same effect as a reduction in the field magnetization; therefore, the effective or net field mmf may be expressed as

$$N_f I_f^{\text{eff}} = N_f I_f^{\text{act}} - (NI)^{\text{ar}} \tag{6.13}$$

where

$\quad N_f$ = number of turns in the shunt field winding

$\quad I_f^{\text{eff}}$ = effective field current

$\quad I_f^{\text{act}}$ = actual field current

$(NI)^{\text{ar}}$ = demagnetizing mmf due to armature reaction

The flux distortion and demagnetization caused by armature reaction are illustrated in Fig. 6.17.

4. *Reduced terminal voltage* in self-excited generators. The terminal voltage is reduced as a result of the preceding three causes. Since the field excitation is supplied by the generator itself, the reduced terminal voltage supplies a reduced field current. This causes a further reduction in the voltage.

The load characteristics of the different types of DC generators are shown in Fig. 6.18. The load characteristic of the separately excited generator is affected mainly by the first three causes in the preceding list; thus, it has a drooping characteristic.

The load characteristic of the self-excited shunt generator is similar to that of the separately excited generator. However, the terminal voltage is lower because of the fourth cause. Moreover, if the generator current is increased beyond a certain level, the terminal voltage collapses to zero.

In a series generator, the field excitation current is the same as the load current. Thus, the field flux and the resulting generated voltage will increase

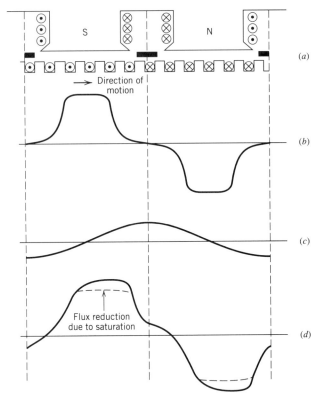

FIGURE 6.17 Flux distribution due to armature reaction: (*a*) armature and field winding mmfs; (*b*) uniform flux distribution due to field mmf; (*c*) flux distribution due to armature mmf; (*d*) resultant flux distribution.

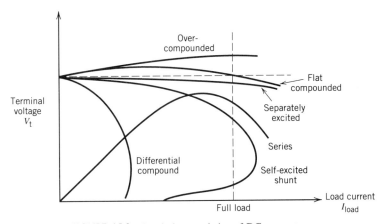

FIGURE 6.18 Load characteristics of DC generators.

with the load up to a certain point, beyond which the effects of saturation will cause the voltage of the series generator to collapse to zero.

In compound generators, the series and shunt fields are either in opposition or aiding each other. In a cumulative-compound generator, the series and shunt fields aid each other. The mmfs of either or both fields may be adjusted to produce a terminal voltage at full-load current that is less than the no-load voltage for an undercompounded generator, or equal to the no-load voltage for a flat-compounded generator, or greater than the no-load voltage for an overcompounded generator.

For a cumulative-compound generator, because of the presence of armature reaction, the effective or net field mmf is given by

$$N_f I_f^{\text{eff}} = N_f I_f^{\text{act}} + N_s I_s - (NI)^{\text{ar}} \tag{6.14}$$

where

N_s = number of turns in the series field winding

I_s = current through the series field winding

In a differential-compound generator, the series field opposes the shunt field, so that the terminal voltage drops a large amount for a small increase in load current. The differential-compound generator can be characterized as a constant-current generator.

6.6.4 Voltage Regulation

The terminal or load characteristics of the different types of DC generators are described in the previous section. A performance measure that gives essentially the same information about the generator is its voltage regulation. Voltage regulation (VR) is defined as follows:

$$\text{Voltage regulation (VR)} = \frac{V_{nl} - V_{fl}}{V_{fl}} \, 100\% \tag{6.15}$$

Voltage regulation gives an approximate description of the terminal characteristic. Positive voltage regulation implies a drooping characteristic, whereas negative voltage regulation implies a rising characteristic. Zero regulation implies a flat characteristic.

EXAMPLE 6.4

The magnetization curve of the 12-kW, 240-V, 1200-rpm, DC machine of Example 6.3 is shown in Fig. 6.19. The machine is operated as a separately

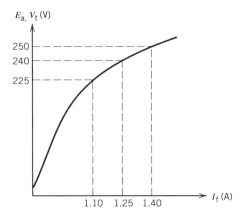

FIGURE 6.19 Magnetization curve of DC machine of Example 6.4.

excited generator, and the field current is adjusted to 1.25 A to obtain a terminal voltage of 240 V at no load. The machine is run at 1200 rpm. The shunt field winding has $N_f = 2500$ turns/pole. The armature and field winding resistances are given as $0.20\ \Omega$ and $200\ \Omega$, respectively.

a. Neglecting armature reaction, determine the terminal voltage at rated load current. Calculate voltage regulation.

b. Assume armature reaction at rated load will cause a reduction of 0.15 A in field current. Determine the terminal voltage at rated load current. Calculate voltage regulation.

c. Assume armature reaction at rated load current will cause a demagnetization of 375 A-t, and determine the field current required to produce a terminal voltage of 240 V at rated load current.

Solution

a. From the magnetization curve shown in Fig. 6.19, for a field current $I_f = 1.25$ A, the generated voltage is $E_a = 240$ V. At rated load, the line or terminal current is equal to the armature current and is given by

$$I_t = I_a = 12,000/240 = 50\ A$$

The terminal voltage is obtained as follows:

$$V_t = E_a - I_a R_a = 240 - (50)(0.20) = 230\ V$$

Therefore, the voltage regulation is given by

$$\text{Voltage regulation} = \frac{V_{nl} - V_{fl}}{V_{fl}} 100\% = \frac{E_a - V_t}{V_t} 100\%$$

$$= \frac{240 - 230}{230} 100\% = 4.4\%$$

b. Due to armature reaction, the effective field current is

$$I_f^{eff} = I_f^{actual} - I_f^{ar} = 1.25 - 0.15 = 1.10 \text{ A}$$

From the magnetization curve, for $I_f = 1.10$ A, the generated voltage is $E_a = 225$ V. Thus,

$$V_t = E_a - I_a R_a = 225 - (50)(0.20) = 215 \text{ V}$$

Therefore,

$$\text{Voltage regulation} = \frac{E_a - V_t}{V_t} 100\%$$

$$= \frac{240 - 215}{215} 100\% = 11.6\%$$

c. At $I_a = 50$ A and $V_t = 240$ V, the generated voltage is calculated as

$$E_a = V_t + I_a R_a = 240 + (50)(0.20) = 250 \text{ V}$$

From the magnetization curve, for a voltage $E_a = 250$ V, the field current is $I_f^{eff} = 1.4$ A.

In the presence of armature reaction, the net mmf is expressed as follows:

$$N_f I_f^{eff} = N_f I_f^{actual} - (N_f I_f)^{ar}$$

Solving for the actual field current yields

$$I_f^{actual} = (1/N_f)[N_f I_f^{eff} + (N_f I_f)^{ar}] = I_f^{eff} + (N_f I_f)^{ar}/N_f$$
$$= 1.4 + 375/2500 = 1.55 \text{ A}$$

EXAMPLE 6.5

The DC generator of Example 6.4 is provided with a series winding. It is operated as a cumulative compound generator, and the terminal voltage is 240 V at both no-load and full-load conditions; that is, there is zero voltage regulation. Assuming a short-shunt connection, determine the number of series turns per pole required. Take the value of the series winding resistance R_s to be 0.02 Ω, and assume that armature reaction causes a demagnetization of 375 A-t.

Solution At rated load conditions, the terminal voltage and current are given by

$$V_t = 240 \text{ V}$$
$$I_s = I_t = 12{,}000/240 = 50 \text{ A}$$

The equivalent circuit of a short-shunt cumulative compound generator is shown in Fig. 6.20.

From the equivalent circuit shown in Fig. 6.20, the generated voltage is found as follows:

$$V_f = V_t + I_s R_s = 240 + (50)(0.02) = 241 \text{ V}$$
$$I_f^{actual} = V_f/R_f = 241/240 = 1.205 \text{ A}$$
$$I_a = I_s + I_f = 50 + 1.205 = 51.205 \text{ A}$$
$$E_a = V_f + I_a R_a = 241 + (51.205)(0.20) = 251.24 \text{ V}$$

From the magnetization curve, for $E_a = 251.24$ V, $I_f^{eff} = 1.41$ A. Also, from Example 6.4, $N_f = 2500$ turns, $(NI)^{ar} = 375$ A-t.

To obtain zero voltage regulation, the total (or net) mmf required at full load must equal the resultant mmf of the shunt and series field windings less the mmf due to armature reaction. Thus,

$$N_f I_f^{eff} = N_f I_f^{actual} + N_s I_s - (NI)^{ar}$$
$$(2500)(1.41) = (2500)(1.205) + N_s(50) - 375$$

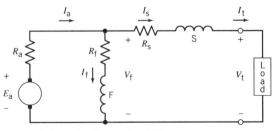

FIGURE 6.20   Equivalent circuit of a short-shunt cumulative compound generator.

Solving for N_s yields

$$N_s = [(2500)(1.41 - 1.205) + 375]/50 = 17.75$$

Say 18 turns/pole.

6.6.5   DC Generator Efficiency

The power flow diagram for a DC shunt generator is shown in Fig. 6.21. As can be seen in this diagram, not all the mechanical power input reaches the load as electrical power output because there are always losses associated with this electromechanical conversion process. Since the generator is assumed to be self-excited, the shunt field winding loss is included with the copper losses in the power flow diagram. If the generator is separately excited, the shunt field copper loss is supplied from a separate electrical source, and it is not included and must be handled separately.

The various losses that occur in the DC shunt generator may be classified as follows:

1. *Electrical or copper losses.* These are the copper losses that occur in the armature and field windings. For a self-excited shunt generator, these losses include

$$\text{Armature copper loss} \quad P_a = I_a^2 R_a$$
$$\text{Shunt field copper loss} \quad P_f = I_f^2 R_f$$

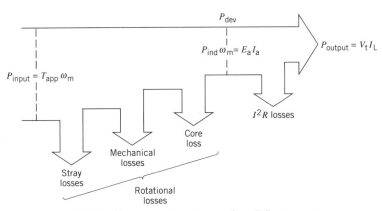

FIGURE 6.21   Power flow diagram for a DC generator.

For a compound generator, another copper loss is present in the series field winding and is added to the armature and shunt field copper losses. The additional copper loss is

$$\text{Series field copper loss} \quad P_s = I_s^2 R_s$$

The armature copper loss depends on load conditions and is typically about 5% when rated current is delivered. The field copper losses are typically from 1% to 2% at rated conditions.

2. *Brush loss.* The electrical loss incurred in the carbon brushes is usually taken equal to $2I_a$ based on the assumption that the total voltage drop across the brushes is about 2 V.

3. *Magnetic or core loss.* These are the hysteresis and eddy current losses occurring in the magnetic circuits of the stator core and poles and the armature core on the rotor.

4. *Mechanical losses.* These are the friction and windage losses. Friction losses include the losses caused by bearing friction and the friction between the brushes and commutator. Windage losses are caused by the friction between rotating parts and air inside the DC machine's casing.

5. *Stray load losses.* These are other losses that cannot be accounted for by the preceding categories.

When the mechanical losses are lumped together with the core loss and stray load loss, they are collectively called rotational loss. Typically, the rotational loss constitutes 3% to 5% of the machine rating, and it is assumed to remain constant for all loading levels.

The DC shunt generator efficiency may be expressed as follows:

$$\eta = \frac{P_{\text{output}}}{P_{\text{input}}} \, 100\% \tag{6.16}$$

The difference between the mechanical input power and the electrical output power constitutes the various losses incurred within the DC shunt generator. Therefore, efficiency may also be expressed as

$$\eta = \frac{P_{\text{input}} - \Sigma P_{\text{losses}}}{P_{\text{input}}} \, 100\%$$

$$= \frac{P_{\text{output}}}{P_{\text{output}} + \Sigma P_{\text{losses}}} \, 100\% \tag{6.17}$$

EXAMPLE 6.6

For the DC compound generator of Example 6.5, the total rotational losses amount to 750 W. Calculate the efficiency when the generator supplies rated current to a load at 240 V.

Solution   From Example 6.5, at rated load conditions,

$$V_t = 240 \text{ V}$$

$$I_t = 50 \text{ A} \qquad R_s = 0.02 \ \Omega$$
$$I_f = 1.205 \text{ A} \qquad R_f = 200 \ \Omega$$
$$I_a = 51.205 \text{ A} \qquad R_a = 0.20 \ \Omega$$

The output power of the generator is

$$P_{output} = V_t I_t = (240)(50) = 12{,}000 \text{ W}$$

The power losses are

$$P_{rotational} = 750 \text{ W}$$
$$I_s^2 R_s = (50)^2(0.02) = 50 \text{ W}$$
$$I_f^2 R_f = (1.205)^2(200) = 290 \text{ W}$$
$$I_a^2 R_a = (51.205)^2(0.20) = 524 \text{ W}$$

The total power input is the sum of the power output and the generator losses.

$$P_{input} = P_{output} + \Sigma(P_{losses}) = 13{,}614 \text{ W}$$

Therefore, the efficiency is

$$\eta = (P_{output}/P_{input})100\% = (12{,}000/13{,}614)100\% = 88\%$$

DRILL PROBLEMS

D6.6   The open-circuit saturation curve of a DC generator driven at rated speed and separately excited is given by

E_a (V)	10	50	100	200	300	350	400	450	500
I_f (A)	0	0.5	1.0	2.0	3.5	4.4	5.4	6.5	8.0

The field winding has a resistance of 62.5 Ω. The armature winding resistance may be assumed to be negligible.

a. The generator is now operated as a shunt generator by connecting the field winding directly across the armature terminals. Determine the generated voltage.

b. What additional resistance must be added in series with the field winding to obtain a generated voltage of 450 V?

c. What additional resistance must be added in series to the field winding to make the total field-circuit resistance equal to the critical value?

D6.7   A separately excited DC generator has the following open-circuit characteristic when running at 1200 rpm.

E_a (V)	40	80	120	160	200	220	240
I_f (A)	0.15	0.30	0.50	0.75	1.05	1.25	1.50

The effect of the armature winding resistance is negligible.

a. The machine is operated as a shunt generator, and is driven at 1200 rpm. Determine the required field-circuit resistance to obtain an open-circuit voltage of 230 V.

b. The operating speed is reduced to 1000 rpm, and the field-circuit resistance remains unchanged. Find the open-circuit generator voltage.

6.7   DC MOTOR PERFORMANCE

The DC motor differs from a DC generator in that the direction of power flow is reversed. In a motor, electrical energy is converted to mechanical energy. DC motors are used where there is a need for variable-speed drives and for traction-type loads.

The performance of a DC motor can be described by using its equivalent circuit, which is shown in Fig. 6.22. The equivalent circuit looks exactly like that of a DC generator except for the direction of the current that is entering the armature.

The generated voltage E_a across the armature has a polarity opposite to the applied voltage V_t. Thus, it is sometimes referred to as *counter emf*, or *back emf*, and is also denoted as E_b. It is given by Eq. 6.8, which is rewritten here as

$$E_a = E_b = K_a \Phi_p \omega_m \qquad (6.18)$$

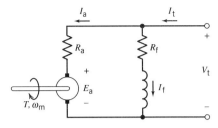

FIGURE 6.22   Equivalent circuit of a DC motor.

The induced or developed torque T_{ind} is given by Eq. 6.9, which is rewritten here as

$$T_{ind} \; = \; T_{dev} \; = \; K_a \Phi_p I_a \qquad (6.19)$$

This torque is related to the electric power converted to mechanical power P_{conv} as follows:

$$\omega_m T_{dev} \; = \; P_{conv} \; = \; E_a I_a \qquad (6.20)$$

6.7.1   Speed Regulation

Whereas with a DC generator, the performance measure of interest is its voltage regulation, in the case of a DC motor the performance measure of interest is its speed regulation. The *speed regulation* (*SR*) is similarly defined as follows:

$$\text{Speed regulation (SR)} \; = \; \frac{n_{nl} - n_{fl}}{n_{fl}} \, 100\% \qquad (6.21)$$

or

$$\text{Speed regulation (SR)} \; = \; \frac{\omega_{nl} - \omega_{fl}}{\omega_{fl}} \, 100\% \qquad (6.22)$$

A motor with zero speed regulation has a full-load speed equal to its no-load speed. Positive speed regulation implies that the motor speed will decrease when the load on its shaft is increased. On the other hand, negative speed regulation implies that the speed will become higher as the load on its shaft becomes higher.

EXAMPLE 6.7

A 220-V shunt motor has an armature resistance of 0.2 Ω and a field resistance of 110 Ω. At no load, the motor runs at 1000 rpm, and it draws a line current of 7 A. At full load, the input to the motor is 11 kW. Consider that the air-gap flux remains fixed at its value at no load; that is, neglect armature reaction.

a. Find the speed, speed regulation, and developed torque at full load.

b. Find the starting torque if the starting armature current is limited to 150% of full-load current.

c. Consider that armature reaction reduces the air-gap flux by 5% when full-load current flows in the armature. Repeat part (a).

Solution

a. Armature reaction is neglected. Referring to Fig. 6.22, the values of the various currents and the generated voltage at no-load conditions are found as follows:

$$I_{t,nl} = 7 \text{ A}$$
$$I_{f,nl} = V_t/R_f = 220/110 = 2 \text{ A}$$
$$I_{a,nl} = I_{t,nl} - I_{f,nl} = 7 - 2 = 5 \text{ A}$$
$$E_{b,nl} = V_t - I_{a,nl}R_a = 220 - (5)(0.2) = 219 \text{ V}$$

Similarly, the currents and the generated voltage at full-load conditions are found as follows:

$$I_{t,fl} = 11,000/220 = 50 \text{ A}$$
$$I_{f,fl} = 220/110 = 2 \text{ A}$$
$$I_{a,fl} = 50 - 2 = 48 \text{ A}$$
$$E_{b,fl} = 220 - (48)(0.2) = 210.4 \text{ V}$$

The no-load speed is $n_{nl} = 1000$ rpm. Since armature reaction is neglected, $\Phi_{nl} = \Phi_{fl}$. Thus, the full-load speed is found as

$$n_{fl} = (E_{b,fl}/E_{b,nl})n_{nl} = (210.4/219)(1000) = 960.7 \text{ rpm}$$
$$\omega_{fl} = 2\pi n_{fl}/60 = 2\pi(960.7)/60 = 100.6 \text{ rad/s}$$

Hence, the speed regulation is computed as

$$\text{Speed regulation} = \frac{n_{nl} - n_{fl}}{n_{fl}} 100\%$$
$$= \frac{1000 - 960.7}{960.7} 100\% = 4.1\%$$

The power developed and torque developed at full load are found as follows:

$$P_{\text{dev,fl}} = E_{b,fl}I_{a,fl} = (210.4)(48) = 10{,}099 \text{ W}$$

$$T_{e,fl} = P_{\text{dev,fl}}/\omega_{fl} = 10{,}099/100.6 = 100.4 \text{ N-m}$$

b. At starting, with a 150% limit for the armature current,

$$I_{a,\text{start}} = 1.50I_{a,fl} = (1.50)(48) = 72 \text{ A}$$

The electromagnetic torque varies directly with the flux and the armature current; thus,

$$\frac{T_{e,\text{start}}}{T_{e,fl}} = \frac{K_a\Phi_{fl}I_{a,\text{start}}}{K_a\Phi_{fl}I_{a,fl}}$$

Since armature reaction is neglected, the flux is assumed to remain constant. Solving for the starting torque yields

$$T_{e,\text{start}} = \frac{I_{a,\text{start}}}{I_{a,fl}} T_{e,fl} = \frac{72}{48}100.4 = 150.6 \text{ N-m}$$

c. The back emf varies directly with the flux and the speed; thus,

$$\frac{E_{b,fl}}{E_{b,nl}} = \frac{K_a\Phi_{fl}n_{fl}}{K_a\Phi_{nl}n_{nl}}$$

At full load, the effect of armature reaction is to reduce the field flux by 5%, that is, $\Phi_{fl} = 0.95\Phi_{nl}$. Solving for the full-load speed yields

$$n_{fl} = \frac{E_{b,fl}}{E_{b,nl}} \frac{\Phi_{nl}}{\Phi_{fl}}n_{nl} = \frac{210.4}{219} \frac{1.0}{0.95}1000 = 1011.3 \text{ rpm}$$

$$\omega_{fl} = 2\pi(1011.3)/60 = 105.9 \text{ rad/s}$$

Therefore, the speed regulation is computed as

$$\text{Speed regulation} = \frac{n_{nl} - n_{fl}}{n_{fl}} 100\%$$

$$= \frac{1000 - 1011.3}{1011.3} 100\% = -1.1\%$$

The power developed $P_{\text{dev,fl}}$ is the same as in part (a), since the generated voltage and the armature current are unchanged. Because the full-load speed has changed, the torque has also changed; thus,

$$T_{e,fl} = P_{dev,fl}/\omega_{fl} = 10,099/105.9 = 95.4 \text{ N-m}$$

DRILL PROBLEMS

D6.8   A DC motor develops a torque of 30 N-m. Determine the electromagnetic torque when the armature winding current is increased by 50% and the flux is reduced by 10%.

D6.9   A shunt motor draws 41 A from a 120-V source when it drives a load at 1750 rpm. The armature and field winding resistances are 0.2 Ω and 120 Ω, respectively. Determine the developed torque.

D6.10   A 240-V shunt motor running at 1750 rpm develops a counter emf of 228 V. It has an armature resistance of 0.15 Ω and a total brush voltage drop of 2 V. Calculate

   a. The armature current at 1750 rpm

   b. The speed when the armature current is 50 A

   c. The speed when the armature current is 25 A

D6.11   The field and armature winding resistances of a 440-V DC shunt machine are 110 Ω and 0.15 Ω, respectively.

   a. Calculate the power developed by the DC machine if it absorbs 22 kW while running as a motor.

   b. Calculate the power developed by the DC machine if it supplies 22 kW while running as a generator.

6.7.2   DC Motor Efficiency

The power flow diagram for a DC compound motor is shown in Fig. 6.23. The electrical power input P_{input} is equal to $V_t I_t = V_t I_L$. The electrical losses consist of the armature winding loss ($I_a^2 R_a$) and the copper losses ($I_f^2 R_f + I_s^2 R_s$) in the field windings.

The motor efficiency is given by

$$\eta = \frac{P_{output}}{P_{input}} \, 100\% \tag{6.23}$$

For the case of a shunt motor, the series field winding is not present or is disconnected. Thus, the field copper loss consists only of the shunt field loss

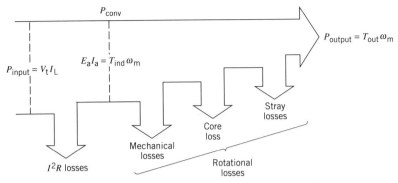

FIGURE 6.23   Power flow diagram for a DC compound motor.

$(I_f^2 R_f)$. The mechanical losses, core losses, and stray load losses are commonly lumped together under rotational losses and are assumed to remain constant at any loading level.

EXAMPLE 6.8

A 10-hp, 220-V, DC shunt motor has armature and field resistances of 0.25 Ω and 220 Ω, respectively. It is supplied by a 220-V source, and it draws a load current of 40 A. The total rotational losses are 450 W. Find the efficiency of the motor.

Solution   When the motor is driving a load, the power input P_{input} is

$$P_{\text{input}} = V_t I_t = (220)(40) = 8800 \text{ W}$$

The field and armature currents are given by

$$I_f = V_t/R_f = 220/220 = 1 \text{ A}$$
$$I_a = I_t - I_f = 40 - 1 = 39 \text{ A}$$

The copper losses P_{Cu} consist of

$$I_a^2 R_a = (39)^2(0.25) = 380 \text{ W}$$
$$I_f^2 R_f = (1)^2(220) = 220 \text{ W}$$

The total copper losses of the motor are

$$P_{\text{Cu}} = 380 + 220 = 600 \text{ W}$$

The power output P_{output} is equal to the difference between the power input and the sum of the total copper losses and the rotational losses P_{rot}.

$$P_{output} = P_{input} - (P_{Cu} + P_{rot}) = 8800 - (600 + 450) = 7750 \text{ W}$$

Therefore,

$$\eta = (P_{output}/P_{input})100\% = (7750/8800)100\% = 88\%$$

DRILL PROBLEMS

D6.12   The series field and armature winding resistances of a 230-V series motor are 0.05 Ω and 0.2 Ω, respectively. The motor draws a current of 20 A while running at 1500 rpm. If the total rotational losses are 400 W, determine the efficiency of the motor.

D6.13   A DC shunt motor is rated at 5 hp, 115 V, 1150 rpm. At rated operating conditions, the efficiency is 85%. The armature circuit resistance is 0.5 Ω, and the field circuit resistance is 115 Ω. Determine the induced voltage at rated operating conditions.

D6.14   A 5-hp, 120-V, 1800-rpm shunt motor is operating at full load and takes a line current of 36 A. Its armature and field resistances are 0.30 Ω and 120 Ω, respectively.

  a. What is the efficiency of this motor?

  b. What is its rotational loss?

6.7.3   Speed-Torque Characteristics

Consider the DC shunt motor whose equivalent circuit is shown in Fig. 6.24.

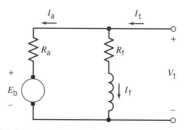

FIGURE 6.24   Equivalent circuit of a DC shunt motor.

From Kirchhoff's voltage law,

$$V_t = E_b + I_a R_a \qquad (6.24)$$

Substituting the expression for the back emf given by Eq. 6.18 into Eq. 6.24 and solving for ω_m yields

$$\omega_m = \frac{V_t - I_a R_a}{K_a \Phi_p} \qquad (6.25)$$

It may be observed that loss of field excitation results in overspeeding for a shunt motor. Thus, care should be taken to prevent the field circuit from getting open.

From Eq. 6.19, the armature current may be expressed as follows:

$$I_a = \frac{T_{dev}}{K_a \Phi_p} \qquad (6.26)$$

Substituting Eq. 6.26 into Eq. 6.25 yields the speed-torque equation of a DC shunt motor.

$$\omega_m = \frac{1}{K_a \Phi_p} V_t - \frac{R_a}{(K_a \Phi_p)^2} T_{dev} \qquad (6.27)$$

If the applied voltage V_t and the flux Φ_p remain constant for any load, the speed will decrease linearly with torque. In an actual machine, however, as the load increases, the flux is reduced because of armature reaction. Since the denominator terms decrease, there is less reduction in speed and speed regulation is improved somewhat. The speed-torque characteristics of DC motors are shown in Fig. 6.25.

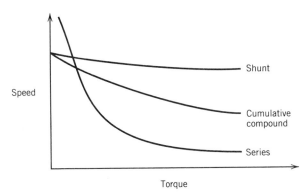

FIGURE 6.25 Speed-torque characteristics of DC motors.

In a compound motor, the magnetic field of the series field winding either aids or opposes the magnetic field of the shunt field winding; thus, the net magnetic flux at the pole is given by

$$\Phi_p = \frac{(mmf)_f \pm (mmf)_s}{R_{path}} \tag{6.28}$$

where

Φ_p = magnetic pole flux

$(mmf)_f$ = shunt field mmf

$(mmf)_s$ = series field mmf

When the mmfs of the two field windings are additive, the speed-torque characteristic is more drooping than that for a shunt motor. When the mmfs are opposing, the speed-torque characteristic lies above that of a shunt motor.

In a series motor, the excitation is provided solely by the series field winding, which is connected in series with the armature. The flux produced is proportional to the armature current. Thus, the torque developed may be written as follows:

$$T_{dev} = K_a\Phi_p I_a = K'I_a^2 \tag{6.29}$$

Therefore, the speed may be expressed as

$$\omega_m = \frac{V_t}{K'I_a} - \frac{R_a}{K'} = \frac{V_t}{K''\sqrt{T_{dev}}} - \frac{R_a}{K'} \tag{6.30}$$

It is seen that the series motor will run at dangerously high speeds at no load. For this reason, a series motor is never started with no load connected to its shaft.

6.7.4   Motor Starting

Consider the DC shunt motor. At starting, the armature is not rotating. Therefore, the counter emf $E_b = K_a\Phi_p\omega_m = 0$. E_b is also called back emf. Hence, the starting current will be dangerously high and is given by

$$I_{a,start} = \frac{V_t - E_b}{R_a} = \frac{V_t}{R_a} \tag{6.31}$$

For the motor of Example 6.7, $I_{a,start}$ will have a value of $(220/0.2) = 1100$ A, which is approximately 23 times the rated current of 48 A. This starting current value is obviously too high. Provision must be made to limit the starting current to prevent damage to the motor. Two times rated current is typically allowed to flow during starting so that sufficient torque will be developed. Two methods of limiting the starting current are as follows:

1. Insert external resistance in the armature circuit.

2. Apply a reduced voltage at starting.

The first method means an additional copper loss, albeit during the starting period only. The second method has the major disadvantage of requiring an expensive variable-voltage supply.

The external resistance R_{ae} is inserted in the armature circuit either manually or automatically. At starting, the armature current is given by

$$I_{a,start} = \frac{V_t}{R_a + R_{ae}} \tag{6.32}$$

As the motor accelerates to a higher speed, the starting resistance is shunted out in steps. When the full starting resistance is shorted out, the motor accelerates to its base speed.

EXAMPLE 6.9

Design a starter for the DC shunt motor of Example 6.7 using an external resistance to be connected in series with the armature as shown in Fig. 6.26.

This resistance is to be cut out in steps so that the armature current does not exceed 200% of full-load armature current. As the motor speeds up, the armature current will drop. As soon as the armature current falls to its full-load value, sufficient resistance is to be cut out so that the current returns to the 200% level. This process is repeated until the full starting resistance is shorted out. The field winding is to be connected directly across the DC supply.

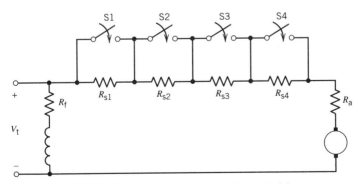

FIGURE 6.26 DC motor starter of Example 6.9.

Solution   From Example 6.7, the full-load or rated armature current is

$$I_{a,rated} = 48\ A$$

The maximum value of starting current is limited to two times rated armature current; therefore,

$$I_{a,start} = 2I_{a,rated} = (2)(48) = 96\ A$$

With the external resistance R_e inserted in series with the armature, Kirchhoff's voltage law yields

$$V_t = E_b + I_a(R_a + R_e)$$

Therefore,

$$R_e = (V_t - E_b)/I_a - R_a$$
$$E_b = V_t - I_a(R_a + R_e)$$

At starting, the counter emf $E_{b0} = 0$. Thus,

$$R_{e0} = (220 - 0)/96 - 0.2 = 2.292 - 0.2 = 2.092\ \Omega$$
$$E_{b1} = 220 - (48)(2.292) = 110\ V$$
$$R_{e1} = (220 - 110)/96 - 0.2 = 1.146 - 0.2 = 0.946\ \Omega$$
$$E_{b2} = 220 - (48)(1.146) = 165\ V$$
$$R_{e2} = (220 - 165)/96 - 0.2 = 0.573 - 0.2 = 0.373\ \Omega$$
$$E_{b3} = 220 - (48)(0.573) = 192.5\ V$$
$$R_{e3} = (220 - 192.5)/96 - 0.2 = 0.286 - 0.2 = 0.086\ \Omega$$
$$E_{b4} = 220 - (48)(0.286) = 206.5\ V$$
$$R_{e4} = (220 - 206.5)/96 - 0.2 = 0.141 - 0.2 = -0.059\ \Omega$$

The negative sign of R_{e4} means that for this step the full value of R_{e3} is shorted out.

The values of resistances that are to be shorted out successively are:

$$R_{s1} = R_{e0} - R_{e1} = 2.092 - 0.946 = 1.146\ \Omega$$
$$R_{s2} = R_{e1} - R_{e2} = 0.946 - 0.373 = 0.573\ \Omega$$
$$R_{s3} = R_{e2} - R_{e3} = 0.373 - 0.086 = 0.287\ \Omega$$
$$R_{s4} = 0.086\ \Omega$$

DRILL PROBLEMS

D6.15 A DC shunt motor is rated 5 kW, 125 V, and 1800 rpm. When the armature is held stationary, a voltage of 5 V applied to the motor terminals will cause a full-load current of 40 A to flow through the armature.

  a. Determine the armature current if rated voltage is applied directly across the motor terminals.

  b. Determine the value of the external resistance that must be connected in series with the armature in order to limit the starting current to twice the rated armature current.

D6.16 A 240-V DC shunt motor has an armature winding resistance of 0.2 ohm. The full-load armature current is 50 A.

  a. Determine the starting current if the motor is connected directly across the 240-V supply. Express this current as a percentage of the full-load value.

  b. Calculate the value of resistance that must be connected in series with the armature circuit to limit the starting current to 150% of full-load value.

6.7.5   Applications of DC Motors

The expression for developed torque T_{dev} of a DC motor is given by Eq. 6.19, which shows that the torque is directly proportional to the product of the field flux and the armature current. This relationship is valid not only for the motor operating under load at steady state but also during starting conditions. This torque characteristic is illustrated in Fig. 6.27a for the shunt, compound, and series motors.

The expression for the speed ω_m of a DC motor is given by Eq. 6.25, which describes the linear relation between speed and armature current and the inverse variation of speed with respect to field flux. The speed characteristics of the shunt, compound, and series motors are shown in Fig. 6.27b.

The DC shunt motor has a relatively constant speed characteristic almost independent of load. It has a speed regulation of about 5% to 10%. Because the field flux changes very little with changes in load levels, the torque developed by the motor is almost directly proportional to the armature current. Therefore, the motor is able to develop a reasonably good starting torque, which is usually limited to less than 250% because of starting current restrictions. DC shunt motors are used primarily for constant-speed applications requiring medium starting torque, such as for driving centrifugal pumps, fans, blowers, conveyors, and machine tools.

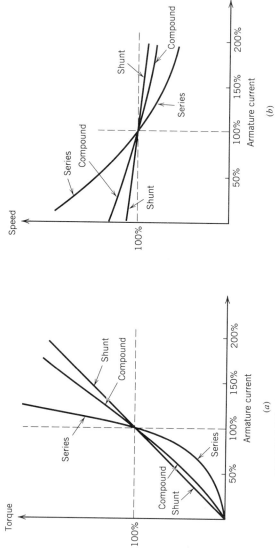

FIGURE 6.27 Speed and torque characteristics of DC motors.

In a DC cumulative compound motor, the flux developed by the series field winding reinforces the flux produced by the shunt field winding. Thus, the torque developed by the compound motor is much higher than that of a DC shunt motor, especially for armature currents above rated value. For the same reason, however, the speed of the compound motor decreases more rapidly with increasing armature current than that of the shunt motor. The speed regulation of the compound motor varies from 15% to 30%, depending on the degree of compounding. Compound motors are used for applications requiring high starting torques and only fairly constant speed and for pulsating loads. They are used to drive conveyors, hoists, compressors, metal-stamping machines, reciprocating pumps, punch presses, crushers, and shears.

The DC series motor derives its flux from its series field, which is connected in series with the armature. Therefore, the torque developed by the motor is directly proportional to the square of the armature current. The speed characteristic of the series motor is described by a large variation in speed from full-load to no-load conditions. This indicates that loads should not be removed completely or reduced to very low levels because of the possibility of the motor "running away." Series motors are used for applications requiring very high starting torques and where varying speed is acceptable. They are used to drive hoists, cranes, and so forth.

*6.8 DC MACHINE DYNAMICS

The previous sections have described the steady-state behavior of DC generators and motors, and the models derived and used are valid only for steady-state conditions. In this section, the dynamic behavior of DC machines is described and transient models are presented. The block diagram representations and the transfer functions of these models are derived. In an introductory course in power engineering, the professor may choose to omit this section.

*6.8.1 Dynamic Equations

A DC machine may be represented by two coupled electrical circuits both containing resistances and inductances as shown in Fig. 6.28. These circuits, the field and the armature, are coupled through the electromagnetic field, which is represented by the generated voltage e_a. The electrical system is coupled to the mechanical system through the electromagnetic torque T_{fld} and an external mechanical torque, which could be an input torque T_S from a prime mover or a load torque T_L. The reference directions of the arrows shown are applicable for modeling the DC generator.

In the following discussion, it is assumed that saturation is negligible and the air-gap flux is directly proportional to the field current i_f. Armature reaction

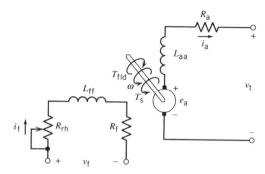

FIGURE 6.28   Schematic representation of a DC generator.

is also assumed negligible; however, armature reaction effects may be added on later as additional field excitation requirements. Hence, the expressions for the electromagnetic torque and generated voltage are given by Eqs. 6.33 and 6.34, respectively.

$$T_{fld} = K_a K_f' i_f i_a$$
$$= K_f i_f i_a \qquad (6.33)$$
$$e_a = K_a K_f' i_f \omega_m$$
$$= K_f i_f \omega_m \qquad (6.34)$$

Here $K_f = K_a K_f'$ is a constant.

*6.8.2   Separately Excited DC Generator

For the DC generator shown in Fig. 6.28, the field excitation is supplied by a separate voltage source. If it is assumed that the field circuit is closed at time $t = 0$, the voltage equation may be written as

$$L_{ff} \frac{di_f}{dt} + R_f i_f = V_f u(t)$$
$$\tau_f \frac{di_f}{dt} + i_f = \frac{V_f}{R_f} u(t) \qquad (6.35)$$

where $\tau_f = L_{ff}/R_f$ is the time constant of the field circuit and $u(t)$ is the unit step function.

The generated voltage is given by Eq. 6.34, assuming that the effect of saturation is negligible. For a speed ω_{m0}, this voltage may be written in terms of the generator constant K_g as follows:

$$e_{a0} = (K_f \omega_{m0}) i_f = K_g i_f \qquad (6.36)$$

When the field current is held constant, as when it has reached steady state, the generated voltage becomes directly proportional to the angular velocity; that is,

$$\frac{e_a}{\omega_m} = \frac{e_{a0}}{\omega_{m0}} \tag{6.37}$$

or

$$e_a = \left(\frac{e_{a0}}{\omega_{m0}}\right)\omega_m \tag{6.38}$$

For the armature circuit, the voltage equation may be written in terms of the generator terminal voltage as

$$L_{aa}\frac{di_a}{dt} + R_a i_a = e_a - V_t$$
$$\tau_a\frac{di_a}{dt} + i_a = \frac{e_a - V_t}{R_a} \tag{6.39}$$

where $\tau_a = L_{aa}/R_a$ is the time constant of the armature circuit.

When the separately excited DC generator is supplying an armature current i_a to an electrical load, the electromagnetic torque T_{fld} may be derived from the developed power P_{dev} as follows:

$$\omega_m T_{fld} = P_{dev} = e_a i_a \tag{6.40}$$

Therefore,

$$T_{fld} = \frac{e_a i_a}{\omega_m}$$
$$= \left(\frac{e_{a0}}{\omega_{m0}}\right)i_a \tag{6.41}$$

The dynamic equation of motion of the DC machine is dependent on the mechanical torque applied to its shaft; thus,

$$J\frac{d\omega_m}{dt} + T_{fld} = T_{shaft}$$
$$J\frac{d\omega_m}{dt} = T_{shaft} - T_{fld} \tag{6.42}$$

where
J = moment of inertia of the rotor and the prime mover
T_{shaft} = mechanical applied torque from the prime mover

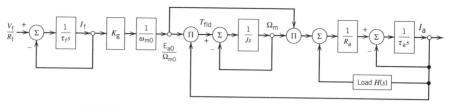

FIGURE 6.29 Block diagram representation of a DC generator.

The differential equations describing the dynamic performance of a separately excited DC generator may be represented with a block diagram. The block diagram representation employing the Laplace-transformed variables is shown in Fig. 6.29.

EXAMPLE 6.10

A 240-V, 50-kW, 1800-rpm, separately excited DC generator has an armature circuit resistance and inductance of 0.10 Ω and 1.0 mH, respectively. The field winding resistance and inductance are 125 Ω and 75 H, respectively. The generator emf constant K_g is 125 V per field ampere at 1800 rpm. The field and armature circuits are initially open, and the prime mover is driving the generator at a constant speed of 1800 rpm.

a. At time $t = 0$, the field circuit is connected to a constant-voltage source of 250 V. Find the expression for the armature terminal voltage as a function of time.

b. In part (a), after the field circuit has reached a steady state, the armature is suddenly connected to a load consisting of a series-connected resistance and inductance of 1.15 Ω and 1.5 mH, respectively. Find the expressions for the armature current, terminal voltage, and electromagnetic torque as functions of time.

Solution

a. The voltage equation for the field circuit is given by Eq. 6.35 as follows:

$$\tau_f \frac{di_f}{dt} + i_f = \frac{V_f}{R_f}u(t) = \frac{250}{150}u(t) = 2.0u(t)$$

where $\tau_f = L_{ff}/R_f = 75/125 = 0.6$ s. Thus, the expression for the field current is given by

$$i_f(t) = 2.0\,(1.0 - e^{-t/\tau_f})u(t)$$
$$= 2.0\,(1.0 - e^{-t/0.60})u(t) \qquad \text{A}$$
$$= 2.0\,(1.0 - e^{-1.667t})u(t)$$

The generated voltage is given by Eq. 6.36 as

$$e_{a0} = K_g i_f = (125)(2.0)(1.0 - e^{-1.667t})u(t) \qquad V$$
$$= 250(1.0 - e^{-1.667t})u(t)$$

Since the generator is initially operating at open circuit, the terminal voltage is equal to the generated voltage. Thus,

$$v_t = e_{a0} = 250(1.0 - e^{-1.667t})u(t) \qquad V$$

b. Since the field circuit has reached a steady-state condition, the generated voltage becomes a constant value $e_a(t) = E_a = 250$ V. The differential equation for the armature circuit takes the form

$$\tau_a \frac{d i_a}{dt} = i_a = \frac{E_a}{R_a + R_L} u(t) = \frac{250}{0.10 + 1.15} u(t) = 200u(t)$$

where

$$\tau_a = \frac{L_{aa} + L_L}{R_a + R_L} = \frac{(1.0 + 1.5) \times 10^{-3}}{0.10 + 1.15} = 2.0 \times 10^{-3} \text{ s}$$

Thus, the expression for the armature current is

$$i_a(t) = 200(1.0 - e^{-t/\tau_a})u(t)$$
$$= 200(1.0 - e^{-t/0.002})u(t) \qquad A$$
$$= 200(1.0 - e^{-500t})u(t)$$

The terminal voltage of the generator can be found as

$$v_t = L_L \frac{d i_a}{dt} + R_L i_a$$
$$= 1.5 \times 10^{-3}(-500)(200)(-e^{-500t}) \qquad V$$
$$+ (1.15)(200)(1.0 - e^{-500t})$$
$$= 230 - 80e^{-500t}$$

The angular velocity is given by

$$\omega_m = \frac{2\pi n}{60} = \frac{2\pi(1800)}{60} = 60\pi \text{ rad/s}$$

Therefore, the electromagnetic torque is

$$T_{fld} = \frac{e_a i_a}{\omega_m} = \frac{(250)(200)(1.0 - e^{-500t})}{60\pi} \quad \text{N-m}$$
$$= 265.2(1.0 - e^{-500t})$$

*6.8.3   Separately Excited DC Motor

The schematic representation of a separately excited DC motor is shown in Fig. 6.30. The field circuit is assumed to have been connected to the separate voltage source V_f for a sufficiently long time that the field current I_f has reached a steady value, that is,

$$I_f = \frac{V_f}{R_{wdg} + R_{rh}} = \frac{V_f}{R_f} \tag{6.43}$$

Assuming that the effects of saturation and armature reaction are negligible, and since the field current I_f is constant, the expressions for the developed electromagnetic torque and induced voltage may be expressed as in Eqs. 6.44 and 6.45, respectively.

$$T_{fld} = K_a K_f' I_f i_a$$
$$= K_f I_f i_a$$
$$= K_m i_a \tag{6.44}$$
$$e_a = K_a K_f' I_f \omega_m$$
$$= K_f I_f \omega_m$$
$$= K_m \omega_m \tag{6.45}$$

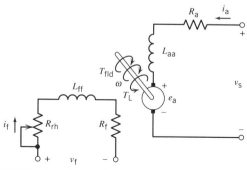

FIGURE 6.30   Schematic representation of a DC motor.

where $K_m = K_f I_f$ = constant and measured in newton-meters per ampere. The motor constant K_m may also be derived from Eq. 6.45 and expressed in volts per radian per second or volt-seconds per radian as follows:

$$K_m = \frac{e_{a0}}{\omega_{m0}} \qquad (6.46)$$

The developed torque is available at the motor shaft for driving a mechanical load. The motor develops just enough torque to balance the required torque of the load and its rotational losses. The induced voltage is in opposition to the applied voltage and thus is also called counter emf or back emf. Because of its reverse polarity, the back emf serves to limit the armature current.

For the armature circuit, a loop voltage equation may be written in terms of the voltage applied to the motor terminals as follows:

$$L_{aa} \frac{d i_a}{dt} + R_a i_a = V_s - e_a$$

$$\tau_a \frac{d i_a}{dt} + i_a = \frac{V_s - e_a}{R_a} \qquad (6.47)$$

where $\tau_a = L_{aa}/R_a$ is the time constant of the armature circuit. Substituting the expression for the counter emf given by Eq. 6.45 into Eq. 6.47 and rearranging the terms, Eq. 6.48 is obtained.

$$\tau_a \frac{d i_a}{dt} + i_a + \left(\frac{K_m}{R_a}\right)\omega_m = \left(\frac{1}{R_a}\right)V_s \qquad (6.48)$$

The dynamic equation of motion of the DC motor is dependent on the mechanical load connected to its shaft. The electromagnetic torque developed by the motor must be equal to the sum of all opposing torques. Thus,

$$T_{fld} = J \frac{d\omega_m}{dt} + T_{loss} + T_{load} \qquad (6.49)$$

where
J = moment of inertia of the rotor and the prime mover

T_{loss} = rotational losses

T_{load} = mechanical load torque connected to the motor shaft

Substituting the expression for the developed electromagnetic torque given in Eq. 6.44 into Eq. 6.49 and rearranging the terms yields

$$-\left(\frac{K_m}{J}\right)i_a + \frac{d\omega_m}{dt} = -\frac{T_{load}}{J} - \frac{T_{loss}}{J} \qquad (6.50)$$

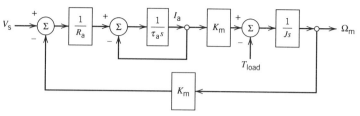

FIGURE 6.31 Block diagram representation of a DC motor.

The block diagram representation of a separately excited DC motor is derived from Eqs. 6.48 and 6.50 and is shown in Fig. 6.31. It may be noted that there are two independent input variables, namely the supply voltage V_s and the load torque T_{load}.

When the response of the motor to changes in the supply voltage is being investigated, the block diagram may be simplified to that of Fig. 6.32. The load torque is typically a function of speed, and it is customary to assume that the load torque is directly proportional to the speed; thus

$$T_{load} = B_{load}\omega_m \tag{6.51}$$

where B_{load} is a proportionality constant.

When the damping torque $T_{load} = B_{load}\omega_m$ is neglected, Fig. 6.32 may be simplified into one simple feedback loop circuit as shown in Fig. 6.33, where $\tau_m = (R_a J)/K_m^2$ is called the inertial time constant.

The overall transfer function may be expressed as follows:

$$\frac{\Omega_m}{V_s/K_m} = \frac{1/(\tau_m \tau_a)}{s(s + 1/\tau_a) + 1/(\tau_m \tau_a)} \tag{6.52}$$

From this overall transfer function, the characteristic equation of the speed response to the voltage input is found as

$$\left(s + \frac{1}{\tau_a}\right)s + \frac{1}{\tau_m \tau_a} = s^2 + s\left(\frac{1}{\tau_a}\right) + \frac{1}{\tau_m \tau_a} = 0 \tag{6.53}$$

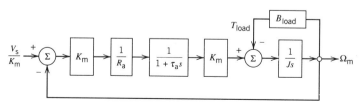

FIGURE 6.32 Block diagram representation of a DC motor.

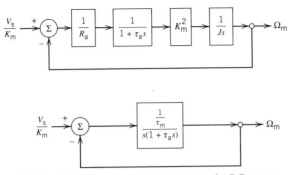

FIGURE 6.33 Simplified block diagrams of a DC motor.

The standard form of the characteristic equation of a second-order system is given by

$$s^2 + 2\alpha s + \omega_n^2 = 0 \qquad (6.54)$$

Comparing Eqs. 6.53 and 6.54, the undamped natural frequency ω_n is

$$\omega_n = \sqrt{\frac{1}{\tau_m \tau_a}} \qquad (6.55)$$

and the damping factor α is

$$\alpha = \frac{1}{2\tau_a} \qquad (6.56)$$

The damping ratio is given by

$$\zeta = \frac{\alpha}{\omega_n} = \frac{1}{2}\sqrt{\frac{\tau_m}{\tau_a}} \qquad (6.57)$$

When the response of the motor to load changes is of interest, the block diagram of Fig. 6.31 may be simplified to that shown in Fig. 6.34.

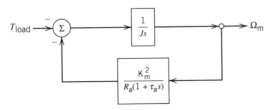

FIGURE 6.34 Simplified block diagram of a DC motor.

The overall transfer function relating the speed response to a change in load torque is

$$\frac{\Omega_m}{T_{load}/J} = -\frac{s + 1/\tau_a}{s(s + 1/\tau_a) + 1/(\tau_m \tau_a)} \tag{6.58}$$

It is seen that the characteristic equation for the speed response with change in load torque has the same undamped natural frequency and damping factor, which are given in Eqs. 6.55 and 6.56, respectively. The negative sign indicates that an additional load torque produces a reduction of speed.

EXAMPLE 6.11

A 250-V, separately excited DC motor has the following parameters:

$$\begin{aligned}
\text{Armature resistance} \quad & R_a = 0.025 \ \Omega \\
\text{Armature inductance} \quad & L_{aa} = 0.006 \ \text{H} \\
\text{Moment of inertia} \quad & J = 25 \ \text{kg-m}^2 \\
\text{Motor constant} \quad & K_m = 2.75 \ \text{N-m/A}
\end{aligned}$$

The motor is connected to a constant 400-V supply. The motor is initially running without load, and the system is at steady state. The no-load armature current is 15 A. The effects of saturation and armature reaction may be neglected.

A constant load torque T_{load} of 1500 N-m is suddenly connected to the shaft of the DC motor. Determine

a. The undamped natural frequency of the speed response
b. The damping factor and damping ratio
c. The initial speed in rpm
d. The initial acceleration
e. The ultimate speed drop

Solution

a. The armature and inertial time constants are given by

$$\tau_a = \frac{L_{aa}}{R_a} = \frac{0.006}{0.025} = 0.24 \ \text{s}$$

$$\tau_m = \frac{JR_a}{K_m^2} = \frac{(25)(0.025)}{(2.75)^2} = 0.083 \ \text{s}$$

The undamped natural frequency is

$$\omega_m = \sqrt{\frac{1}{\tau_m \tau_a}} = \sqrt{\frac{1}{(0.083)(0.24)}} = 7.10 \text{ rad/s}$$

b. The damping factor and damping ratio are

$$\alpha = \frac{1}{2\tau_a} = \frac{1}{(2)(0.24)} = 2.083$$

$$\zeta = \frac{\alpha}{\omega_n} = \frac{2.08}{7.10} = 0.29$$

c. At time $t = 0$, at the instant the load is suddenly added, the induced emf, or counter emf, is given by

$$E_{a0} = V_s - I_{a0}R_a = 400 - (15)(0.025) = 399.6 \text{ V}$$

Therefore, the initial motor speed is

$$\omega_m = \frac{E_{a0}}{K_m} = \frac{399.6}{2.75} = 145.3 \text{ rad/s}$$

$$= (145.3)\left(\frac{60}{2\pi}\right) = 1388 \text{ rpm}$$

d. The initial acceleration, assuming losses are negligible, is found by using Eq. 6.50. Thus,

$$\alpha = \frac{d\omega_m}{dt} = \frac{-T_{load} + K_m I_{a0}}{J}$$

$$= \frac{-1500 + (2.75)(15)}{25} = -58.35 \text{ rad/s}^2$$

e. The ultimate drop in speed is found by application of the final-value theorem of Laplace transforms to Eq. 6.58. Thus,

$$\Delta\omega_m = \lim_{s \to \infty} \left[s \frac{-(1/J)(s + 1/\tau_a)}{s(s + 1/\tau_a) + 1/(\tau_m \tau_a)} \frac{\Delta T_{load}}{s} \right]$$

$$= -\frac{\tau_m}{J} \Delta T_{load} = -\frac{0.083}{25}(1500) = -4.98 \text{ rad/s}$$

$$= -4.98\left(\frac{60}{2\pi}\right) = -47.6 \text{ rpm}$$

REFERENCES

1. Bergseth, F. R., and S. S. Venkata. *Introduction to Electric Energy Devices.* Prentice Hall, Englewood Cliffs, N.J., 1987.
2. Chapman, Stephen J. *Electric Machinery Fundamentals.* McGraw-Hill, New York, 1985.
3. Del Toro, Vincent. *Electric Machines and Power Systems.* Prentice-Hall, Englewood Cliffs, N.J., 1985.
4. Fitzgerald, A. E., Charles Kingsley, and Stephen Umans. *Electric Machinery.* 4th ed. McGraw-Hill, New York, 1983.
5. Hubert, Charles I. *Electric Machines.* Merrill, New York, 1991.
6. Kosow, Irving L. *Electric Machinery and Transformers.* Prentice Hall, Englewood Cliffs, N.J., 1991.
7. Macpherson, George and Robert D. Laramore. *An Introduction to Electrical Machines and Transformers.* 2nd ed. Wiley, New York, 1990.
8. Nasar, Syed A. *Electric Energy Conversion and Transmission.* Macmillan, New York, 1985.
9. Ramshaw, Raymond, and R. G. van Heeswijk. *Energy Conversion Electric Motors and Generators.* Saunders College Publishing, Philadelphia, 1990.
10. Sen, P. C. *Principles of Electrical Machines and Power Electronics.* Wiley, New York, 1989.
11. Wildi, Theodore. *Electrical Machines, Drives, and Power Systems.* 2nd ed. Prentice Hall, Englewood Cliffs, N.J., 1991.

PROBLEMS

6.1 A six-pole DC machine has an armature connected as a lap winding. The armature has 48 slots with four conductors per slot. The armature is rotated at 600 rpm, and the flux per pole is 30 mWb. Calculate the induced voltage.

6.2 A four-pole DC generator has a wave-wound armature containing 384 armature conductors. The generator is driven at 1180 rpm and generates a voltage of 480 V. What is the flux per pole?

6.3 A six-pole, 1200-rpm, DC generator has 48 armature slots with four conductors per slot. The flux per pole is 20 mWb, and each armature conductor has a maximum current-carrying capacity of 40 A.

a. Compare the terminal voltages, currents, and power rating for a lap winding.

b. Repeat part (a) for a wave winding.

6.4 A DC generator has a flux per pole of 125 mWb. The generator has six poles, and it is rotated at 1200 rpm.

a. Determine the induced voltage if the armature has 48 conductors connected as a lap winding.

b. Determine the induced voltage if the armature has 48 conductors connected as a wave winding.

6.5 A shunt-connected generator has four poles, and its lap-connected armature has 576 conductors. The armature and field resistances are $0.10\ \Omega$ and $100\ \Omega$, respectively. The flux per pole is 30 mWb. The generator supplies 3.5 kW to a load connected to its terminals at a voltage of 120 V. Determine the generator speed.

6.6 A separately excited generator has six poles with a flux per pole of 25 mWb. The armature is lap wound and has 620 conductors. The generator supplies a certain load at 240 V, and at this load the armature copper loss is 600 W. The generator is driven at a speed of 1120 rpm. Assuming that the total brush contact voltage drop is 2 V, calculate the current and power delivered by the generator.

6.7 A 50-kW, 240-V DC shunt generator has an armature resistance of $0.10\ \Omega$, a field circuit resistance of $120\ \Omega$, and a total brush voltage drop of 2 V. The generator delivers rated current at rated speed and rated voltage. Calculate the following:

   a. Load current

   b. Field current

   c. Armature current

   d. Armature induced voltage

6.8 A 10-kW, 120-V DC shunt generator has an armature circuit resistance of $0.2\ \Omega$ and a field circuit resistance of $240\ \Omega$. The generator delivers rated current at rated voltage and rated speed. Assume a total brush voltage drop of 2 V. Calculate (a) the armature current and (b) the armature induced voltage.

6.9 A compound generator has armature, shunt field, and series field winding resistances of $0.2\ \Omega$, $200\ \Omega$, $0.1\ \Omega$, respectively. The generator induced voltage is 255 V, and the terminal voltage is 240 V. The generator is connected for long-shunt compound operation.

   a. Calculate the power supplied to a load.

   b. Repeat part (a) if the generator is reconnected for short-shunt compound operation.

6.10 The shunt field current of a 60-kW, 125-V, DC generator has to be increased from 4.0 A on no load to 5.0 A at full load to produce flat compounding. Each field pole has 1500 turns.

   a. Calculate the number of series field turns per pole, assuming short-shunt connection.

   b. Repeat part (a) assuming long-shunt connection.

6.11 The open-circuit characteristic of a DC shunt generator is given by

E_a (V)	60	120	170	210	240	265	285
I_f (A)	1	2	3	4	5	6	7

The generator operates at a speed of 1200 rpm.

a. Plot the open-circuit characteristic.

b. From the open-circuit characteristic, find the maximum field-circuit resistance in order that the self-excited shunt generator will build up.

c. Determine the value the no-load voltage will build up to for a field-circuit resistance of 48 Ω and a speed of 1200 rpm.

6.12 A separately excited DC generator has the following open-circuit characteristic when driven at 1200 rpm.

E_a (V)	100	200	300	400	500	600	700
I_f (A)	0.50	1.00	1.75	2.75	4.00	5.50	7.50

The field winding resistance is 120 Ω, and the armature winding resistance is 0.2 Ω. Assume a total brush voltage drop of 4 V. Armature reaction effects may be neglected. The machine is driven at 1200 rpm as a shunt-excited generator.

a. Determine the terminal voltage on no load.

b. At full load when the armature winding current is 80 A, determine the terminal voltage.

6.13 A DC machine is rated 10 kW, 250 V, 1750 rpm and has armature and field winding resistances of 0.2 Ω and 125 Ω, respectively. The machine is self-excited and is driven at 1750 rpm. The data for the magnetization curve are

E_a (V)	15	50	100	150	188	212	250	275	305	330
I_f (A)	0	0.1	0.2	0.3	0.4	0.5	0.75	1.0	1.5	2.0

a. Determine the generated voltage with no field current.

b. Determine the critical field circuit resistance.

c. Determine the resistance of the field rheostat if the no-load terminal voltage is 250 V.

d. Determine the value of the no-load generated voltage if the generator is driven at 1500 rpm and the rheostat is short-circuited.

e. Determine the speed at which the generator is to be driven such that the no-load generated voltage is 200 V with the rheostat short-circuited.

6.14 The self-excited DC machine in Problem 6.13 delivers rated load when driven at 1750 rpm. The rotational loss is 450 W. Neglect the effects of armature reaction. Calculate

a. The generated voltage

b. The developed torque

  c. The field current

  d. The efficiency of the generator

6.15 A DC shunt generator is rated 20 kW, 220 V, and 1800 rpm. It has armature and field winding resistances of 0.1 Ω and 110 Ω, respectively. The data for the magnetization curve at 1800 rpm are

E_a (V)	5	25	60	120	170	200	215	225	240
I_f (A)	0.0	0.1	0.25	0.5	0.75	1.0	1.25	1.5	2.0

The machine is connected as a self-excited generator.

  a. Determine the maximum generated voltage.

  b. The generator delivers full-load current at rated voltage with a field current of 2 A. Determine the resistance of the field rheostat.

  c. Determine the electromagnetic power and torque developed at full-load condition.

  d. Determine the armature reaction effect in equivalent field amperes (I_f^{ar}) at full load.

6.16 The shunt generator of Problem 6.15 is connected as a long-shunt compound generator.

  a. Draw a schematic diagram for the generator connection.

  b. Determine the number of turns per pole of the series field winding needed to make the no-load and full-load terminal voltages equal to the rated voltage. The series field winding resistance is 0.05 Ω, and the shunt field winding has $N_f = 1200$ turns/pole.

6.17 A 220-V DC shunt motor has armature and field winding resistances of 0.15 Ω and 110 Ω, respectively. The motor draws a line current of 5 A while running on no load. When driving a load, the motor runs at 1100 rpm and draws 48 A of line current. Calculate the no-load speed.

6.18 A 10-hp, 230-V, 1150-rpm, four-pole, DC shunt motor has a total of 596 conductors arranged in a wave winding having two parallel paths. The armature circuit resistance is 0.15 Ω. When the motor delivers rated output power at rated speed, the motor draws a line current of 38 A and a field current of 2 A. Assume that the effects of armature reaction are negligible. Compute (a) the flux per pole and (b) the developed torque.

6.19 A 10-hp, 230-V, 1200-rpm DC series motor draws a current of 36 A when delivering rated output at its rated speed. The armature and series field winding resistances are 0.20 Ω and 0.1 Ω, respectively. Assume that the magnetization curve is linear and that the effects of armature reaction are negligible.

  a. Find the speed of this motor when it is taking a current of 24 A.

  b. Determine the developed torque of this motor for the load conditions of part (a).

6.20 A 20-hp, 250-V, 1100-rpm DC shunt motor drives a load that requires a constant torque regardless of the speed of operation. The armature circuit resistance is 0.10 Ω. When this motor delivers rated power, the armature current is 65 A.

  a. If the flux is reduced to 75% of its original value, find the new value of armature current.

  b. What is the new speed for the conditions of part (a)?

6.21 A shunt motor develops a total torque of 250 N-m at rated load. The field flux decreases by 15%, and at the same time the armature current increases by 40%. Calculate the new value of torque.

6.22 A 220-V DC shunt motor has an armature resistance of 0.2 Ω and a rated armature current of 40 A. Assume a total brush voltage drop of 3 V. Calculate

  a. The generated voltage

  b. The power developed by the armature in watts

  c. The mechanical power developed in horsepower

6.23 A 120-V shunt motor has armature and field winding resistances of 0.10 Ω and 120 Ω, respectively, and a total brush voltage drop of 2 V. The motor operates at rated load and draws a line current of 41 A at a speed of 200 rad/s. Calculate

  a. The field current and armature current

  b. The counter emf

  c. The developed power in kW

  d. The developed torque in N-m

6.24 A 25-hp, 240-V shunt motor has an armature resistance of 0.20 Ω and a brush voltage drop of 4 V. The field circuit resistance is 120 Ω. At no load, the motor draws 14 A and has a speed of 1700 rpm. At full load, the motor draws 82 A. Calculate

  a. The motor speed at full load

  b. The speed regulation

  c. The mechanical power developed at full load

6.25 A four-pole DC motor has a flux per pole of 10 mWb. The armature has 600 conductors, which are connected so that there are four parallel paths between the brushes. The armature resistance is 0.50 Ω, including the effect of brush drop. The machine is connected to a source of 120 V, and the machine is loaded so that it takes rated armature current of 50 amperes. Neglect the effect of armature reaction. Calculate

  a. The motor speed

  b. The torque developed

  c. The speed regulation of the motor

6.26 In the DC motor of Problem 6.25, the effect of armature reaction, which was previously ignored, is to reduce the main field flux by 15% under the load of 50 A. All other data remain the same as in Problem 6.25. Determine the speed and horsepower output of the motor.

6.27 A 10-hp, 230-V series motor has a line current of 37 A and a rated speed of 1200 rpm. The armature and series field resistances are 0.4 Ω and 0.2 Ω, respectively. The total brush voltage drop is 2 V. Calculate the following:

   a. Speed at a line current of 20 A

   b. No-load speed when the line current is 1 A

   c. Speed at 150% rated load when the line current is 60 A and the series field flux is 125% of the full-load flux.

6.28 A DC series motor is rated 230 V, 12 hp, and 1200 rpm. It is connected to a 230-V supply, and it draws a current of 40 A while rotating at 1200 rpm. The armature and series field winding resistances are 0.25 Ω and 0.1 Ω, respectively.

   a. Determine the power and torque developed by the motor.

   b. Determine the speed, torque, and power if the motor draws 20 amperes.

6.29 A 40-hp, 230-V DC shunt motor has armature and field winding resistances of 0.2 Ω and 115 Ω, respectively. At no-load and rated voltage, the speed is 1200 rpm and the motor draws a line current of 5 A. If load is applied to the motor, its speed drops to 1100 rpm. Assume that the effects of armature reaction are negligible. At this load, determine

   a. The armature current and the line current

   b. The developed torque

   c. The horsepower output assuming the rotational losses are constant at 350 W.

6.30 A 250-V shunt motor delivers 15 kW of power at the shaft at 1200 rpm while drawing a line current of 75 A. The field and armature resistances are 250 Ω and 0.10 Ω, respectively. Assuming a contact voltage drop per brush of 1 V, calculate (a) the torque developed by the motor and (b) the motor efficiency.

6.31 A 220-V DC shunt motor has armature and field winding resistances of 0.15 Ω and 110 Ω, respectively. The motor draws a line current of 5 A while running at 1200 rpm on no load. When driving a load, the input to the motor is 12 kW. Calculate

   a. The speed of the motor

   b. The developed torque

   c. The efficiency of the motor at this load

6.32 A 230-V, DC shunt motor has an armature circuit resistance of 0.25 Ω and a field circuit resistance of 115 Ω. At full load the armature draws a current of 38 A and the speed is measured at 1050 rpm. Neglect saturation.

   a. Find the developed torque in newton-meters.

   b. The field rheostat is adjusted so that the resulting field circuit resistance is 144 Ω. Find the new operating speed assuming the developed torque and armature current remain constant.

   c. Assuming that rotational losses amount to 600 W, calculate the efficiency in part (b).

6.33   A 30-hp, 1150-rpm, six-pole shunt motor has an efficiency of 87% when operating at rated load from a 240-V supply. The motor has armature and field resistances of 0.10 Ω and 120 Ω, respectively. Determine

   a. The mechanical power developed
   b. The developed torque
   c. The output shaft torque

6.34   A 20-hp, 240-V, four-pole shunt motor operates at rated load and 1700 rpm and an efficiency of 88%. It has armature and field resistances of 0.2 Ω and 240 Ω, respectively. Determine

   a. The power input to the motor
   b. The mechanical power developed
   c. The shaft torque

6.35   A DC shunt motor is rated 10 hp, 250 V and is connected to a 230-V source. The armature resistance is 0.2 Ω; the field circuit resistance is 125 Ω, and the total rotational losses are 380 W. The motor is driving a load, and it draws an armature current of 30 A while running at a speed of 1700 rpm. Calculate

   a. The generated voltage
   b. The value of the load torque
   c. The efficiency of the motor

6.36   A 220-V shunt motor delivers 20 hp to a load connected to its shaft on full load at 1150 rpm. The motor full-load efficiency is 85%. The armature and field winding resistances are 0.15 Ω and 110 Ω, respectively. Determine

   a. The starting resistance such that the starting line current does not exceed twice the full-load current
   b. The starting torque with the starting resistance computed in part (a) inserted in the armature circuit

6.37   A 240-V, DC shunt motor has an armature winding resistance of 0.2 ohm. The full-load armature current is 50 A. Find the value of the resistance to be connected in series with the armature circuit so that the speed is 75% of the rated speed if full-load current flows.

6.38   An automatic starter is to be designed for a 10-hp, 230-V shunt motor. The resistance of the armature circuit is 0.20 Ω, and the resistance of the field circuit is 115 Ω. The field winding is connected directly across the 230-V source. When operated at no load and rated voltage, the armature current is 4 A and the motor runs at a speed of 1180 rpm. When the motor is delivering rated output, the armature current is 37 A. The resistance in series with the armature is to be adjusted automatically so that during the starting period the armature current is allowed to rise up to but not exceed twice the rated value. As soon as the current falls to rated value, sufficient series resistance is to be cut out (short-circuited) to allow the current to increase once more. This process is repeated until all the series resistance has been cut out.

a. Determine the total resistance of the starter.

b. Calculate the resistance that should be cut out at each step in the starting process.

6.39 A 20-kW, 220-V, 1800-rpm shunt motor has a total armature circuit resistance of 0.4 ohm. The resistance of the shunt field circuit is 220 ohms. At starting, the maximum allowable armature current is twice the rated value. Determine the steps of the starting resistor and the speed attained by the armature at each step. Neglect the effect of armature reaction.

*6.40 A DC shunt generator is shown in Fig. 6.35. The generator is driven at constant speed. The armature and field winding resistances are 0.2 Ω and 50 Ω, respectively, and the armature and field winding inductances are 10 mH and 25 H, respectively. The generator terminal voltage is passed through a low-pass filter with R $= 1$ Ω and L $= 1$ H. Determine the transfer function that relates the output voltage V_0 to the input voltage V_i. The generator constant is $K_g = 120$ V per field ampere.

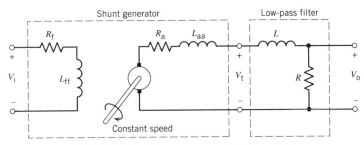

FIGURE 6.35 Shunt generator of Problem 6.40.

*6.41 A separately excited DC motor is shown in Fig. 6.36. The motor drives a mechanical load connected to its shaft. The motor is operated with constant field current. The armature resistance is 0.5 Ω, and the armature inductance is negligible. The other parameters of the motor are $K_m = 2$ V/(rad/s) and $K_t = 1.5$ N-m/A. The equivalent inertia of the mechanical load and the motor is $J = 10$ kg-m^2.

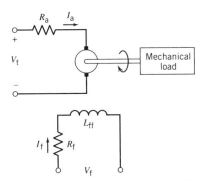

FIGURE 6.36 DC motor of Problem 6.41.

a. Show the block diagram representation of the system.

b. Find the transfer function that relates the speed in rad/s to the voltage applied to the armature circuit.

c. Let the applied voltage be $V_t = 200u(t)$, where $u(t)$ is the unit step function. Find the expression for the motor speed as a function of time t.

d. Find the steady-state speed of the motor.

Seven

Synchronous Machines

7.1 INTRODUCTION

A synchronous machine is an AC machine in which alternating current flows in the armature windings and DC excitation is supplied to the field winding. The armature windings are designed to carry large currents at high voltages; therefore, they are located on the stator. The field winding is excited by smaller currents at a lower voltage; thus, it is placed on the rotor.

A synchronous machine can be operated as a generator or as a motor, just like other rotating machines. However, it is different from the others in that its operating speed is constant. This speed is called *synchronous speed*. The synchronous speed is related to the frequency of the stator currents and the number of magnetic poles of the rotor. This relationship is expressed in the following equation (See Chapter 5, Section 5.3.2):

$$pn_s = 120f \qquad (7.1)$$

where

n_s = synchronous speed

p = number of poles

f = frequency

In the United States, the frequency is fixed at 60 Hz. Therefore, the type of rotor and the required number of poles are basically dependent on the particular application, that is, on the speed rating. Thus, for high-speed turbogenerators that operate at either 1800 rpm or 3600 rpm, the number of poles required is four poles or two poles, respectively. Because of their high speeds, these synchronous machines have nonsalient, or round, or cylindrical, rotors. A cylindrical-rotor synchronous machine is illustrated in Fig. 7.1.

221

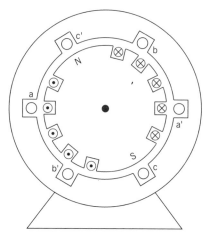

FIGURE 7.1   Cylindrical-rotor synchronous machine.

On the other hand, low-speed synchronous machines like those in hydroelectric power plants have several pairs of poles and allow the use of salient-pole, or projecting-pole, rotors. A salient-pole synchronous machine is illustrated in Fig. 7.2.

DRILL PROBLEMS

D7.1   At what speed must a six-pole synchronous generator be run to generate 50-Hz voltage?

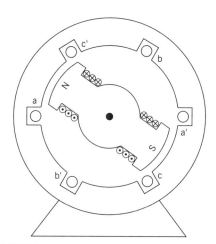

FIGURE 7.2   Salient-pole rotor synchronous machine.

D7.2 Determine the number of poles required for an alternator driven by a prime mover having a speed of 1200 rpm to generate AC at a frequency of 60 Hz.

D7.3 Determine the frequency required to operate a 16-pole, 600-V synchronous motor at 375 rpm.

7.2   ROUND-ROTOR SYNCHRONOUS MACHINES

When the cylindrical-rotor synchronous machine shown in Fig. 7.1 is operated as a generator, the induced voltages expressed in phasor form, with the phase a voltage chosen as reference phasor, are given by Eqs. 7.2, 7.3, and 7.4 for phases a, b, and c, respectively.

$$\mathbf{E}_{an} = E_{rms} \underline{/0°} \tag{7.2}$$

$$\mathbf{E}_{bn} = E_{rms} \underline{/-120°} \tag{7.3}$$

$$\mathbf{E}_{cn} = E_{rms} \underline{/-240°} \tag{7.4}$$

Equations 7.2–7.4 give the voltages that can be measured at the generator terminals when the stator windings are open. These voltages are due to the flux produced by the rotor or field current.

The expression for phase voltage E_{rms} has been derived in Chapter 5 and is rewritten here as Eq. 7.5.

$$E_{rms} = 4.44 K_w f N_a \Phi_p \tag{7.5}$$

where
K_w = machine stator winding factor
f = frequency
N_a = stator winding number of turns per phase
Φ_p = flux per pole

DRILL PROBLEMS

D7.4 A three-phase, 60-Hz synchronous generator operating at no load has a generated voltage of 620 V at rated frequency. If the pole flux is decreased by 15% and the speed is increased by 10%, determine (a) the induced voltage and (b) the frequency.

D7.5 A three-phase, eight-pole, 900-rpm, wye-connected synchronous generator has 120 turns per phase and a stator winding factor of 0.90. A voltage of 2400 V is measured across the machine terminals on no load. Determine the flux per pole.

7.2.1 Equivalent Circuit of a Round-Rotor Machine

The voltage E_a is the internal generated voltage produced across one phase of the synchronous generator when its terminals are open, or at no-load conditions. However, this voltage E_a is not the voltage V_t that is measured at the terminals when the generator is supplying stator current to an electrical load.

When no load is connected to the generator terminals, the rotor magnetic field F_r induces the internal voltage E_a. Since there is no stator current at no load, the terminal voltage V_t is equal to E_a.

Consider a lagging power factor load connected to the generator terminals. A stator current I_a will flow that lags the internal voltage E_a. This current flowing through the stator, or armature, windings produces a synchronously rotating field F_s at the same angular speed as the rotor magnetic field F_r. The stator magnetic field induces a second voltage E_s in the stator windings. Because E_s has been produced by the armature current, it is called armature reaction voltage.

The two magnetic fields F_s and F_r are rotating at the same angular velocity. Hence, the respective induced voltages E_s and E_a have the same angular frequency. Therefore, they can be combined or added as phasors to give the resultant voltage E_t.

$$E_t = E_a + E_s \tag{7.6}$$

This resultant voltage can also be thought of as the internal voltage induced by the net magnetic field F_t in the air gap, which is the sum of the stator and rotor magnetic fields:

$$F_t = F_r + F_s \tag{7.7}$$

The phasor voltages and currents and the various magnetic fields are illustrated in Fig. 7.3.

The armature reaction voltage E_{ar} is directly proportional to the amount of stator current flowing. It is 90° behind the stator current I_a. Thus, E_{ar} may be expressed as a voltage drop:

$$E_{ar} = -jX_{ar}I_a \tag{7.8}$$

where X_{ar} is a proportionality constant. Therefore, the net or resultant voltage may be expressed as follows.

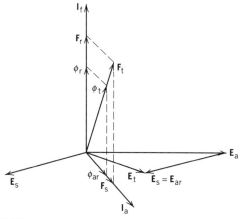

FIGURE 7.3   Synchronous generator voltages and magnetic fields.

$$\mathbf{E}_t = \mathbf{E}_a - jX_{ar}\mathbf{I}_a \tag{7.9}$$

Equation 7.9 may be recognized as Kirchhoff's voltage equation for the circuit of Fig. 7.4.

In an actual physical synchronous generator, the net or resultant magnetic field present in the air gap is realistically not linked completely by the stator windings. The portion of the magnetic flux that does not link the windings is referred to as the leakage flux ϕ_{al}. This leakage flux leads to a voltage drop across what is called leakage reactance X_{al}. In addition, the stator windings inherently contain resistances that give rise to an armature resistance drop. Thus, the overall equivalent circuit of the synchronous generator may be presented as shown in Fig. 7.5.

It is customary to add the leakage reactance to the reactance due to armature reaction to form what is referred to as *synchronous reactance* X_s. Thus,

$$X_s = X_{ar} + X_{al} \tag{7.10}$$

Finally, the equivalent circuit of the synchronous generator is presented in Fig. 7.6. This equivalent circuit is on a per-phase basis; the voltages are given

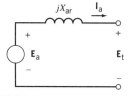

FIGURE 7.4   Internal equivalent circuit.

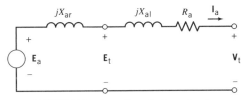

FIGURE 7.5 Overall equivalent circuit.

in line-to-neutral volts, and the resistance and reactance are in ohms per phase. In the per-phase representation, the line current is equal to the phase current.

The phasor diagram illustrating the relationships among the different phasors is shown in Fig. 7.7 for a synchronous generator supplying a lagging power factor load.

DRILL PROBLEMS

D7.6 A three-phase, 10-kVA, 208-V, four-pole, wye-connected synchronous generator has a synchronous reactance of 2 Ω per phase and negligible armature resistance. The generator is connected to a three-phase, 208-V infinite bus. Neglect rotational losses.

a. The field current and the mechanical input power are adjusted so that the synchronous generator delivers 6 kW at 0.85 lagging power factor. Determine the excitation voltage and the angle δ.

b. The mechanical input power is kept constant but the field current is adjusted to make the power factor unity. Determine the percent change in the field current with respect to its value to part (a).

D7.7 A three-phase, 500-kVA, 12-kV, wye-connected synchronous generator has an armature resistance of 1.5 Ω and a synchronous reactance of 36 Ω. At a certain field current, the generator delivers rated kVA to a load at 12 kV and 0.866 lagging power factor. For the same field excitation and the same kVA load but at a leading power factor of 0.9, what will the terminal voltage be?

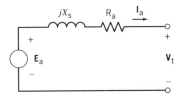

FIGURE 7.6 Equivalent circuit of a synchronous generator.

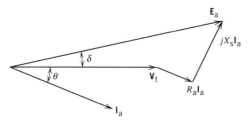

FIGURE 7.7   Phasor diagram for a round-rotor synchronous generator.

7.2.2   Open-Circuit and Short-Circuit Characteristics

Two basic characteristics of the synchronous machine are of interest. These are the open-circuit characteristic and the short-circuit characteristic, and they are discussed ahead.

Open-Circuit Characteristic   The relationship described by Eq. 7.5 is plotted in Fig.7.8. This plot is referred to as the magnetization curve, or saturation curve, or *open-circuit characteristic (OCC)* of the synchronous machine.

The open-circuit characteristic is derived experimentally by driving the synchronous machine at synchronous speed and measuring the terminal voltage (line-to-line) on no load, or at open circuit, for various values of field current. The straight line drawn tangent to the lower portion of the OCC is called the air-gap line; it gives the value of generated voltage if saturation is not present.

Short-Circuit Characteristic   Consider a synchronous generator initially operating at steady state and rated voltage at open circuit. Suppose a short circuit is suddenly applied to its terminals. A transient condition would ensue, and the stator current would rise to a high value. After a while the transients would

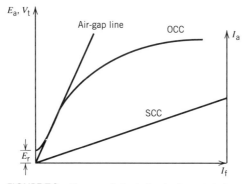

FIGURE 7.8   Open- and short-circuit characteristics.

die down and the stator short-circuit current would settle to a new steady-state value.

If readings of short-circuit current are taken and plotted for different values of field current, the plot described is called the *short-circuit characteristic (SCC)* of the synchronous machine. A typical SCC exhibiting its linear feature is shown in Fig. 7.8 together with the OCC.

Calculation of Synchronous Reactance from OCC and SCC   The open-circuit and short-circuit characteristics of the synchronous machine can be used to determine the value of its synchronous reactance. It is seen from Fig. 7.9 that at short-circuit conditions, the stator current is

$$\mathbf{I}_{a,sc} = \frac{\mathbf{E}_a}{R_a + jX_s} \qquad (7.11)$$

In most synchronous machines, the armature resistance is negligible compared to the synchronous reactance. If the value of R_a is set equal to zero, the value of X_s may be found from Eq. 7.11. This value of unsaturated synchronous reactance is found from Fig. 7.10 by reading the line-to-line voltage from the air-gap line and dividing by the current read from the SCC corresponding to the field current that produces the air-gap voltage; thus,

$$X_{s,unsat} = \frac{E_{a,ag}/\sqrt{3}}{I_{a,sc}} = \frac{V_{0a}/\sqrt{3}}{I_{0'b}} \qquad (7.12)$$

The saturated synchronous reactance may also be found from Fig. 7.10 by taking the rated terminal voltage (line-to-line) measured on the OCC and dividing by the current read from the SCC corresponding to the field current that produces rated terminal voltage. Thus,

$$X_{s,sat} = \frac{V_{t,rated}/\sqrt{3}}{I_{a,sc}} = \frac{V_{0c}/\sqrt{3}}{I_{0'b}} \qquad (7.13)$$

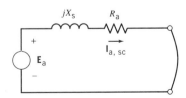

FIGURE 7.9   Short-circuited synchronous generator.

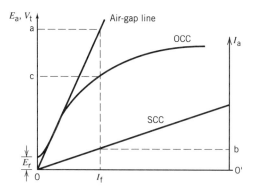

FIGURE 7.10 Calculation of synchronous reactance from open-and short-circuit characteristics.

EXAMPLE 7.1

A three-phase, wye-connected, two-pole synchronous generator is rated at 300 kVA, 480 V, 60 Hz, and 0.8 PF lagging. The open- and short-circuit characteristics are given in Table 7.1.

a. Determine the unsaturated synchronous reactance.

b. Determine the saturated synchronous reactance at the rated conditions.

Solution The open-circuit and short-circuit characteristics of this machine are plotted in Fig. 7.11.

a. The unsaturated synchronous reactance is found by using the line-to-line voltage measured from the air-gap line and the short-circuit current

Table 7.1 Generator Characteristics of Example 7.1

I_f (A)	OCC (V_{LL})	AG line (V_{LL})	SCC (A)
1.0	120		
2.0	240		
3.0	340		
4.0	430		
5.0	480	600	360
6.0	520		
7.0	540		
8.0	550		
9.0	555		
10.0	560		

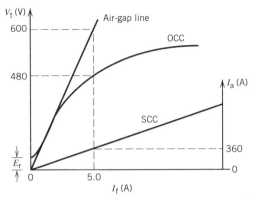

FIGURE 7.11 Open- and short-circuit characteristics of machine of Example 7.1.

corresponding to a field current $I_f = 5.0$ A; thus,

$$X_{s,unsat} = (600/\sqrt{3})/360 = 0.962 \ \Omega \text{ per phase}$$

b. The saturated synchronous reactance is found by using the rated terminal voltage (line-to-line) as measured from the OCC and the short-circuit current corresponding to a field current $I_f = 5.0$ A; thus,

$$X_{s,sat} = (480/\sqrt{3})/360 = 0.770 \ \Omega \text{ per phase}$$

7.2.3   Voltage Regulation

Just as in transformers and DC machines, a measure of the performance of a synchronous generator is its voltage regulation, which is defined as

$$\text{Voltage regulation} = \frac{V_{nl} - V_{fl}}{V_{fl}} 100\% \qquad (7.14)$$

where

V_{nl} = voltage at open-circuit, or no-load condition

V_{fl} = voltage at rated, or full-load, condition

The full-load voltage V_{fl} is the same as the terminal voltage V_t, and V_{nl} is equal to the corresponding generated voltage E_a. Thus, voltage regulation may also be expressed as

$$\text{Voltage regulation} = \frac{E_a - V_t}{V_t} 100\% \qquad (7.15)$$

EXAMPLE 7.2

Calculate the percent voltage regulation for a three-phase, wye-connected, 20-MVA, 13.8-kV synchronous generator operating at full-load, or rated, conditions and 0.8 power factor lagging. The synchronous reactance is 8 Ω per phase, and the armature resistance can be neglected.

Solution The rated voltage of 13.8 kV is normally given as a line-to-line voltage. The per-phase terminal voltage is taken as reference phasor; thus,

$$\mathbf{V_t} = (13.8 \times 10^3/\sqrt{3})\underline{/0°} = 7967\underline{/0°} \text{ V} \quad \text{(line-to-neutral)}$$

At rated operating conditions and 0.8 power factor lagging, the stator current is found as

$$\mathbf{I_a} = \frac{S_{\text{rated}}}{\sqrt{3}\ V_{\text{rated}}} \underline{/-\cos^{-1} \text{PF}}$$

$$= \frac{20 \times 10^3}{13.8\sqrt{3}} \underline{/-\cos^{-1} 0.8} = 836.7\underline{/-36.9°} \text{ A}$$

The generated voltage is computed as follows:

$$\mathbf{E_a} = \mathbf{V_t} + \mathbf{I_a}(R_a + jX_s)$$
$$= 7967\underline{/0°} + (836.7\underline{/-36.9°})(0 + j8)$$
$$= 13,125\underline{/24.1°} \text{ V} \quad \text{(line-to-neutral)}$$

Since the stator current $\mathbf{I_a} = 0$ at no load, $|\mathbf{V_{nl}}| = |\mathbf{E_a}|$. Therefore, the percent voltage regulation is computed as follows:

$$\text{Voltage regulation} = \frac{E_a - V_t}{V_t}100\%$$
$$= \frac{13,125 - 7967}{7967}100\% = 64.7\%$$

DRILL PROBLEMS

D7.8 A three-phase, 1000-kVA, 12-kV, wye-connected synchronous generator supplies 750 kW at 12 kV and 0.8 lagging power factor load. The synchronous reactance is 30 Ω per phase, and the armature resistance is negligible. Calculate the voltage regulation.

D7.9   A three-phase, 1500-kVA, 13.2-kV, wye-connected synchronous generator has an armature resistance of 0.5 Ω and a synchronous reactance of 9.0 Ω. The generator is supplying rated load at rated voltage.

    a. Calculate the generated voltage at unity power factor, at 0.8 PF lagging, and at 0.8 leading power factor.

    b. Calculate voltage regulation for each of the loads specified for part (a).

7.2.4   Power-Angle Characteristic of a Round-Rotor Machine

The maximum average power that a synchronous machine can deliver is determined by the maximum mechanical torque that can be applied without loss of synchronism with the external system to which the synchronous machine is connected. In this section, an expression for the power supplied by the synchronous generator is derived in terms of the parameters of the machine equivalent circuit and of the system.

For the purpose of this analysis, the external system is represented by an inductive reactance X_e in series with an ideal voltage source $\mathbf{V_e}$. The synchronous machine is represented by its excitation voltage $\mathbf{E_a}$ in series with its synchronous reactance X_s. The armature resistance is assumed to be negligible. Figure 7.12 shows the per-phase circuit representation.

The per-phase terminal voltage is taken as the reference phasor:

$$\mathbf{V_t} = V_t \underline{/0°} \tag{7.16}$$

The generated voltage is expressed as

$$\mathbf{E_a} = E_a \underline{/\delta} \tag{7.17}$$

The angle δ is called the *power angle*. For a synchronous generator, δ is always a positive angle; that is, the generated voltage leads the terminal voltage by the angle δ.

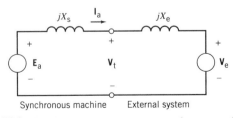

FIGURE 7.12   Synchronous generator connected to external system.

The stator current is found as

$$\mathbf{I}_a = \frac{\mathbf{E}_a - \mathbf{V}_t}{jX_s} \tag{7.18}$$

The complex power delivered by the synchronous generator to the external system is given by

$$\mathbf{S} = P + jQ = 3\mathbf{V}_t\mathbf{I}_a^* \tag{7.19}$$

Substituting Eq. 7.18 into Eq. 7.19, and simplifying, yields

$$P + jQ = 3V_t\underline{/0^\circ}\left(\frac{E_a\underline{/-\delta} - V_t\underline{/0^\circ}}{-jX_s}\right)$$

$$= 3\left(\frac{E_aV_t}{X_s}\underline{/-\delta + 90^\circ} - j\frac{V_t^2}{X_s}\right) \tag{7.20}$$

The real part and the imaginary part of Eq. 7.20 represent the three-phase real power P and the three-phase reactive power Q, respectively.

$$P = 3\frac{E_aV_t}{X_s}\sin\delta \tag{7.21}$$

$$Q = 3\left(\frac{E_aV_t}{X_s}\cos\delta - \frac{V_t^2}{X_s}\right) \tag{7.22}$$

Equation 7.21 is called the *power-angle characteristic* for the synchronous machine. Its plot is called the *power-angle curve,* and it is shown in Fig. 7.13.

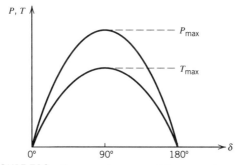

FIGURE 7.13   Power-angle curve and torque-angle curve.

The *maximum power* P_{max} delivered to the external system occurs when $\delta = 90°$; that is,

$$P_{max} = 3\frac{E_a V_t}{X_s} \qquad (7.23)$$

The maximum torque T_{max} that can be applied to the shaft of the synchronous generator without stepping out of synchronism, also called the *pull-out torque*, is related to the maximum power P_{max} by

$$T_{max} = \frac{P_{max}}{\omega_m} \qquad (7.24)$$

Equations 7.23 and 7.24 give the maximum power and maximum torque, respectively, that a synchronous generator can deliver before it steps out of synchronism. Another important parameter to consider is the synchronous generator MVA rating, which specifies the maximum power that the generator can deliver to an electrical load, continuously without overheating, at a specified voltage and power factor. Thus, the real power (MW) output of the generator is limited to a maximum value equal to its rated MVA at rated voltage and unity power factor.

A synchronous generator is said to be *overexcited* if it delivers, or supplies, reactive power to the load. It is said to be *underexcited* if it receives, or absorbs, reactive power. An overexcited synchronous generator operates at a lagging power factor. The underexcited generator operates at a leading power factor.

EXAMPLE 7.3

A 25-kVA, 230-V, three-phase, four-pole, 60-Hz, wye-connected synchronous generator has a synchronous reactance of 1.5 Ω/phase and negligible stator resistance. The generator is connected to an infinite bus (of constant voltage magnitude and constant frequency) at 230 V and 60 Hz.

a. Determine the excitation voltage E_a when the machine is delivering rated kVA at 0.8 power factor lagging.

b. The field excitation current I_f is increased by 20% without changing the power input from the prime mover. Find the stator current I_a, power factor, and reactive power Q supplied by the machine.

c. With the field excitation current I_f as in part (a), the input power from the prime mover is increased very slowly. What is the steady-state limit? Determine stator current I_a, power factor, and reactive power Q.

Solution

a. The terminal voltage is taken as reference phasor. Thus,

$$\mathbf{V_t} = (230/\sqrt{3})\underline{/0°} = 132.8\underline{/0°} \text{ V} \quad \text{(line-to-neutral)}$$

The stator current is obtained as follows:

$$\mathbf{I_a} = \frac{25,000}{230\sqrt{3}}\underline{/-\cos^{-1}0.8} = 62.8\underline{/-36.9°} \text{ A}$$

The excitation voltage is calculated as follows:

$$\begin{aligned}
\mathbf{E_a} &= \mathbf{V_t} + \mathbf{I_a}(R_a + jX_s) \\
&= 132.8\underline{/0°} + (62.8\underline{/-36.9°})(0 + j1.5) \\
&= 203.8\underline{/21.7°} \text{ V} \quad \text{(line-to-neutral)}
\end{aligned}$$

Therefore, the line-to-line excitation voltage magnitude is

$$E_a = 203.8\sqrt{3} = 353 \text{ V} \quad \text{(line-to-line)}$$

The power angle is the phase angle by which $\mathbf{E_a}$ leads $\mathbf{V_t}$, and it is given by

$$\delta = 21.7°$$

b. The excitation voltage magnitude is increased by 20%; that is,

$$E_a' = 1.20E_a = (1.2)(203.8) = 244.6\text{V} \quad \text{(line-to-neutral)}$$

Since the input power from the prime mover remains unchanged, $P' = P$. Therefore,

$$\begin{aligned}
3(E_a'V_t/X_s)\sin\delta' &= 3(E_aV_t/X_s)\sin\delta \\
244.6\sin\delta' &= 203.8\sin 21.7°
\end{aligned}$$

Solving for the new power angle yields

$$\delta' = 17.9°$$

Hence, the stator current is calculated as follows:

$$\mathbf{I}_a = \frac{\mathbf{E}_a - \mathbf{V}_t}{jX_s}$$

$$= \frac{244.6\,\underline{/17.9^\circ} - 132.8\,\underline{/0^\circ}}{j1.5} = 83.4\,\underline{/-53^\circ}\ \text{A}$$

The power factor is

$$PF = \cos 53^\circ = 0.60 \text{ lagging}$$

The reactive power is found as follows:

$$Q = 3V_tI_a \sin\theta = 3(132.8)(83.4)\sin 53^\circ = 26.5 \text{ kVAR}$$

Alternatively, the reactive power may be obtained by using Eq.7.22.

$$Q = 3\left[\frac{(244.6)(132.8)}{1.5}\cos 17.9^\circ - \frac{(132.8)^2}{1.5}\right] = 26.5 \text{ kVAR}$$

c. With the field excitation current as in part (a), the excitation voltage magnitude is

$$E_a = 203.8\text{V} \quad \text{(line-to-neutral)}$$

The steady-state limit is the maximum power P_{max} of the generator, and it occurs at $\delta = 90^\circ$. Therefore,

$$P_{max} = 3E_aV_t/X_s = 3(203.8)(132.8)/1.5 = 54.13 \text{ kW}$$

The maximum power condition is depicted in Fig. 7.14.

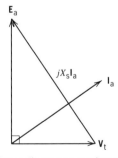

FIGURE 7.14   Phasor diagram at maximum power condition.

Based on the phasor diagram shown in Fig. 7.14, the stator current is computed as follows:

$$
\begin{aligned}
\mathbf{I}_a &= \frac{\mathbf{E}_a - \mathbf{V}_t}{jX_s} \\
&= \frac{203.8 \underline{/90^\circ} - 132.8 \underline{/0^\circ}}{j1.5} = 162.2 \underline{/33.1^\circ} \ \text{A}
\end{aligned}
$$

The power factor is given by

$$
PF = \cos(\underline{/\mathbf{V}_t} - \underline{/\mathbf{I}_a}) = \cos(0^\circ - 33.1^\circ) = 0.84 \ \text{leading}
$$

The reactive power is given by

$$
Q = 3V_t I_a \sin\theta = 3(132.8)(162.2)\sin(-33.1^\circ) = -35.3 \ \text{kVAR}
$$

Alternatively, the reactive power may be found by using Eq. 7.22.

$$
Q = 3\left[\frac{(203.8)(132.8)}{1.5} \cos 90^\circ - \frac{(132.8)^2}{1.5} \right] = -35.3 \ \text{kVAR}
$$

DRILL PROBLEMS

D7.10 A three-phase, six-pole, 60-Hz synchronous generator has a synchronous reactance of 4 Ω per phase and a terminal voltage of 2300 V. The field current is adjusted so that the excitation voltage is 2300 V at a power angle of 15°. Find

    a. Stator current

    b. Power factor

    c. Output power

    d. Torque required to drive the machine

D7.11 A three-phase, 12-kV, 60-Hz, wye-connected generator has a synchronous reactance of 15 Ω per phase and negligible armature resistance. For a given field current, the open-circuit voltage is 13 kV.

    a. Calculate the maximum power developed by the generator.

    b. Determine the armature current and power factor for the maximum power condition.

7.2.5   Efficiency

A power flow diagram for a synchronous generator is shown in Fig. 7.15. The input consists of the mechanical power supplied to the machine shaft from the prime mover, and the output is the AC electrical power delivered to the load. The DC electrical power input to the field circuit for field excitation has not been included in Fig. 7.15 because this power is supplied from a separate DC source.

The losses of the synchronous generator are classified in the same manner as in DC machines. Copper losses are present in the three stator windings ($3I_a^2 R_a$). Since the field circuit is excited from a separate DC source, the field winding copper losses ($I_f^2 R_f$) are not included. Mechanical losses consist of bearing friction and windage losses. Hysteresis and eddy current losses constitute the core losses. Other unaccounted losses are grouped under stray losses and are usually assumed to be equal to 1% of the machine power output.

The efficiency of the synchronous generator is defined as

$$\eta = \frac{P_{\text{output}}}{P_{\text{input}}} 100\%$$

$$= \frac{P_{\text{output}}}{P_{\text{output}} + \Sigma(\text{losses})} 100\%$$

$$= \frac{P_{\text{input}} - \Sigma(\text{losses})}{P_{\text{input}}} 100\% \tag{7.25}$$

7.3   SALIENT-POLE SYNCHRONOUS MACHINES

The salient-pole synchronous machine has a nonuniform air gap, as discussed in Chapter 5. Hence, the equivalent circuit and power-angle characteristics derived for nonsalient-pole, or round-rotor, machines have to be modified before they can be applied to salient-pole machines. These modifications are discussed in this section.

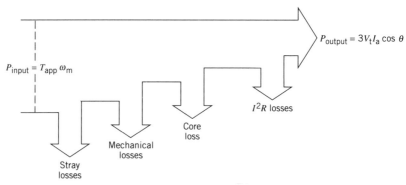

FIGURE 7.15   Power flow diagram for a synchronous generator.

7.3.1 Equivalent Circuit of a Salient-Pole Machine

Consider a salient-pole synchronous machine whose armature resistance can be assumed to be negligible. The flux path through the protruding poles is called the direct axis, and the flux path through the large air gap is called the quadrature axis. Because of the much shorter air gap between the stator and the salient poles, the flux along the direct axis encounters lower reluctance. The flux along the quadrature axis, on the other hand, encounters much higher reluctance. Hence the principles and formulas that have been derived for the round-rotor synchronous machine cannot be used for the salient-pole machine.

The two-reactance theory is used to describe the operation of salient-pole machines. It takes into account the difference in the reluctances in the direct-axis and quadrature-axis flux paths. The stator current I_a is resolved into two mutually perpendicular components: the direct-axis component I_d is along the axis of the rotor salient pole, and the quadrature-axis component I_q is in quadrature to I_d. As shown in the phasor diagram of Fig. 7.16, I_a is the phasor sum of I_d and I_q:

$$I_a = I_d + I_q \qquad (7.26)$$

Corresponding to the d axis and q axis and associated with each current component, a reactance is defined. These reactances are called *direct-axis reactance* X_d and *quadrature-axis reactance* X_q and are associated with I_d and I_q, respectively. Values of X_d and X_q are available from the manufacturer of synchronous machines. Hence, the equivalent circuit looks as shown in Fig. 7.17.

Based on the phasor diagram of Fig. 7.16 and the equivalent circuit of Fig. 7.17, the voltage-current relationship for a salient-pole machine is written as

$$E_a = V_t + jX_dI_d + jX_qI_q \qquad (7.27)$$

Equations 7.26 and 7.27 are used to analyze a salient-pole machine.

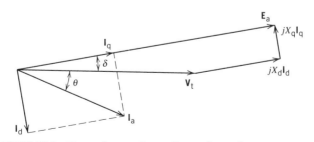

FIGURE 7.16 Phasor diagram for a salient-pole synchronous generator.

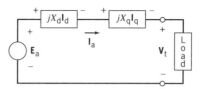

FIGURE 7.17   Equivalent circuit of a salient-pole synchronous generator.

EXAMPLE 7.4

A 75-MVA, 13.8-kV, three-phase, eight-pole, 60-Hz salient-pole synchronous machine has the following d-axis and q-axis reactances: $X_d = 1.0$ pu and $X_q = 0.6$ pu. The synchronous generator is delivering rated MVA at rated voltage and 0.866 power factor lagging. Choose a power base of 75 MVA and a voltage base of 13.8 kV. Compute the excitation voltage E_a.

Solution   The following calculations are performed in per unit using a power base of 75 MVA and a voltage base of 13.8 kV. The per-unit terminal voltage $\mathbf{V}_t$ is taken as reference phasor; thus

$$\mathbf{V}_t = 1.0 \underline{/0°}$$

At rated conditions and 0.866 PF lagging, the per-unit stator current is given by

$$\mathbf{I}_a = 1.0 \underline{/-30°}$$

Refer to the phasor diagram of Fig. 7.18. The expression for $\mathbf{E}_a$ is given by Eq. 7.27 and may be rewritten as follows:

$$
\begin{aligned}
\mathbf{E}_a &= \mathbf{V}_t + jX_d\mathbf{I}_d + jX_q\mathbf{I}_q + (jX_q\mathbf{I}_d - jX_q\mathbf{I}_d) \\
&= \mathbf{V}_t + jX_q(\mathbf{I}_q + \mathbf{I}_d) + j(X_d - X_q)\mathbf{I}_d \\
&= \mathbf{V}_t + jX_q\mathbf{I}_a + j(X_d - X_q)\mathbf{I}_d \\
&= \mathbf{E}' + j(X_d - X_q)\mathbf{I}_d
\end{aligned}
\tag{7.28}
$$

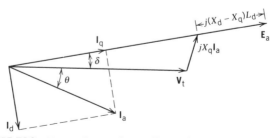

FIGURE 7.18   Phasor diagram for a salient-pole synchronous generator.

where

$$\mathbf{E'} = \mathbf{V_t} + jX_q\mathbf{I_a} \qquad (7.29)$$

From Fig. 7.18, it may be seen that the phasor $j(X_d - X_q)\mathbf{I_d}$ lies parallel to $\mathbf{E_a}$. Therefore, $\mathbf{E'} = \mathbf{V_t} + jX_q\mathbf{I_a}$ must be in parallel with $\mathbf{E_a}$. This implies that the phase angle of $\mathbf{E'}$ is equal to the phase angle of $\mathbf{E_a}$, which is δ. Hence,

$$\mathbf{E'} = 1.0\,\underline{/0^\circ} + (j0.6)(1.0\,\underline{/-30^\circ}) = 1.40\,\underline{/21.8^\circ} \text{ pu}$$

Therefore, angle δ is equal to 21.8°.

The angle between $\mathbf{E_a}$ and $\mathbf{I_a}$ is found as follows:

$$(\delta + \theta) = 21.8^\circ + 30^\circ = 51.8^\circ$$

This angle is used to resolve $\mathbf{I_a}$ into its components:

$$\mathbf{I_d} = [I_a \sin(\delta + \theta)]\,\underline{/\delta - 90^\circ}$$
$$= [1.0 \sin 51.8^\circ]\,\underline{/21.8^\circ - 90^\circ} = 0.786\,\underline{/-68.2^\circ} \text{ pu}$$
$$\mathbf{I_q} = [I_a \cos(\delta + \theta)]\,\underline{/\delta}$$
$$= [1.0 \cos 51.8^\circ]\,\underline{/21.8^\circ} = 0.618\,\underline{/21.8^\circ} \text{ pu}$$

Substituting the values of $\mathbf{I_d}$ and $\mathbf{I_q}$ into Eq. 7.27 yields

$$E_a = 1.0\,\underline{/0^\circ} + (j1.0)(0.786\,\underline{/-68.2^\circ}) + (j0.6)(0.618\,\underline{/21.8^\circ})$$
$$= 1.714\,\underline{/21.8^\circ} \text{ pu}$$

Alternatively, $\mathbf{E_a}$ may be calculated using the known value of $\mathbf{E'}$ and the fact that these phasors are in phase with $j(X_d - X_q)\mathbf{I_d}$. Therefore, the magnitudes of the phasors $\mathbf{E'}$ and $j(X_d - X_q)\mathbf{I_d}$ add directly. Thus,

$$E_a = E' + [(X_d - X_q)I_d]$$
$$= 1.40 + (1.0 - 0.6)(0.786) = 1.714 \text{ pu}$$

Therefore, the excitation voltage phasor is found as follows:

$$\mathbf{E_a} = E_a\underline{/\delta} = 1.714\,\underline{/21.8^\circ} \text{ pu} = 23.6\,\underline{/21.8^\circ} \text{ kV}$$

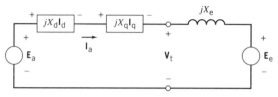

FIGURE 7.19 Equivalent circuit of a synchronous generator connected to an external system.

7.3.2 Power-Angle Characteristic of a Salient-Pole Machine

The derivation of the expression for the power generated by a salient-pole synchronous machine is similar to that of a round-rotor synchronous machine. Consider a salient-pole synchronous generator connected to an external power system. The per-phase equivalent circuit is shown in Fig. 7.19. The relationships among the phasors $\mathbf{V_t}$, $\mathbf{E_a}$, $\mathbf{I_a}$, $\mathbf{I_d}$, and $\mathbf{I_q}$ are shown in the phasor diagram of Fig. 7.20.

The generated voltage is taken as reference phasor; thus,

$$\mathbf{E_a} = E_a \underline{/0^\circ} \tag{7.30}$$

$$\mathbf{V_t} = V_t \underline{/-\delta} \tag{7.31}$$

From the phasor diagram of Fig. 7.20, the expressions for the magnitudes of the d-axis and q-axis components of the current are found as follows:

$$I_q = \frac{V_t \sin \delta}{X_q} \tag{7.32}$$

$$I_d = \frac{E_a - V_t \cos \delta}{X_d} \tag{7.33}$$

The expression for the complex power delivered by the synchronous generator to the external system is given by

$$\mathbf{S} = P + jQ = 3\mathbf{V_t}\mathbf{I_a^*} = 3V_t\underline{/-\delta}(I_q\underline{/0^\circ} + I_d\underline{/-90^\circ})^* \tag{7.34}$$

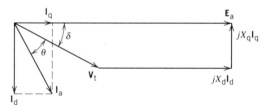

FIGURE 7.20 Phasor diagram for a salient-pole synchronous generator.

Substituting Eqs. 7.32 and 7.33 into Eq. 7.34, and simplifying, yields the expressions for the real power P and reactive power Q.

$$P = 3\left[\frac{E_a V_t}{X_d} \sin\delta + \frac{V_t^2}{2}\left(\frac{1}{X_q} - \frac{1}{X_d}\right)\sin 2\delta\right] \qquad (7.35)$$

$$Q = 3\left[\frac{E_a V_t}{X_d}\cos\delta - V_t^2\left(\frac{\sin^2\delta}{X_q} + \frac{\cos^2\delta}{X_d}\right)\right] \qquad (7.36)$$

On the right-hand side of Eq. 7.35, the first term is identical to the expression for the power delivered by a round-rotor synchronous generator. The second term represents the effects of generator saliency, and it is called the *reluctance power*. The plot of Eq. 7.35 is the power-angle curve for a salient-pole synchronous generator. It is shown in Fig. 7.21.

The direct-axis reactance X_d is larger than the quadrature-axis reactance X_q. When the salient-pole machine approaches a round rotor, the values of X_d and X_q will both approach the value of X_s. When this substitution is made, Eqs. 7.35 and 7.36 will reduce to Eqs. 7.21 and 7.22, respectively.

EXAMPLE 7.5

For the 75-MVA, 13.8-kV synchronous generator of Example 7.4, the excitation voltage E_a and the terminal voltage V_t are kept constant at their respective values of 1.714 pu and 1.0 pu. Choose a power base of 75 MVA and a voltage base of 13.8 kV. Find the maximum power and maximum kVA that the generator can deliver.

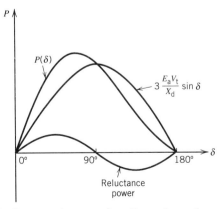

FIGURE 7.21   Power-angle curve of a salient-pole synchronous generator.

Solution

a. The per-unit real power that the generator delivers is found by using Eq. 7.35.

$$P = \frac{(1.714)(1.0)}{1.0} \sin\delta + \frac{(1.0)^2}{2}\left(\frac{1}{0.6} - \frac{1}{1.0}\right)\sin 2\delta$$
$$= 1.714\sin\delta + 0.333\sin 2\delta$$

For maximum power, differentiate P with respect to δ, and set the derivative equal to zero; thus,

$$\frac{dP}{d\delta} = 1.714\cos\delta + 0.666\cos 2\delta$$

By applying a trigonometric identity on $\cos 2\delta$ and simplifying,

$$1.333\cos^2\delta + 1.714\cos\delta - 0.666 = 0.0$$

This equation is a quadratic equation in terms of $\cos\delta$ and can be solved by using the quadratic formula. Thus,

$$\cos\delta = 0.313 \quad \text{or} \quad \cos\delta = -1.60$$

The second solution is obviously extraneous; therefore,

$$\delta = \cos^{-1} 0.313 = 71.8°$$

To check whether $\delta = 71.8°$ yields a maximum, the second derivative is evaluated:

$$\frac{d^2P}{d\delta^2} = -1.714\sin\delta - 1.333\sin 2\delta$$
$$= -1.714\sin 71.8° - 1.333\sin[(2)(71.8°)] = -2.42$$

Since $d^2P/d\delta^2 < 0$, P is maximum at $\delta = 71.8°$, and the maximum power P_{max} is computed as follows:

$$P_{max} = 1.714\sin 71.8° + 0.333\sin[(2)(71.8°)] = 1.826 \text{ pu} = 137 \text{ MW}$$

b. The reactive power delivered at $\delta = 71.8°$ is obtained by using Eq. (7.36).

$$Q = \left[\frac{(1.714)(1.0)}{1.0} \cos 71.8° - (1.0)^2 \left(\frac{\sin^2 71.8}{0.6} + \frac{\cos^2 71.8°}{1.0} \right) \right]$$
$$= -1.066 \text{ pu} = -80 \text{ MVAR}$$

The total MVA is computed as follows:

$$S = \sqrt{(1.826)^2 + (-1.066)^2} = 2.114 \text{ pu} = 158 \text{ MVA}$$

DRILL PROBLEMS

D7.12 A three-phase, 100-MVA, 12-kV, 60-Hz, salient-pole synchronous machine has direct-axis and quadrature-axis reactances of 1.0 pu and 0.7 pu, respectively. The stator resistance may be neglected. The machine is connected to an infinite bus and delivers 72 MW at 0.8 power factor lagging.

  a. Use V_t as reference phasor, and draw the phasor diagram. Determine the excitation voltage and the power angle.

  b. Determine the maximum power the synchronous generator can supply if the field circuit becomes open. Determine the machine current and power factor for this condition.

D7.13 A 50-kVA, 480-V, three-phase, wye-connected, salient-pole synchronous generator runs at full load at 0.9 leading power factor. The per-phase direct-axis and quadrature-axis reactances are 1.5 Ω and 1.0 Ω, respectively. The armature resistance is negligible. Calculate (a) the excitation voltage and (b) the power angle.

7.4 GENERATOR SYNCHRONIZATION

An individual synchronous generator supplying power to an impedance load acts as a voltage source whose frequency is determined by the speed of rotation of its prime mover and whose voltage is determined by its excitation system. The major disadvantage of such an operating practice is that anytime the generator is out of order, or is under maintenance, the supply of electricity to the load is interrupted.

The electricity supply systems of industrialized countries have hundreds of synchronous generators operating in parallel. These generators are interconnected by a network of transmission lines and substations. The main reasons for interconnection are reliability of service, economy of power system operation, and improved operating efficiency of the individual generators.

The point of connection of the generator to the power system is called an *infinite bus*. The infinite bus can be represented as a voltage source of constant magnitude and constant frequency. When a synchronous generator is connected to a large power system, the frequency and rms voltage at the generator terminals are fixed by the power system.

The process of properly connecting a synchronous generator in parallel with the other generators in the power system, or to the infinite bus, is called *synchronization*. In order to synchronize properly, the following conditions have to be satisfied:

1. The magnitude of the terminal voltage of the incoming generator must be the same as the voltage at the point of interconnection with the power system or infinite bus.

2. The frequency of the incoming generator must be the same as the frequency of the power system or infinite bus.

3. The generator must have the same phase sequence as the infinite bus.

4. The phase angles of corresponding phases of the incoming generator and the power system must be equal.

To verify that these conditions for connecting the incoming generator in parallel with the infinite bus are satisfied, a set of three synchronizing lamps may be used. The schematic diagram for the connection of these lamps for synchronization in a laboratory setup is shown in Fig. 7.22.

The field rheostat of the generator is adjusted to vary the field current until the generator voltage V_2 becomes equal to, or slightly greater than, the infinite bus voltage V_1. If the phase sequences of the generator and the infinite

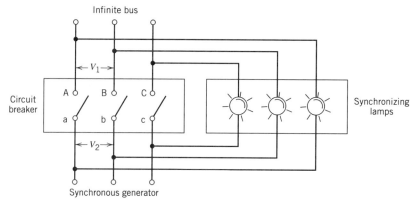

FIGURE 7.22 Schematic diagram for synchronization using synchronizing lamps.

bus are different, the three lamps will brighten up alternately. To correct for this improper condition, any two of the three connections to the synchronous generator are interchanged. If the phase sequence is correct, the lamps are all bright or are all dark at the same time. If the frequencies are slightly different, the three lamps will brighten or darken at the same time. The speed of the prime mover of the synchronous generator is adjusted so that the generator's frequency is the same as that of the infinite bus, at which time all lamps stay dark. When all four conditions are satisfied simultaneously, the circuit breaker is closed, and the generator is now operating in parallel with the rest of the synchronous machines of the system.

* 7.5 DYNAMICS OF THE PRIME MOVER–GENERATOR SYSTEM

The previous discussions have dealt with modeling and analyzing a synchronous machine under steady-state conditions. In steady state, the electromagnetic torque and mechanical torque balance each other. The machine operates at the synchronous speed determined by the number of poles and the frequency of the power system, which is normally 60 Hz.

When the load on the synchronous generator changes, that is, when the power demand either increases or decreases, the machine will either decelerate or accelerate temporarily. Hence, the dynamics of the rotor come into the picture. There is now a need to apply Newton's second law of motion. Application of Newton's second law results in

$$J \frac{d\omega_m}{dt} = T_m - T_e \qquad (7.37)$$

where

J = moment of inertia of the rotor (N-m-s^2)

ω_m = speed of the rotor (rad/s)

$d\omega_m/dt$ = angular acceleration of the rotor

T_m = shaft mechanical torque (N-m)

T_e = electromagnetic torque (N-m)

The angular acceleration may also be expressed in terms of the second derivative of the angular displacement:

$$\frac{d\omega_m}{dt} = \frac{d^2\theta_m}{dt^2} \qquad (7.38)$$

The angle θ_m represents the angular position of the rotor with respect to some stationary reference frame. It is convenient to measure the rotor angular position with respect to a synchronously rotating reference frame instead of a stationary frame.

$$\theta_m(t) = \omega_{m,s} t + \delta_m(t) \tag{7.39}$$

where

$\omega_{m,s}$ = synchronous angular velocity of the rotor

δ_m = rotor angular position with respect to the synchronously rotating reference frame

For a synchronous generator with p poles, the electrical radian frequency ω and the rotor angle δ are given by

$$\omega_m(t) = \left(\frac{2}{p}\right)\omega(t)$$

$$\delta_m(t) = \left(\frac{2}{p}\right)\delta(t) \tag{7.40}$$

Substituting Eqs. 7.39 and 7.40 into Eqs. 7.37 and 7.38 and simplifying,

$$\frac{2}{p}J\frac{d^2\delta}{dt^2} = T_m - T_e \tag{7.41}$$

Equation 7.41 is known as the *swing equation* and is used to solve for the electromechanical dynamics of the synchronous machine. It should be noted that Eq. 7.41 is written for generator action. The electromagnetic torque T_e is found as follows:

$$T_e = 3\frac{E_a V_t}{\omega_m X_s}\sin\delta = K_s\sin\delta \tag{7.42}$$

where

$$K_s = 3\frac{E_a V_t}{\omega_m X_s} \tag{7.43}$$

The quantity K_s is called synchronous torque coefficient. For dynamic analysis, K_s can be taken as a constant.

EXAMPLE 7.6

Express Eq. 7.41 in terms of power.

Solution Multiply both sides of Eq. 7.41 by $\omega_m = (2/p)\omega$, where $\omega = 2\pi f$ and $f = 60$ Hz. On the left-hand side of the equation, let the synchronous rotor velocity ω_s represent the approximate value of ω_m. Thus,

$$\frac{2}{p}\omega_s J \frac{d^2\delta}{dt^2} = \omega_m T_m - \omega_m T_e \qquad (7.44)$$

The first term on the right-hand side of Eq. 7.44 is the mechanical power P_m, and the second term is given by Eq. 7.42. Therefore, Eq. 7.44 may also be written as

$$\frac{2}{p}\omega_s J \frac{d^2\delta}{dt^2} = P_m - 3\frac{E_a V_t}{X_s}\sin\delta \qquad (7.45)$$

7.6 SYNCHRONOUS MOTOR PERFORMANCE

The equivalent circuit and torque equations derived for a synchronous generator also apply to synchronous motors. Therefore, analyzing the performance of a motor parallels the analysis of generator performance. The per-phase equivalent circuit of the synchronous motor is shown in Fig. 7.23.

For phasor analysis purposes, the terminal voltage V_t is usually taken as reference, and the positive direction of stator current is into the motor. The phasor diagram of Fig. 7.24 applies to a lagging power factor current.

EXAMPLE 7.7

A three-phase, 208-V synchronous motor has a synchronous reactance of 1.0 Ω per phase and negligible armature resistance. The motor draws 50 kVA at 0.8 power factor leading. Calculate (a) the stator current I_a and (b) the excitation voltage E_a.

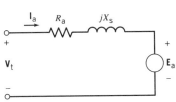

FIGURE 7.23 Equivalent circuit of a synchronous motor.

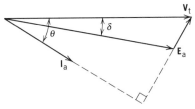

FIGURE 7.24 Phasor diagram of a synchronous motor.

Solution

a. Take the terminal voltage $\mathbf{V}_t$ as reference phasor:

$$\mathbf{V}_t = (208/\sqrt{3})\underline{/0^\circ} = 120\underline{/0^\circ} \text{ V} \quad \text{(line-to-neutral)}$$

The stator current is computed as

$$\mathbf{I}_a = \frac{50,000}{208\sqrt{3}}\underline{/\cos^{-1} 0.8} = 138.8\underline{/36.9^\circ} \text{ A}$$

b. The excitation voltage is found as follows:

$$\begin{aligned}
\mathbf{E}_a &= \mathbf{V}_t - (R_a + jX_s)\mathbf{I}_a \\
&= 120\underline{/0^\circ} - (0 + j1.0)(138.8\underline{/36.9^\circ}) \\
&= 231.7\underline{/-28.6^\circ} \text{ V} \quad \text{(line-to-neutral)}
\end{aligned}$$

Therefore, the line-to-line excitation voltage is

$$E_a = 231.7\sqrt{3} = 401 \text{ V} \quad \text{(line-to-line)}$$

Now suppose that the field current is increased. This increased field excitation increases the generated voltage E_a. However, the real power supplied by the motor remains the same because the mechanical load torque did not change and the rotational speed is not affected by the increased I_f. When the machine losses are neglected, the expression for the real power delivered by the motor is

$$P = 3\frac{V_t E_a'}{X_s}\sin\delta' = 3\frac{V_t E_a}{X_s}\sin\delta \tag{7.46}$$

Alternatively, the real power may be expressed as

$$P = 3V_t I_a'\cos\theta' = 3V_t I_a\cos\theta \tag{7.47}$$

It is evident from Eqs. 7.46 and 7.47 that the expressions $E_a\sin\delta$ and $I_a\cos\theta$ must be constant. Of course, when the field current is increased, E_a increases but only in such a way that its projection $E_a\sin\delta$ remains constant. This is illustrated in Fig. 7.25. Similarly, the stator current I_a is constrained to change so as to have a constant projection $I_a\cos\theta$. It may be seen that as the

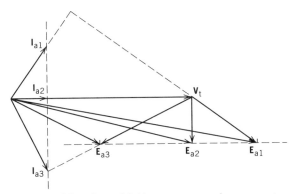

FIGURE 7.25   Effect of field current on synchronous motor.

field current and E_a continue to increase, the magnitude of the stator current initially decreases and then increases again.

At low values of E_a, the stator current lags the terminal voltage. The motor acts as an inductive load and is said to be underexcited. As the field current is increased, the stator current becomes less and less lagging and will later become in phase with the voltage. At this point, the motor is operating at unity power factor and is said to be *normally excited*. As the field current is increased further, the stator current becomes leading. The motor becomes a source of reactive power, and it is said to be overexcited. The variation of the stator current as the field current is changed is plotted in Fig. 7.26. Because of its shape, the plot is called the *V-curve* of the synchronous motor.

In Fig. 7.26, each V-curve corresponds to a different power level. At the vertex of any V-curve, the stator current is at its minimum for that power level and the motor operates at unity power factor. For field currents less than the value corresponding to minimum stator current, the motor operates at a lagging

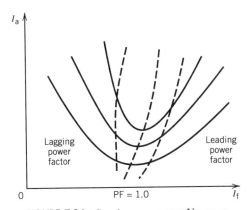

FIGURE 7.26   Synchronous motor V-curves.

power factor and it absorbs reactive power as an underexcited motor. For field currents greater than the value for minimum stator current, the motor operates at a leading power factor, and it supplies reactive power to the external system and is said to be overexcited.

EXAMPLE 7.8

A three-phase, 200-hp, 480-V, six-pole, 60-Hz synchronous motor has a synchronous reactance of 1.5 Ω/phase. Neglect all losses. The motor is connected to an infinite bus, and it delivers its rated horsepower at unity power factor. Determine the pullout torque, that is, the maximum torque the motor can deliver without losing synchronism.

Solution   Take the line-to-neutral terminal voltage $\mathbf{V}_t$ as reference phasor:

$$\mathbf{V}_t = (480/\sqrt{3}) \underline{/0^\circ} = 277.1 \underline{/0^\circ} \text{ V}$$

At unity power factor, the stator current is in phase with the applied voltage; thus,

$$\mathbf{I}_a = \frac{(200)(746)}{\sqrt{3}\ 480} \underline{/0^\circ} = 179.46 \underline{/0^\circ} \text{ A}$$

Then the excitation voltage is given by

$$\mathbf{E}_a = \mathbf{V}_t - (R_a + jX_s)\mathbf{I}_a = 277.1 \underline{/0^\circ} - (0 + j1.5)(179.46 \underline{/0^\circ})$$
$$= 386.3 \underline{/-44.2^\circ} \text{ V}\quad \text{(line-to-neutral)}$$

Since the synchronous motor is assumed to be lossless,

$$T_{max}\omega_m = P_{max}$$

The maximum power occurs at a power angle $\delta = 90^\circ$.

$$P = 3\frac{V_t E_a}{X_s}\sin 90^\circ = 3\frac{(277.1)(386.3)}{1.5} = 214 \text{ kW}$$

The mechanical angular speed is found from the radian frequency as

$$\omega_m = (2/p)\omega = (2/6)(2\pi 60) = 125.7 \text{ rad/s}$$

Finally,

$$T_{max} = P_{max}/\omega_m = (214 \times 10^3)/125.7 = 1702.5 \text{ N-m}$$

The power flow diagram for a synchronous motor is shown in Fig. 7.27. The input power to the motor is electrical and is equal to $3V_tI_a\cos\theta$, where θ is the angle between the $\mathbf{V}_t$ and $\mathbf{I}_a$ phasors. The mechanical power developed, P_{dev}, is equal to $T_{ind}\omega_m$. It may be noted that the DC electrical power for field excitation is not included in Fig. 7.27 because it is supplied by a separate DC source.

The copper losses consist of the stator losses of $3I_a^2R_a$. The field winding losses are not included in this diagram because the field winding is excited by a separate DC source. The rotational losses include core losses, consisting of hysteresis and eddy current losses, and mechanical losses, consisting of friction and windage losses. The output power P_{output} is equal to $T_{load}\omega_m$, where T_{load} is the output torque. When the motor is operated at rated conditions, the output mechanical power P_{output} is equal to the motor rating, which is usually given in horsepower.

EXAMPLE 7.9

A 200-hp, 2300-V, three-phase, 60-Hz, cylindrical-rotor motor has a synchronous reactance of 12Ω per phase and negligible armature resistance. When it is delivering its rated output, the motor's efficiency is 90% and its power angle δ is 17°. Determine

a. The excitation voltage $\mathbf{E}_a$

b. The stator current $\mathbf{I}_a$ and power factor

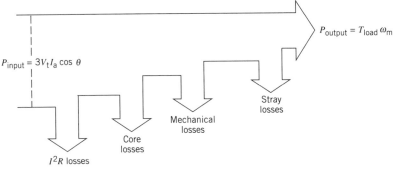

FIGURE 7.27 Power flow diagram for a synchronous motor.

Solution

a. The line-to-neutral terminal voltage is taken as reference phasor; thus,

$$\mathbf{V_t} = (2300/\sqrt{3})\underline{/0°} = 1328\underline{/0°} \text{ V}$$

From the power flow diagram, the input power is the same as the mechanical power developed, since the stator copper loss is negligible. Hence,

$$P_{input} = \frac{P_{output}}{\eta} = 3\frac{V_t E_a}{X_s}\sin\delta$$

$$= \frac{(200)(746)}{0.90} = 3\frac{1328 E_a}{12}\sin 17°$$

Thus, the line-to-neutral excitation voltage is found as

$$E_a = \frac{(200)(746)(12)}{(0.90)(3)(1328)(\sin 17°)} = 1708 \text{ V}$$

Therefore, the line-to-line excitation voltage is

$$E_a = 1708\sqrt{3} = 2958\text{V} \quad \text{(line-to-line)}$$

b. Since the armature resistance is negligible, the stator current is given by

$$\mathbf{I_a} = \frac{\mathbf{V_t} - \mathbf{E_a}}{jX_s}$$

$$= \frac{1328\underline{/0°} - 1708\underline{/-17°}}{j12} = 48.8\underline{/31.4°} \text{ A}$$

The power factor is

$$PF = \cos(\underline{/V_t} - \underline{/I_a}) = \cos(0° - 31.4°) = 0.85 \text{ leading}$$

DRILL PROBLEMS

D7.14 A three-phase, 6-kV, wye-connected synchronous motor has a synchronous reactance of 12 Ω per phase and negligible armature resistance. Calculate the induced voltage when the motor takes 1000 kVA at rated voltage and

a. 0.8 power factor lagging

b. Unity power factor

c. 0.8 power factor leading

D7.15 A three-phase, 200-hp, 2400-V, 60-Hz, wye-connected, cylindrical-rotor synchronous motor has a synchronous reactance of 12 Ω per phase and negligible armature resistance. The motor draws 150 kW at a power angle of 18 electrical degrees. Determine

    a. The excitation voltage

    b. The line current

    c. The power factor

D7.16 A three-phase, 2400-V, 60-Hz, 8-pole, wye-connected synchronous motor has 5 Ω per phase synchronous reactance and negligible stator resistance. The motor is connected to a 2400-V infinite bus, and it draws 120 amperes at 0.8 power factor lagging. Neglect rotational losses.

    a. Determine the output power.

    b. Calculate the maximum power.

    c. Determine the torque, stator current, and power factor for the maximum power condition.

D7.17 The synchronous reactance of a synchronous motor is 10 Ω per phase, and its armature resistance is negligible. The input power is 1500 kW, and the induced voltage is 4600 V. If the terminal voltage is 4160 V, determine (a) the armature current and (b) the power factor.

REFERENCES

1. Bergseth, F. R., and S. S. Venkata. *Introduction to Electric Energy Devices.* Prentice Hall, Englewood Cliffs, N. J., 1987.

2. Brown, David, and E. P. Hamilton III. *Electromechanical Energy Conversion.* Macmillan, New York, 1984.

3. Chapman, Stephen J. *Electric Machinery Fundamentals.* 2nd ed. McGraw-Hill, New York, 1991.

4. Del Toro, Vincent. *Electric Machines and Power Systems.* Prentice-Hall, Englewood Cliffs, N.J., 1985.

5. Fitzgerald, A. E., Charles Kingsley, and Stephen Umans. *Electric Machinery.* 5th ed. McGraw-Hill, New York, 1990.

6. Kosow, Irving L. *Electric Machinery and Transformers.* Prentice Hall, Englewood Cliffs, N. J., 1991.

7. Macpherson, George, and Robert D. Laramore. *An Introduction to Electrical Machines and Transformers.* 2nd ed. Wiley, New York, 1990.

8. Nasar, Syed A. *Electric Energy Conversion and Transmission*. Macmillan, New York, 1985.

9. Ramshaw, Raymond, and R. G. van Heeswijk. *Energy Conversion Electric Motors and Generators*. Saunders College Publishing, Philadelphia, 1990.

10. Sen, P. C. *Principles of Electrical Machines and Power Electronics*. Wiley, New York, 1989.

11. Wildi, Theodore. *Electrical Machines, Drives, and Power Systems*. 2nd ed. Prentice Hall, Englewood Cliffs, N. J., 1991.

PROBLEMS

7.1 From 2 poles to 10 poles, calculate the prime mover speeds in rpm and rad/s required to generate AC at a frequency of 60 Hz.

7.2 Calculate the frequency produced by a prime mover turning a 10-pole synchronous generator at 800 rpm.

7.3 A three-phase, 50-hp, 2300-V, 60-Hz synchronous motor is operating at 900 rpm. Determine the number of poles in the rotor.

7.4 Determine the speed of an eight-pole synchronous motor operating from a three-phase, 50-Hz, 4160-V system.

7.5 A three-phase, eight-pole, 60-Hz synchronous generator has a 50-mWb flux per pole. The stator winding factor is 0.90. The armature has 104 turns per phase. Calculate the induced voltage.

7.6 A three-phase, 60-Hz synchronous generator operating at no load has an induced voltage of 4300 V at rated frequency. The pole flux is increased by 10% and the rotor speed is increased by 5%. Determine (a) the induced voltage and (b) frequency.

7.7 The open-circuit voltage of a 60-Hz generator is 12 kV at a field current of 8 A. The synchronous generator is operated at 50 Hz and a field current of 3 A. Neglect saturation. Calculate the open-circuit voltage.

7.8 A three-phase, wye-connected synchronous generator is rated at 150 kVA and 2300 V. The generator delivers rated current to a load at rated terminal voltage and 0.8 PF lagging. When the load is removed, the terminal voltage rises to 3500 V. Assume that armature resistance is negligible.

    a. Calculate the synchronous reactance of the generator.

    b. Draw a phasor diagram.

7.9 A synchronous generator is connected to a 13.8-kV infinite bus. It has a synchronous reactance of 7.5 Ω per phase, and the armature resistance is negligible. The generator delivers a real power output of 50 MW and a reactive power output of 30 MVAR to the infinite bus.

    a. Determine the excitation voltage and angle.

b. Draw a phasor diagram indicating the terminal voltage, the excitation voltage, the armature current, and the voltage drop across the synchronous reactance.

7.10 A three-phase, 100-kVA, 240-V, 60-Hz, six-pole, wye-connected synchronous generator is supplying a load of 80 kVA at 230 V and 0.866 power factor lagging. The armature has a synchronous impedance of $0.1 + j0.5$ Ω per phase. Determine the following:

    a. Armature current

    b. Excitation voltage

    c. Power angle

    d. Input shaft torque (neglecting losses)

7.11 A cylindrical-rotor synchronous generator has a per-unit synchronous reactance of 1.0 and a negligible armature resistance. The generator supplies rated kVA to a load at a terminal voltage of 1.0 per unit and a leading power factor of 80%. Determine the excitation voltage.

7.12 The following readings are taken from the results of open-circuit and short-circuit tests on a three-phase, 10-MVA, 12-kV, two-pole, 60-Hz cylindrical-rotor generator driven at synchronous speed.

Field current (A)	150	180
Armature current, short-circuit test (A)	400	480
Line voltage, open-circuit test (kV)	11.2	12.0
Line voltage, air-gap line (kV)	13.5	15.0

The armature resistance is measured separately as 0.10 Ω per phase.

    a. Determine the unsaturated synchronous reactance in ohms per phase and per unit.

    b. Calculate the saturated synchronous reactance in ohms per phase and per unit.

7.13 A three-phase, 50-MVA, 12-kV, wye-connected, 60-Hz synchronous generator has a resistance of 0.10 Ω per phase and a synchronous reactance of 4 Ω per phase. The generator delivers rated load current at rated voltage and 0.9 power factor leading. Calculate the voltage regulation.

7.14 A three-phase, 1000-kVA, 4160-V, wye-connected synchronous generator has an armature resistance and synchronous reactance of $1.5 + j25$ Ω per phase. Determine the voltage regulation if the power factor is

    a. 0.8 leading

    b. Unity

    c. 0.8 lagging

7.15 The synchronous generator described in Problem 7.12 delivers rated kVA to a load at rated voltage and 0.8 power factor lagging.

a. Compute the field current required.

b. Additional points on the open-circuit characteristic are given in the following tabulation. Find the voltage regulation.

Field current (A)	200	250	300	350
Line voltage (kV)	12.5	13.5	14.3	14.8

7.16 A three-phase, 2400-V, 60-Hz, six-pole, wye-connected synchronous generator is connected to an infinite bus. The generator is delivering 500 kW at a power angle of 30°. The stator has a synchronous reactance of 10 Ω per phase and negligible armature resistance. Determine the following:

a. Input torque to the generator

b. Excitation voltage

c. Armature current and power factor

d. The reactive power delivered

7.17 A three-phase, 2000-kVA, 12-kV, 1800-rpm synchronous generator has a synchronous reactance of 20 Ω per phase and negligible armature resistance.

a. The field current is adjusted to obtain the rated terminal voltage at open circuit. Determine the excitation voltage.

b. A short circuit occurs across the machine terminals. Find the stator current.

c. The synchronous machine is connected to an infinite bus. The generator delivers its rated current at 0.8 power factor lagging. Determine the excitation voltage.

d. Calculate the maximum power the synchronous machine can deliver for the excitation current of part (c).

7.18 Loss data for the synchronous generator of Problem 7.12 are as follows:

Open-circuit core loss at 12 kV = 75 kW
Short-circuit load loss at 480 A = 60 kW
Friction and windage = 65 kW
Field winding resistance = 0.35 Ω

Compute the efficiency at rated load and 0.8 power factor lagging.

7.19 A salient-pole synchronous generator has a direct-axis synchronous reactance of 1.0 per unit and a quadrature-axis reactance of 0.6 per unit. Neglect saturation. The generator delivers full-load current at rated terminal voltage and 0.866 lagging power factor.

a. Draw the phasor diagram.

b. Determine the excitation voltage.

7.20 The salient-pole synchronous machine of Problem 7.12 operates as a generator, and it delivers 0.8 pu of power at 1.0 pu voltage and a power factor of 0.8 leading. Determine (a) the excitation voltage and (b) the power angle.

7.21 A three-phase, 60-MVA, 12-kV, 60-Hz, salient-pole synchronous machine has d-axis and q-axis reactances of 1.2 pu and 0.6 pu, respectively, and negligible armature resistance. The machine is connected to an infinite bus at 12 kV, and the field current is adjusted to make the excitation voltage equal to the terminal voltage.

    a. Determine the maximum power that the machine can supply.

    b. Find the stator current and power factor at this maximum power condition.

    c. Draw the phasor diagram.

7.22 A salient-pole synchronous generator has d-axis and q-axis synchronous reactances of 1.60 pu and 1.20 pu, respectively. The generator is connected to an infinite bus through an external reactance of 0.20 pu. The generator delivers its rated output power at 0.8 power factor lagging to the infinite bus.

    a. Draw a phasor diagram showing the bus voltage, the armature current, the generator terminal voltage, the excitation voltage, and the rotor angle.

    b. Calculate the rotor angle in degrees.

    c. Compute the per-unit terminal and excitation voltages.

7.23 The induced voltage of a synchronous motor is 4160 V. It lags behind the terminal voltage by 30°. If the terminal voltage is 4000 V, determine the operating power factor. The per-phase armature reactance is 6 Ω, and the armature resistance is negligible.

7.24 A three-phase, wye-connected synchronous generator is operating at 80% power factor leading. The synchronous reactance is 2.5 Ω per phase, and the armature resistance is negligible. The armature current is 20 A. The terminal voltage is 440 V.

    a. Find the excitation voltage and the power angle.

    b. Repeat part (a) when the armature current is increased to 40 A while the PF is maintained at 80% leading.

    c. Draw a phasor diagram describing the conditions of both parts (a) and (b).

7.25 A three-phase, 1000-hp, cylindrical-rotor, wye-connected synchronous motor has a negligible armature winding resistance and a synchronous reactance of 38 Ω per phase. The motor receives a constant power of 850 kW at 12 kV. The motor has a full-load current of 48 A. Determine the excitation voltage.

7.26 A three-phase, 2400-V, 60-Hz, four-pole, wye-connected synchronous motor has a synchronous reactance of 16 Ω and negligible armature resistance. The excitation is adjusted so that the induced voltage is 2400 V. The motor drives a load connected to its shaft, and the stator current is 80 A. Calculate

    a. The power angle

    b. The input power

    c. The developed torque

7.27 An overexcited synchronous motor is connected across a 250-kVA inductive load of 0.6 lagging power factor. The motor takes 20 kW while running on no load. Calculate the kVA rating of the motor in order to raise the overall power factor of the motor-inductive load combination to 0.95 lagging.

7.28 A small industrial plant has a total electrical load of 300 kW at 0.6 lagging power factor. A 50-hp pump is to be installed. A synchronous motor operating at 0.8 PF leading is selected to drive the pump. Neglect all losses in the synchronous motor. Calculate (a) the new total load real and reactive powers and (b) the resultant power factor.

7.29 A three-phase, 500-hp, 2400-V, 60-Hz, six-pole synchronous motor has a synchronous reactance of 12 Ω per phase and is assumed to be lossless.

   a. Find the motor speed.

   b. Determine the maximum possible torque when the motor operates at rated load conditions and unity power factor.

   c. Repeat part (b) when the motor operates at 0.8 PF leading.

7.30 A three-phase, 2300-V, wye-connected synchronous motor has a synchronous impedance of $0.05 + j1.25$ Ω per phase. The motor draws its full-load current of 500 A at unity power factor and a field current of 6 A. Find the field current when the motor takes 400 A at 0.8 power factor leading.

7.31 A three-phase, 2300-V, 60-Hz, cylindrical-rotor synchronous motor has a synchronous reactance of 10 Ω per phase and negligible armature resistance. The motor delivers 250 hp at a power angle of 20 electrical degrees, and the efficiency is 90%. Determine

   a. The excitation voltage

   b. The stator current

   c. The power factor

7.32 A three-phase, 13.2-kV, 60-Hz, wye-connected, synchronous motor has an armature resistance of 2 Ω per phase and a synchronous reactance of 30 Ω per phase. When the motor delivers 1500 hp, it takes a current of 80 A at a leading power factor and the efficiency is 90%.

   a. Determine the power factor.

   b. Calculate the excitation voltage.

7.33 A three-phase, 20-kVA, 480-V, wye-connected synchronous machine operates as a motor, and it draws rated current at rated voltage and 0.8 lagging power factor. The total iron, friction, and field copper losses are 1500 W, and the armature resistance is 2.5 Ω per phase. Determine the efficiency of the motor.

7.34 A three-phase, 4160-V, wye-connected, cylindrical-rotor synchronous motor has a synchronous reactance of 8 Ω per phase and negligible armature resistance. The combined rotational losses (friction and windage plus core loss) amount to 5 kW. The highest excitation voltage possible is 4350 V. The motor delivers an output of 400 hp to a mechanical load connected to its shaft.

a. Calculate the stator current at maximum excitation.

b. Compute the smallest excitation voltage for which the motor will remain in synchronism.

7.35 A three-phase, 1000-hp, 2300-V, wye-connected synchronous motor operates at rated voltage and rated frequency. The motor delivers rated power at 0.8 power factor leading, an efficiency of 92%, and a power angle of 25°. The synchronous reactance is 8 Ω per phase, and the armature resistance is negligible. Determine (a) the line current and (b) the excitation voltage.

7.36 A three-phase, 2000-hp, 13.2-kV, 60-Hz, six-pole, wye-connected, cylindrical-rotor synchronous motor operates at rated load, 0.85 power factor leading, and an efficiency of 94%. The synchronous reactance per phase is 32 Ω, and the armature resistance is negligible. Determine the following:

a. Rated torque

b. Armature current

c. Excitation voltage and power angle

d. Pullout torque

7.37 A synchronous motor delivers rated kVA at rated voltage and leading power factor. Let the power angle between the excitation voltage phasor and the terminal voltage phasor be denoted by δ and the power factor angle denoted by θ. Draw the phasor diagram corresponding to this load condition, and show the following relationship.

$$\tan \delta = \frac{I_a X_q \cos \theta + I_a R_a \sin \theta}{V_t + I_a X_q \sin \theta - I_a R_a \cos \theta}$$

Consider θ to be negative when the stator current $\mathbf{I}_a$ lags the terminal voltage $\mathbf{V}_t$.

7.38 A three-phase, salient-pole synchronous machine has d-axis and q-axis reactances of 1.4 pu and 0.6 pu, respectively. The armature resistance is negligible. The machine operates as a synchronous motor and draws 0.8 pu of power at 1.0 per-unit voltage and 0.866 power factor leading.

a. Draw the phasor diagram.

b. Determine the excitation voltage and power angle δ.

c. Determine the power due to field excitation and the power due to the saliency of the machine.

Eight

Induction Motors

8.1 INTRODUCTION

Just like DC machines and synchronous machines, the induction machine may be used as a generator or as a motor. Because their performance cannot compare with that of synchronous machines, induction generators have not been very popular. In recent years, however, induction generators have found use in wind power plants. Because of its wide use and popularity, the induction motor is called the workhorse of the power industry. This chapter will describe the principles of operation and performance analysis of three-phase and single-phase induction motors.

An induction motor is an AC machine in which alternating current is supplied to the stator armature windings directly and to the rotor windings by induction or transformer action from the stator. Hence, it has also been called a rotating transformer. Its stator windings are similar to the stator windings of synchronous machines. However, the rotor of the induction motor may be either of two types:

a. A *wound rotor* carries three windings similar to the stator windings. The terminals of the rotor windings are connected to insulated slip rings mounted on the rotor shaft. Carbon brushes bearing on these rings make the rotor terminals available to the user of the machine. For steady-state operation, these terminals are shorted.

b. A *squirrel-cage rotor* consists of conducting bars embedded in slots in the rotor magnetic core, and these bars are short-circuited at each end by conducting end rings. The rotor bars and the rings are shaped like a squirrel cage, hence the name squirrel-cage rotor.

262

Most induction motors have squirrel-cage rotors. From a modeling point of view, however, the two types of rotors are similar.

The three balanced alternating voltages applied to the stator cause balanced stator currents to flow. As shown in Chapter 5, these currents produce a rotating mmf that can be represented as a rotating magnetic field. This rotating magnetic field induces voltages in the rotor windings, by Faraday's law. These induced voltages, in turn, cause balanced currents to flow in the short-circuited rotor. These rotor currents then produce a rotor mmf, which can also be represented as a rotating magnetic field. The interaction of these two rotating magnetic fields produces an electromagnetic torque T_e, which is used to turn a mechanical load T_m. At steady state, when the motor losses are neglected, T_m and T_e are equal.

The speed of rotation of the rotor magnetic field, when viewed from a stationary position on the stator, is equal to the speed of rotation of the stator magnetic field, which is the synchronous speed n_s. However, the rotor speed n_r is different from the synchronous speed. If n_r were equal to n_s, there would be no variation of flux linkage in the rotor, or there would be no net flux cutting; hence, no voltage would be induced in the rotor. Therefore, rotor speed has to be less than synchronous speed. The difference in speed is represented by the *slip s*, which is defined as follows.

$$s = \frac{n_s - n_r}{n_s} \qquad (8.1)$$

where

n_r = rotor speed (rpm)

n_s = $120f/p$ = synchronous speed (rpm)

f = frequency

p = number of poles

The difference between the speed of the rotor magnetic field and the speed of the rotor expressed in revolutions per minute is called *slip rpm* and is equal to $(n_s - n_r)$. Therefore, the frequency of the rotor currents is given by

$$f_r = (n_s - n_r)\frac{p}{120} = s\left(\frac{pn_s}{120}\right) \qquad (8.2)$$

Since $(pn_s/120)$ is equal to the frequency of the stator currents, Eq. 8.2 can also be written as

$$f_r = sf \qquad (8.3)$$

For steady-state operation, the slip has a normal range of values between 1% and 5%.

EXAMPLE 8.1

A three-phase, 10-hp, 208-V, 60-Hz, four-pole, wye-connected induction motor delivers rated output power at a slip of 5%. Determine the following:

a. Synchronous speed
b. Rotor speed and slip rpm at the rated load
c. Rotor frequency at the rated load
d. Speed of stator rotating magnetic field
e. Speed of the rotor rotating magnetic field
  (i) relative to the rotor
  (ii) relative to the stator
  (iii) relative to the stator magnetic field

Solution

a. The synchronous speed of this motor is

$$n_s = 120f/p = (120)(60)/4 = 1800 \text{ revolutions/minute (rpm)}$$

b. At rated output power, the slip $s = 0.05$. Thus, the rotor speed, which is also the actual speed of the motor, is given by

$$n = n_r = (1 - s)n_s = (1 - 0.05)(1800) = 1710 \text{ rpm}$$

Thus, the slip rpm is found as follows.

$$\text{Slip rpm} = n_s - n_r = 1800 - 1710 = 90 \text{ rpm}$$

Alternatively,

$$\text{Slip rpm} = sn_s = (0.05)(1800) = 90 \text{ rpm}$$

c. The rotor frequency of this motor is given by

$$f_r = sf = (0.05)(60) = 3 \text{ Hz}$$

d. The speed of rotation of the stator magnetic field is equal to the synchronous speed $n_s = 1800$ rpm.

e. The speed of rotation of the rotor magnetic field is
  (i) Relative to the rotor:

$$n_2 = sn_s = 90 \text{ rpm}$$

(ii) Relative to the stator:

$$n + n_2 = n + sn_s = n_s = 1800 \text{ rpm}$$

(iii) Relative to the stator magnetic field = 0 rpm

DRILL PROBLEMS

D8.1 A six-pole induction motor runs at 1158 rpm when it is connected to a 60-Hz source. Determine the synchronous speed and the percent slip.

D8.2 A three-phase, 60-Hz induction motor runs at 1192 rpm at no load and at 1120 rpm at full load. Determine (a) the number of poles and (b) the slip at rated load.

D8.3 A three-phase, 60-Hz, 12-pole, wye-connected induction motor has a full load slip of 5%. Calculate

a. Full-load speed
b. Synchronous speed
c. Slip rpm

D8.4 A three-phase, 208-V, eight-pole, 60-Hz induction motor operates at a slip of 5%. Determine the following:

a. Speed of the stator and rotor magnetic fields
b. Speed of the rotor
c. Slip speed of the rotor
d. Rotor frequency

D8.5 A four-pole, 60-Hz induction motor drives a load at 1740 rpm. Determine the following:

a. Slip
b. Speed of the stator field with respect to the stator
c. Speed of the stator field with respect to the rotor
d. Speed of the rotor field with respect to the stator
e. Speed of the rotor field with respect to the rotor

8.2 EQUIVALENT CIRCUIT OF A THREE-PHASE INDUCTION MOTOR

An equivalent circuit is invaluable in the performance analysis of the three-phase induction motor. Therefore, the equivalent circuit, as well as approximate equivalent circuits, are developed in this section. When the parameters of the equivalent circuits are not available, standard tests for determining these parameters are performed; these tests are described in the following section.

8.2.1 Development of Equivalent Circuit

The general form of the equivalent circuit for a three-phase induction motor can be derived from the equivalent circuit of a three-phase transformer. The induction motor can be thought of as a three-phase transformer whose secondary, or the rotor, is short-circuited and is revolving at the motor speed. Because the motor normally operates at balanced conditions, only a single-phase equivalent circuit is necessary.

When balanced three-phase currents flow in both stator and rotor windings, the resultant synchronously rotating air-gap flux wave induces balanced three-phase voltages in both stator windings and rotor windings. The stator induced voltage has a frequency equal to the frequency f of the applied voltage, while the rotor induced voltage has a frequency f_r given by Eq. 8.3.

Consider the stator first. The applied voltage per phase across the stator terminals is equal to the sum of the stator induced voltage per phase, plus the voltage drop across the stator winding resistance, plus the voltage drop across the stator leakage reactance due to the leakage flux, which links only the stator winding. Mathematically, in phasor form, the relationship may be expressed as

$$\mathbf{V}_1 = \mathbf{E}_1 + R_1\mathbf{I}_1 + jX_1\mathbf{I}_1$$
$$= \mathbf{E}_1 + (R_1 + jX_1)\mathbf{I}_1 \tag{8.4}$$

where

$\mathbf{V}_1$ = stator terminal voltage per phase

$\mathbf{E}_1$ = stator induced voltage per phase

$\mathbf{I}_1$ = stator current

R_1 = stator winding resistance

X_1 = stator leakage reactance

The magnetic core can be modeled as a parallel combination of a resistance R_c, to account for hysteresis and eddy current losses, and a reactance X_m, to account for the magnetizing current required to produce the air-gap magnetic flux. The magnetizing current in an induction motor is much larger than that in a transformer because of the presence of the air gap in a motor.

Next, a model of the rotor is developed. Let E_2 denote the rotor induced voltage at standstill, that is, $s = 1.0$. At standstill, the induction motor may be viewed as a transformer with an air gap, and the stator per-phase induced voltage E_1 is related to the rotor per-phase induced voltage E_2 by the turns ratio (N_1/N_2); that is,

$$E_1 = \left(\frac{N_1}{N_2}\right)E_2 \tag{8.5}$$

where

N_1 = number of turns in the stator winding

N_2 = number of turns in the rotor winding

The voltage induced in the rotor of the induction motor is directly proportional to the relative motion of the rotor and the synchronously rotating air-gap magnetic field. When the induction motor is rotating at a speed n, or a slip s, the rotor induced voltage E_{2s} is equal to the induced voltage at standstill E_2 multiplied by the slip. In the short-circuited rotor circuit, the induced voltage E_{2s} appears as a voltage drop across the rotor resistance and leakage reactance. The rotor resistance does not depend on the slip. However, the rotor leakage reactance does, and is equal to $X_r = 2\pi f_r L_r = s2\pi f L_r$, where L_r is the leakage inductance of the rotor winding due to flux linking the rotor winding only. Thus, the rotor induced voltage at slip s may be expressed mathematically as follows:

$$
\begin{aligned}
E_{2s} = sE_2 &= R_r I_r + j(2\pi f_r L_r)I_r \\
&= R_r I_r + js(2\pi f L_r)I_r \\
&= R_r I_r + jsX_2' I_r
\end{aligned}
\tag{8.6}
$$

where

E_{2s} = rotor induced voltage at slip s

E_2 = rotor induced voltage at standstill ($s = 1.0$)

I_r = rotor phase current

R_r = rotor resistance per phase

$X_2' = 2\pi f L_r$ = rotor leakage reactance per phase at standstill

Dividing both sides of Eq. 8.6 by the slip s and referring rotor quantities to the stator side, as in a transformer, yields

$$\mathbf{E}_2 = \left(\frac{R_r}{s} + jX_2'\right)\mathbf{I}_r$$

$$\left(\frac{N_1}{N_2}\right)\mathbf{E}_2 = \left(\frac{N_1}{N_2}\right)^2\left(\frac{R_r}{s} + jX_2'\right)\left(\frac{N_2}{N_1}\right)\mathbf{I}_r \qquad (8.7)$$

$$\mathbf{E}_1 = \left(\frac{R_2}{s} + jX_2\right)\mathbf{I}_2$$

where

$\mathbf{E}_1 = (N_1/N_2)\mathbf{E}_2 = $ rotor induced voltage referred to the stator

$\mathbf{I}_2 = (N_2/N_1)\mathbf{I}_r = $ rotor current referred to the stator

$R_2 = (N_1/N_2)^2 R_r = $ rotor resistance referred to the stator

$X_2 = (N_1/N_2)^2 X_2' = $ rotor leakage reactance referred to the stator

The stator circuit represented by Eq. 8.4 and the rotor circuit represented by Eq. 8.7 are at the same frequency f of the applied voltage. Therefore, these stator and rotor circuits can be joined together and combined with the model of the magnetic core into the per-phase equivalent circuit of the induction motor, which is shown in Fig. 8.1.

EXAMPLE 8.2

A three-phase, 25-hp, 440-V, 60-Hz, four-pole, induction motor has the following impedances referred to the stator in Ω/phase.

$$R_1 = 0.50 \qquad R_2 = 0.35$$
$$X_1 = 1.20 \qquad X_2 = 1.20 \qquad X_m = 25$$

The combined rotational losses (mechanical and core losses) amount to 1250 W, and they are assumed to remain constant. For a rotor slip of 2.5% at rated

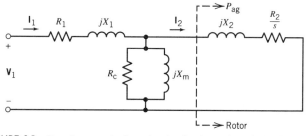

FIGURE 8.1 Per-phase equivalent circuit of a three-phase induction motor.

voltage and rated frequency, find

  a. The motor speed
  b. The stator current
  c. The power factor
  d. The efficiency of the motor

Solution   Since the core loss is lumped with the rotational losses, the resistance R_c is neglected. Thus, the per-phase equivalent circuit of the induction motor appears as shown in Fig. 8.2.

  a. The synchronous speed is

$$n_s = 120f/p = (120)(60)/4 = 1800 \text{ rpm}$$
$$\omega_s = 2\pi n_s/60 = 2\pi(1800)/60 = 188.5 \text{ rad/s}$$

The motor speed is found from the given slip.

$$n = (1 - s)n_s = (1 - 0.025)1800 = 1755 \text{ rpm}$$
$$\omega = 2\pi n/60 = 2\pi(1755)/60 = 183.8 \text{ rad/s}$$

  b. The impedance of the rotor referred to the stator is

$$Z_2 = R_2/s + jX_2 = 0.35/0.025 + j1.20 = 14.0 + j1.20 \ \Omega$$

Therefore, the input impedance is

$$Z_{in} = R_1 + jX_1 + (Z_2)(jX_m)/(Z_2 + jX_m)$$
$$= 0.50 + j1.20 + (14.0 + j1.20)(j25)/[14.0 + j(1.20 + 25)]$$
$$= 10.42 + j7.64 = 12.92 \ \underline{/36.3°} \ \Omega$$

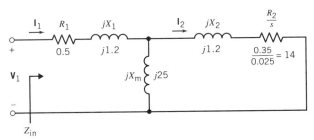

FIGURE 8.2   Equivalent circuit for motor of Example 8.2.

The terminal voltage per phase is taken as reference phasor; thus,

$$\mathbf{V}_1 = (440/\sqrt{3})\underline{/0^\circ} = 254.0\underline{/0^\circ} \text{ V} \quad \text{(line-to-neutral)}$$

The stator current is found as follows.

$$\mathbf{I}_1 = \mathbf{V}_1/Z_{\text{in}} = (254.0\underline{/0^\circ})/(12.92\underline{/36.3^\circ}) = 19.66\underline{/-36.3^\circ}$$

c. The power factor PF is found as follows.

$$\text{PF} = \cos 36.3^\circ = 0.806 \text{ lagging}$$

d. The power input to the motor is given by

$$P_{\text{input}} = 3V_1I_1\text{PF} = 3(254.0)(19.66)(0.806) = 12{,}075 \text{ W}$$

The stator copper loss SCL is given by

$$\text{SCL} = 3I_1^2R_1 = 3(19.66)^2(0.50) = 580 \text{ W}$$

The rotor current is computed as follows:

$$\mathbf{I}_2 = \{j25/[14.0 + j(25 + 1.2)]\}\mathbf{I}_1$$

$$= [j25/(14.0 + j26.2)]\left[19.66\underline{/-36.3^\circ}\right] = 16.54\underline{/-8.2^\circ} \text{ A}$$

Hence, the rotor copper loss RCL is

$$\text{RCL} = 3I_2^2R_2 = 3(16.54)^2(0.35) = 287 \text{ W}$$

The output power is found as follows.

$$P_{\text{out}} = P_{\text{input}} - \text{SCL} - \text{RCL} - P_{\text{rot}}$$
$$= 12{,}075 - 580 - 287 - 1250 = 9958 \text{ W}$$

Therefore, the efficiency is

$$\eta = P_{\text{out}}/P_{\text{input}} = (9958/12{,}075)100\% = 82.5\%$$

An approximate equivalent circuit for an induction motor is derived by moving the shunt elements, R_c and X_m in parallel, representing the core to the motor terminals. This simplification introduces little error but greatly reduces

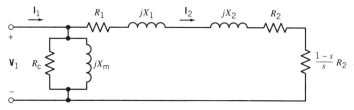

FIGURE 8.3   Approximate equivalent circuit of an induction motor.

the computational effort. This approximate equivalent circuit is shown in Fig. 8.3. Also shown in the figure is the equivalent rotor resistance R_2/s, which has been decomposed into R_2 and $R_2[(1 - s)/s]$. The first resistance component, R_2, represents the rotor copper loss, and the second component represents the power developed by the motor.

EXAMPLE 8.3

A three-phase, 220-V induction motor has the following data:

$$R_1 = 0.20 \ \Omega \qquad R_2 = 0.15 \ \Omega$$
$$X_1 = 0.50 \ \Omega \qquad X_2 = 0.30 \ \Omega$$

The core effect can be neglected. The motor operates at 3% slip. If the total losses are 1000 W, determine the following:

a. Stator current

b. Power factor

c. Efficiency of the motor

Solution

a. With the effects of the core neglected, the approximate per-phase equivalent circuit of the three-phase induction motor reduces to that shown in Fig. 8.4.

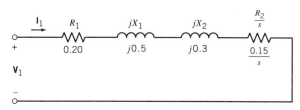

FIGURE 8.4   Equivalent circuit of motor of Example 8.3.

The terminal voltage per phase is taken as reference phasor; thus,

$$\mathbf{V}_1 = (220/\sqrt{3})\underline{/0°} = 127\underline{/0°} \text{ V} \quad \text{(line-to-neutral)}$$

The stator current is found as follows:

$$\mathbf{I}_1 = \mathbf{V}_1/[R_1 + R_2/s + j(X_1 + X_2)]$$
$$= (127\underline{/0°})/[0.2 + 0.15/0.03 + j(0.5 + 0.3)]$$
$$= 24.14\underline{/-8.75°} \text{ A}$$

b. The input power factor is given by

$$PF = \cos(\underline{/\mathbf{V}_1} - \underline{/\mathbf{I}_1}) = \cos 8.75° = 0.988 \text{ lagging}$$

c. The input power is given by

$$P_{\text{input}} = 3V_1 I_1 PF = 3(127)(24.14)(0.988) = 9090 \text{ W}$$

The output power is computed as follows.

$$P_{\text{output}} = P_{\text{input}} - \text{losses} = 9090 - 1000 = 8090 \text{ W}$$

Therefore, the efficiency is

$$\eta = P_{\text{output}}/P_{\text{input}} = (8090/9090)100\% = 89\%$$

DRILL PROBLEMS

D8.6 A three-phase, 440-V, 60-Hz, four-pole, wye-connected induction motor has the following per-phase parameters, which are referred to the stator:

$$R_1 = 0.10 \ \Omega \qquad X_1 = 0.4 \ \Omega$$
$$R_2 = 0.15 \ \Omega \qquad X_2 = 0.4 \ \Omega \qquad X_m = 12 \ \Omega$$

The motor core loss is 2000 W, and the friction and windage losses amount to 1500 W. At a slip of 5%, determine the following:

a. Input current and power factor
b. Power input

c. Power output

d. Efficiency of the motor

D8.7 A three-phase, 30-hp, 480-V, four-pole, 60-Hz induction motor has the following equivalent circuit parameters in Ω per phase referred to the stator:

$$R_1 = 0.25 \qquad R_2 = 0.20$$
$$X_1 = 1.30 \qquad X_2 = 1.20 \qquad X_m = 35$$

The total core, friction, and windage losses may be assumed constant at 1250 W. The motor is connected directly to a 440-V source, and it runs at a slip of 3.5%. Compute the following:

a. Motor speed

b. Input current and power factor

c. Shaft output torque

d. Efficiency of the motor

8.2.2 Determination of Parameters from Tests

To determine the parameters of the equivalent circuit of the three-phase induction motor, it is subjected to tests similar to the open-circuit and short-circuit tests for three-phase transformers.

No-Load Test Like the open-circuit test on a transformer, this test is performed to obtain the shunt parameters of the motor, which represent the magnetizing current and its core loss. The *no-load test* is taken at rated frequency, and the voltage applied to the motor is rated voltage.

When the motor is running at no load, the slip is close to zero; therefore, $n \approx n_s$. Hence, the equivalent rotor resistance (R_2/s) is large. Referring to the approximate equivalent circuit of Fig. 8.2, the rotor impedance containing this large resistance is in parallel with the shunt magnetizing reactance X_m; the parallel combination is approximately equal to jX_m. Thus, the equivalent circuit of Fig. 8.2 reduces to the simple series equivalent circuit shown in Fig. 8.5.

When the motor operates at rated voltage and rated frequency, the combined rotational losses including friction and windage loss, hysteresis and eddy current loss, and stray load loss are assumed to remain constant at any load. This constant value is the value of the rotational loss at no load and is found as follows.

$$P_{\text{rot}} = P_{\text{nl}} - 3I_{\text{nl}}^2 R_1 \tag{8.8}$$

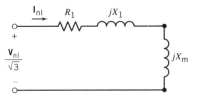

FIGURE 8.5   Approximate equivalent circuit for no-load test.

The three-phase induction motor is considered to be wye connected. Thus, the per-phase no-load resistance R_{nl} is found as

$$R_{nl} = \frac{P_{nl}}{3I_{nl}^2} \qquad (8.9)$$

where

P_{nl} = total power input at no load

I_{nl} = stator current per phase at no load

It may be noted that R_{nl} is different from R_1 because the former includes the effects of the rotational losses and the no-load copper loss.

The per-phase no-load impedance is computed as

$$Z_{nl} = \frac{V_{nl}}{\sqrt{3}I_{nl}} = R_{nl} + jX_{nl} \qquad (8.10)$$

where V_{nl} is the line-to-line terminal voltage at no load. Therefore, the per-phase no-load reactance is computed as follows:

$$X_{nl} = \sqrt{Z_{nl}^2 - R_{nl}^2} \qquad (8.11)$$

From the approximate equivalent circuit of Fig. 8.5, the apparent reactance is seen to be

$$X_{nl} = X_1 + X_m \qquad (8.12)$$

DC Test   The stator resistance R_1 may be assumed to be equal to its DC value. To find this value, a DC voltage is applied to two stator terminals of the motor and the current and applied voltage are measured. The stator resistance is computed as

$$R_1 = \frac{1}{2}\frac{V_{DC}}{I_{DC}} \qquad (8.13)$$

Blocked-Rotor Test   In this test, the rotor of the induction motor is blocked so that it cannot rotate; therefore, $s = 1$. Thus, the three-phase motor appears like a short-circuited three-phase transformer. Similarly, a reduced three-phase voltage is applied to the stator such that rated current will flow through the windings.

During normal running conditions of the induction motor, the rotor frequency is proportional to the slip. Therefore, when the performance of the motor is being investigated at, or near, rated loads (at low values of slip), the *blocked-rotor test* should be taken at a lower frequency. A test frequency of 25% of rated frequency is recommended by the Institute of Electrical and Electronics Engineers (IEEE).

The blocked-rotor resistance is found as follows:

$$R_{bl} = \frac{P_{bl}}{3I_{bl}^2} \qquad (8.14)$$

where

$P_{bl} = $ total power input at blocked rotor

$I_{bl} = $ stator current per phase at blocked rotor

The blocked-rotor impedance at the test frequency f_{test} is given by

$$Z_{bl} = \frac{V_{bl}}{\sqrt{3}I_{bl}} \qquad (8.15)$$

The blocked-rotor reactance at the test frequency f_{test} is computed as

$$X_{bl,test} = \sqrt{Z_{bl}^2 - R_{bl}^2} \qquad (8.16)$$

The blocked-rotor reactance computed at the test frequency is corrected to rated frequency by multiplying $X_{bl,test}$ by the ratio (f_{rated}/f_{test}):

$$X_{bl} = \left(\frac{f_{rated}}{f_{test}}\right)X_{bl,test} \qquad (8.17)$$

If the exciting current component of the stator current is neglected, the equivalent circuit of the induction motor shown in Fig. 8.2 reduces to that shown in Fig. 8.6 for the blocked-rotor conditions.

With the shunt magnetizing reactance neglected in the approximate equivalent circuit of Fig. 8.6, it is seen that the blocked-rotor resistance and frequency-corrected reactance are related to the series parameters as follows.

$$R_{bl} = R_1 + R_2 \qquad (8.18)$$

$$X_{bl} = X_1 + X_2 \qquad (8.19)$$

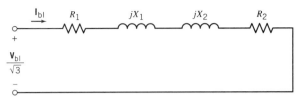

FIGURE 8.6   Approximate equivalent circuit for blocked-rotor test.

The value of the stator resistance R_1 is determined from the DC test. Thus, the rotor resistance R_2 referred to the stator is computed as

$$R_2 = R_{bl} - R_1 \qquad (8.20)$$

There is no simple way to apportion the blocked-rotor reactance between the stator leakage reactance X_1 and the rotor leakage reactance X_2 referred to the stator. However, the performance of the induction motor is only slightly affected by the relative distribution of X_{bl} between X_1 and X_2. Thus, it may be assumed that

$$X_1 = X_2 = \tfrac{1}{2}X_{bl} \qquad (8.21)$$

Hence, the value of the shunt magnetizing reactance is computed as

$$X_m = X_{nl} - X_1 \qquad (8.22)$$

EXAMPLE 8.4

A three-phase, 5-hp, 208-V, four-pole, 60-Hz induction motor is subjected to a no-load test at 60 Hz, a blocked-rotor test at 15 Hz, and a DC test. The following data are obtained.

	No Load	Blocked Rotor	DC
Voltage (V)	208	35	20
Current (A)	4	12	25
Power (W)	250	450	

Determine the parameters of the equivalent circuit and the combined rotational losses of the motor.

Solution   From the DC test, the stator resistance is found as

$$R_1 = \tfrac{1}{2}(20/25) = 0.40 \ \Omega$$

From the no-load test, the combined rotational loss is computed as

$$P_{rot} = 250 - 3(4)^2(0.4) = 230.8 \text{ W}$$

The no-load impedance parameters are

$$R_{nl} = 250/[3(4)^2] = 5.2 \ \Omega$$
$$Z_{nl} = 208/[\sqrt{3}(4)] = 30.0 \ \Omega$$
$$X_{nl} = [(30.0)^2 - (5.2)^2]^{1/2} = 29.5 \ \Omega$$

From the blocked-rotor test, the reduced-frequency parameters are computed as follows:

$$R_{bl} = 450/[3(12)^2] = 1.04 \ \Omega$$
$$Z_{bl} = 35/[\sqrt{3}(12)] = 1.68 \ \Omega$$
$$X_{bl,test} = [(1.68)^2 - (1.04)^2]^{1/2} = 1.32 \ \Omega$$

The rotor resistance referred to the stator side is computed as

$$R_2 = 1.04 - 0.40 = 0.64 \ \Omega$$

The blocked-rotor reactance referred to rated frequency is

$$X_{bl} = (60/15)(1.32) = 5.28 \ \Omega$$

This value of the blocked-rotor reactance is equally divided between stator and rotor leakage reactances; thus,

$$X_1 = X_2 = \tfrac{1}{2}(5.28) = 2.64 \ \Omega$$

Finally, the magnetizing reactance is found by subtracting X_1 from X_{nl}:

$$X_m = 29.5 - 2.64 = 26.9 \ \Omega$$

DRILL PROBLEMS

D8.8 A three-phase, 25-hp, 208-V, six-pole, 60-Hz, wye-connected induction motor is tested with the following results:

	No Load (at 60 Hz)	Blocked Rotor (at 15 Hz)	DC
Voltage (V)	208	25	15
Current (A)	24	66	70
Power (W)	1400	2300	

Determine the parameters of the equivalent circuit of the motor.

D8.9 The following data were obtained from no-load and blocked-rotor tests at rated frequency and a DC test on a three-phase, 30-hp, 460-V, 60-Hz, four-pole, wye-connected induction motor:

	No Load	Blocked Rotor	DC
Voltage (V)	440	95	20
Current (A)	15	52	55
Power (W)	4000	6200	

Determine the parameters of the equivalent circuit and the rotational losses of the motor.

8.3 PERFORMANCE ANALYSIS OF AN INDUCTION MOTOR

When an induction machine is running at no load, the slip is close to zero. Hence, the equivalent rotor resistance R_2/s is infinitely large. This large resistance value results in a very small flow of current I_2. Thus, the electromagnetic torque assumes a small value, just enough to overcome the combined rotational losses consisting of the friction and windage and core losses.

When a mechanical load is connected to the motor shaft, the initial reaction of the motor is for its speed to decrease. Thus, the slip s increases, the equivalent rotor resistance (R_2/s) decreases, and the rotor current I_2 increases. Therefore, the electromagnetic torque increases and, when it becomes equal to the sum of the load torque and rotational losses, the motor will continue to run at a steady-state speed whose value will be less than the no-load speed.

The equivalent circuit of Fig. 8.1 and the power flow diagram of Fig. 8.7 can be used to analyze the steady-state performance of induction motors. In Fig. 8.7, $\omega_s = (2/p)\omega$, where $\omega = 2\pi f = 120\pi$ rad/s.

The power input to the induction motor is expressed as

$$P_{\text{input}} = 3P_1 = 3V_1 I_1 \cos\theta_1 \qquad (8.23)$$

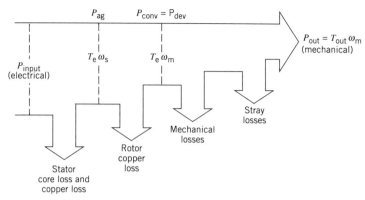

FIGURE 8.7 Power flow diagram for an induction motor.

The stator copper loss SCL is given by

$$\text{SCL} = P_{\text{cu1}} = 3I_1^2 R_1 \tag{8.24}$$

The equivalent rotor resistance R_2/s represents the power transferred across the air gap from the stator to the rotor. The expression for the *air-gap power* is given by

$$P_{\text{ag}} = P_{\text{input}} - \text{SCL} - P_{\text{core}}$$

$$= 3I_2^2\left(\frac{R_2}{s}\right) \tag{8.25}$$

This air-gap power may be decomposed into two components. The first is the rotor copper loss RCL, and it is expressed as

$$\text{RCL} = P_{\text{cu2}} = 3I_2^2 R_2 \tag{8.26}$$

The other component of the air-gap power is P_{conv}, the power converted from electrical to mechanical form. It is also called the *developed power* P_{dev}, and it is given by

$$P_{\text{conv}} = P_{\text{dev}} = P_{\text{ag}} - \text{RCL}$$

$$= 3I_2^2 R_2\left(\frac{1-s}{s}\right) \tag{8.27}$$

$$= P_{\text{ag}}(1-s)$$

The power output available at the shaft of the motor is found by subtracting the rotational loss from the power developed:

$$P_{\text{output}} = P_{\text{dev}} - P_{\text{rot}} \tag{8.28}$$

Finally, the efficiency of the induction motor is calculated as

$$\eta = \frac{P_{output}}{P_{input}} 100\% \qquad (8.29)$$

EXAMPLE 8.5

A three-phase, 25-hp, 230-V, 60-Hz induction motor draws 60 A from the source at 0.866 lagging power factor. The motor losses include the following:

Stator copper loss $P_{cu1} = 850$ W
Magnetic core loss $P_{core} = 450$ W
Rotor copper loss $P_{cu2} = 1050$ W
Rotational loss $P_{rot} = 500$ W

Find the following:

    a. Air-gap power P_{ag}
    b. Slip s
    c. Mechanical power developed P_{dev}
    d. Output power
    e. Efficiency of the motor

Solution

    a. The terminal voltage per phase is taken as reference phasor; thus,

$$\mathbf{V_t} = (230/\sqrt{3})\underline{/0°} = 132.8\underline{/0°} \text{ V} \quad \text{(line-to-neutral)}$$

The power input to the motor is found as

$$\mathbf{P_{input}} = 3\mathbf{V_t I_1}PF = 3(132.8)(60)(0.866) = 20,700 \text{ W}$$

Therefore, the power transferred across the air gap is calculated as

$$P_{ag} = P_{input} - P_{core} - P_{cu1} = 20,700 - 450 - 850 = 19,400 \text{ W}$$

    b. The power transferred across the air gap P_{ag} may also be expressed, in terms of the rotor copper loss, as

$$P_{ag} = 3I_2^2 R_2/s = P_{cu2}/s$$

Thus, the slip is

$$s = P_{cu2}/P_{ag} = 1050/19,400 = 0.054$$

c. The power developed is given by

$$P_{dev} = P_{ag} - P_{cu2} = 19,400 - 1050 = 18,350 \text{ W}$$

d. The power output of the motor is given by

$$P_{output} = P_{dev} - P_{rot} = 18,350 - 500 = 17,850 \text{ W}$$

e. Therefore, the efficiency of the motor is found as

$$\eta = (P_{output}/P_{input})100\% = (17,850/20,700)100\% = 86.2\%$$

EXAMPLE 8.6

A three-phase, 50-hp, 480-V, 60-Hz, four-pole, induction motor has the following data:

$$R_1 = 0.100 \ \Omega/\text{phase} \qquad X_1 = 0.35 \ \Omega/\text{phase}$$
$$R_2 = 0.125 \ \Omega/\text{phase} \qquad X_2 = 0.40 \ \Omega/\text{phase}$$

Stator core losses and mechanical (friction and windage) losses are 1200 W and 900 W, respectively. At no load, the motor draws 18 A at 0.09 lagging power factor. When the motor is operated at $s = 0.025$, determine the following:

a. Line current and power factor
b. Electromagnetic torque
c. Output power
d. Efficiency of the motor

Solution

a. The approximate per-phase equivalent circuit is shown in Fig. 8.8. The terminal voltage per phase is taken as reference phasor; thus,

$$\mathbf{V}_1 = (480/\sqrt{3})\underline{/0°} = 277.1\underline{/0°} \text{ V} \quad \text{(line-to-neutral)}$$

At no load, the rotor current is approximately zero; $\mathbf{I}_2 \approx 0.0$. Therefore, the magnetizing current $\mathbf{I}_m$ is equal to the no-load stator current $\mathbf{I}_{1,nl}$; that is,

$$\mathbf{I}_m = \mathbf{I}_{1,nl} = 18 \underline{/-\cos^{-1} 0.09} = 18 \underline{/-84.8°} \text{ A}$$

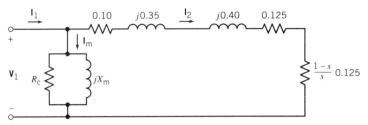

FIGURE 8.8 Equivalent circuit for motor of Example 8.6.

At a slip $s = 0.025$, the rotor current is found as

$$\mathbf{I}_2 = \mathbf{V}_1/[R_1 + (R_2/s) + j(X_1 + X_2)]$$
$$= (277.1 \underline{/0°})/[0.100 + (0.125/0.025) + j(0.35 + 0.40)]$$
$$= 53.8 \underline{/-8.4°} \text{ A}$$

Therefore, the stator current is computed as

$$\mathbf{I}_1 = \mathbf{I}_2 + \mathbf{I}_m = 53.8 \underline{/-8.4°} + 18 \underline{/-84.8°} = 60.5 \underline{/-25.2°} \text{ A}$$

The power factor is found as

$$\text{PF} = \cos(\underline{/\mathbf{V}_1} - \underline{/\mathbf{I}_1}) = \cos 25.2° = 0.90 \text{ lagging}$$

b. The synchronous speed is given by

$$n_s = 120f/p = (120)(60)/4 = 1800 \text{ rpm}$$
$$\omega_s = 2\pi n_s/60 = 2\pi(1800)/60 = 188.5 \text{ rad/s}$$

The air-gap power is computed as

$$P_{ag} = 3I_2^2 R_2/s = 3(53.8)^2(0.125)/0.025 = 43{,}417 \text{ W}$$

Hence, the electromagnetic torque developed is determined from the air-gap power; that is,

$$T_e = P_{ag}/\omega_s = 43{,}417/188.5 = 230 \text{ N-m}$$

c. The power developed by the motor is given by

$$P_{dev} = (1 - s)P_{ag} = (1 - 0.025)43{,}417 = 42{,}332 \text{ W}$$

Therefore, the output power is found as

$$P_{output} = P_{dev} - P_{rot} = 42,332 - 900 = 41,432 \text{ W}$$

d. The power input to the motor is given by

$$P_{input} = P_{ag} + P_{cu1} + P_{core} = P_{ag} + 3I_1^2 R_1 + P_{core}$$
$$= 43,417 + 3(60.5)^2(0.10) + 1200 = 45,715 \text{ W}$$

Therefore, the efficiency is

$$\eta = P_{output}/P_{input} = (41,432/45,715)100\% = 90.6\%$$

DRILL PROBLEMS

D8.10 A three-phase, six-pole, 60-Hz induction motor is operating at a speed of 1152 rpm. The power input to the motor is 44 kW, the rotational losses are 500 W, and the stator copper loss is 1600 W. Find the following:

   a. Slip
   b. Air-gap power
   c. Rotor copper loss
   d. Developed torque and developed horsepower
   e. Output torque and output horsepower

D8.11 A three-phase, four-pole, 60-Hz, wye-connected wound-rotor induction motor is rated at 15 hp and 208 V. Its equivalent circuit parameters are

$$R_1 = 0.25 \ \Omega \qquad R_2 = 0.15 \ \Omega$$
$$X_1 = 0.50 \ \Omega \qquad X_2 = 0.50 \ \Omega \qquad X_m = 18 \ \Omega$$

The combined rotational losses consisting of the friction and windage plus core losses amount to 300 W. For a slip of 4%, find the following:

   a. Line current
   b. Air-gap power
   c. Power converted from electrical to mechanical form
   d. Efficiency of the motor

D8.12 A four-pole, 60-Hz induction motor is rated at 40 hp and 440 V. The motor drives a load at 1710 rpm. The core loss and friction and windage loss are 450 W and 250 W, respectively. The motor parameters are given in Ω/phase as follows:

$$R_1 = 0.15 \qquad R_2 = 1.20$$
$$X_1 = 0.75 \qquad X_2 = 0.75 \qquad X_m = 25$$

Determine the following:

    a. Line current and power factor

    b. Real and reactive power input

    c. Air-gap power

    d. Mechanical power and torque developed

    e. Shaft horsepower and torque

    f. Efficiency of the motor

8.4 TORQUE-SPEED CHARACTERISTICS

The torque-speed characteristics of three-phase induction motors can be analyzed from the approximate equivalent circuit of Fig. 8.3 and the power flow diagram of Fig. 8.7.

8.4.1 Starting Torque

At starting, the slip is unity $[s = (n_s - n_r)/n_s = n_s/n_s = 1]$. The equivalent circuit of Fig. 8.3 reduces to that of Fig. 8.9.

The starting electromagnetic torque $T_{e,start}$ is given by

$$T_{e,start} = P_{ag,start}/\omega_s \tag{8.30}$$

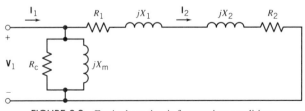

FIGURE 8.9 Equivalent circuit for starting conditions.

The power transferred across the air gap at starting is

$$P_{ag,start} = 3(I_{2,start})^2 \left(\frac{R_2}{s}\right) = 3(I_{2,start})^2 R_2 \tag{8.31}$$

where the rotor current $I_{2,start}$ is given by

$$I_{2,start} = \frac{V_1}{R_1 + R_2 + j(X_1 + X_2)} \tag{8.32}$$

Taking the magnitude of $I_{2,start}$ and substituting it in Eqs. 8.30 and 8.31, the starting torque is obtained.

$$T_{e,start} = \frac{3V_1^2 R_2}{\omega_s[(R_1 + R_2)^2 + (X_1 + X_2)^2]} \tag{8.33}$$

As may be seen from Eq. 8.32, the starting current is large compared to rated load current. This is because R_1, R_2, X_1, and X_2 are small and V_1 is the rated terminal voltage. In order to limit this starting current, for larger motors such as those rated above 5 hp, an applied voltage smaller than rated voltage is used to start the induction motor.

EXAMPLE 8.7

The three-phase, 25-hp, 440-V induction motor of Example 8.2 has the following impedances in Ω per phase referred to the stator.

$$R_1 = 0.50 \quad R_2 = 0.35$$
$$X_1 = 1.20 \quad X_2 = 1.20 \quad X_m = 25$$

The motor is operated at rated voltage and rated frequency.

a. What is the starting torque?

b. When the rotor resistance is doubled, what is the new value of starting torque?

Solution The approximate equivalent circuit at starting is shown in Fig. 8.10.

a. The terminal voltage per phase is taken as reference phasor; thus,

$$V_1 = (440/\sqrt{3})\underline{/0°} = 254.0\underline{/0°} \text{ V} \quad \text{(line-to-neutral)}$$

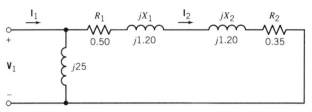

FIGURE 8.10 Approximate equivalent circuit for motor of Example 8.7.

By using Eq. 8.33, the starting torque is calculated as follows.

$$T_{e,\text{start}} = \frac{3(254.0)^2(0.35)}{188.5[(0.50 + 0.35)^2 + (1.20 + 1.20)^2]} = 55.4 \text{ N-m}$$

b. When rotor resistance is doubled, $R_2 = 2(0.35) = 0.70\ \Omega$, the new value of starting torque is found as follows:

$$T_{e,\text{start}} = \frac{3(254.0)^2(0.70)}{188.5[(0.50 + 0.70)^2 + (1.20 + 1.20)^2]} = 99.8 \text{ N-m}$$

8.4.2 Torque Versus Speed

The torque-speed characteristic of an induction motor is studied in terms of its torque-versus-slip relationship. The slip and speed are related through Eq. 8.1.

From the equivalent circuit of Fig. 8.3, the expression for the electromagnetic torque T_e is derived as

$$T_e = \frac{P_{\text{ag}}}{\omega_s} \tag{8.34}$$

where

$$P_{\text{ag}} = 3I_2^2\left(\frac{R_2}{s}\right) \tag{8.35}$$

$$I_2 = \frac{V_1}{|(R_1 + R_2/s) + j(X_1 + X_2)|}$$

$$= \frac{V_1}{\sqrt{(R_1 + R_2/s)^2 + (X_1 + X_2)^2}} \tag{8.36}$$

Substituting Eqs. 8.35 and 8.36 into Eq. 8.34, the expression for the torque is obtained.

$$T_e = \frac{3V_1^2(R_2/s)}{\omega_s[(R_1 + R_2/s)^2 + (X_1 + X_2)^2]}$$ (8.37)

Equation 8.37 can be further simplified as follows:

$$T_e = \frac{3V_1^2 R_2 s}{\omega_s[(R_1 s + R_2)^2 + (X_1 + X_2)^2 s^2]}$$ (8.38)

In order to plot Eq. 8.38, the maximum torque $T_{e,max}$ and the slip s_{max} at which it occurs have to be determined. The maximization theorem from calculus is applied; that is, take the derivative of T_e with respect to s, set the derivative to zero, and solve for s. The solution yields $s = s_{max}$, where

$$s_{max} = \frac{R_2}{\sqrt{R_1^2 + (X_1 + X_2)^2}}$$ (8.39)

It can be shown that the second derivative of T_e with respect to s is negative. Therefore, T_e is at its maximum value at s_{max}. Substituting the value of s_{max} given by Eq. 8.39 into Eq. 8.38 and simplifying, the maximum torque $T_{e,max}$ is obtained as follows:

$$T_{e,max} = \frac{3V_1^2}{2\omega_s[R_1 + \sqrt{R_1^2 + (X_1 + X_2)^2}]}$$ (8.40)

The synchronous speed is given by

$$\omega_s = \left(\frac{2}{p}\right)2\pi f = \frac{4\pi f}{p}$$ (8.41)

With the starting torque $T_{e,start}$, the maximum torque $T_{e,max}$, and s_{max} known, the torque-speed curve can now be plotted, and it is shown in Fig. 8.11.

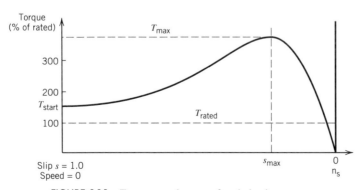

FIGURE 8.11 Torque-speed curve of an induction motor.

EXAMPLE 8.8

The three-phase induction motor of Example 8.6 is connected to a three-phase source at 480 V. Determine the following:

  a. Slip and motor speed at which maximum torque occurs
  b. Rotor current at this slip
  c. Maximum torque

Solution   From Example 8.6, the machine constants are given as follows:

$$R_1 = 0.100 \ \Omega \qquad X_1 = 0.35 \ \Omega$$
$$R_2 = 0.125 \ \Omega \qquad X_2 = 0.40 \ \Omega$$

The terminal voltage per phase is taken as reference phasor; thus,

$$\mathbf{V}_1 = (480/\sqrt{3})\underline{/0^\circ} = 277.1\underline{/0^\circ} \ \text{V} \quad \text{(line-to-neutral)}$$

The synchronous speed is

$$n_s = 120f/p = (120)(60)/4 = 1800 \ \text{rpm}$$

  a. By using Eq. 8.39, the slip at maximum torque is found as

$$s_{max} = \frac{0.125}{\sqrt{(0.10)^2 + (0.35 + 0.40)^2}} = 0.165$$

  Therefore, the speed at maximum torque is

$$n_{max} = (1 - s_{max})n_s = (1 - 0.165)1800 = 1503 \ \text{rpm}$$

  b. The magnetizing branch can be assumed to be negligible. Thus, the rotor current is given by

$$I_{2,max} = \frac{V_1}{\sqrt{(R_1 + R_2/s_{max})^2 + (X_1 + X_2)^2}}$$
$$= \frac{277.1}{\sqrt{(0.10 + 0.125/0.165)^2 + (0.35 + 0.40)^2}} = 243.2 \ \text{A}$$

  c. The maximum torque is found by using Eq. 8.40. Thus,

$$T_{e,max} = \frac{3(277.1)^2}{2(188.5)[0.10 + \sqrt{(0.10)^2 + (0.35 + 0.40)^2}]} = 713 \ \text{N-m}$$

EXAMPLE 8.9

The three-phase, 25-hp, 440-V induction motor of Example 8.2 is operated at rated voltage and rated frequency. Determine the following:

a. Slip at maximum torque
b. Maximum torque
c. Value of slip when the rotor resistance is doubled
d. Value of maximum torque when rotor resistance is doubled

Solution

a. The approximate per-phase equivalent circuit is shown in Fig. 8.12. The rated terminal voltage per phase is

$$V_1 = 440/\sqrt{3} = 254.0 \text{ V} \quad \text{(line-to-neutral)}$$

Therefore, the slip s_{max} at maximum torque is given by

$$
\begin{aligned}
s_{max} &= \frac{R_2}{\sqrt{R_1^2 + (X_1 + X_2)^2}} \\
&= \frac{0.35}{\sqrt{(0.50)^2 + (1.20 + 1.20)^2}} = 0.143
\end{aligned}
$$

b. The rotor current $I_{2,max}$ at s_{max} is computed as

$$
\begin{aligned}
I_{2,max} &= \frac{V_1}{\sqrt{(R_1 + R_2/s_{max})^2 + (X_1 + X_2)^2}} \\
&= \frac{254.0}{\sqrt{(0.50 + 0.35/0.143)^2 + (1.20 + 1.20)^2}} = 66.8 \text{ A}
\end{aligned}
$$

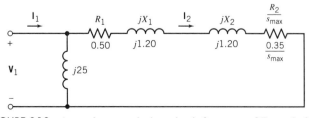

FIGURE 8.12 Approximate equivalent circuit for motor of Example 8.9.

Therefore, the maximum torque $T_{e,max}$ is calculated as

$$T_{e,max} = \frac{3I_{2,max}^2(R_2/s_{max})}{\omega_s}$$

$$= \frac{3(66.8)^2(0.35/0.143)}{188.5} = 174 \text{ N-m}$$

c. When the rotor resistance is doubled, the new value of slip s_{max} at maximum torque is given by

$$s_{max} = \frac{2(0.35)}{\sqrt{(0.50)^2 + (1.20 + 1.20)^2}} = 0.286$$

d. When the rotor resistance is doubled, the new value of the rotor current $I_{2,max}$ is found by using the value of s_{max} computed in part (c). Thus,

$$I_{2,max} = \frac{254.0}{\sqrt{(0.50 + 0.70/0.286)^2 + (1.20 + 1.20)^2}} = 66.8 \text{ A}$$

Therefore, the new value of maximum torque $T_{e,max}$ is found as

$$T_{e,max} = \frac{3I_{2,max}^2(R_2/s_{max})}{\omega_s}$$

$$= \frac{3(66.8)^2(0.70/0.286)}{188.5} = 174 \text{ N-m}$$

DRILL PROBLEMS

D8.13   A three-phase, 20-hp, 440-V, six-pole, 60-Hz, wye-connected induction motor has a starting torque of 98 N-m and a full-load torque of 72 N-m. Calculate (a) the starting torque when the applied voltage is reduced to 300 V and (b) the applied voltage in order to develop a starting torque equal to the full-load torque.

D8.14   For the motor in Problem D8.11, determine (a) the slip at pullout (maximum) torque and (b) the pullout torque.

D8.15   A three-phase, 10-hp, 230-V, four-pole, 60-Hz, wye-connected induction motor develops its full-load torque at a slip of 4.5% when operating at 230 V and 60 Hz. The per-phase equivalent circuit impedances of the motor are

$$R_1 = 0.35\ \Omega \qquad X_m = 15.0\ \Omega$$
$$X_1 = 0.50\ \Omega \qquad X_2 = 0.50\ \Omega$$

The mechanical and core losses are assumed to be negligible. Determine the following:

a. Rotor resistance R_2

b. Slip at maximum torque

c. Maximum torque

d. Rotor speed at maximum torque

e. The starting torque of this motor

8.5 SINGLE-PHASE INDUCTION MOTORS

One of the most common types of residential and commercial loads is the single-phase induction motor. These single-phase motors are rated at less than one horsepower (1 horsepower = 746 watts). Many different designs are available. In the home, these motors are most commonly used as fans and refrigerator compressors.

Compared to a three-phase induction motor, the fractional-horsepower motor is much simpler in construction but much more difficult to analyze. Much of the design of the single-phase motor has been done by building and testing prototype motors until the desired characteristics are obtained. Many different types of single-phase motors have been built in order to match the varying torque requirements of various appliances. Differing cost is another reason for the availability of numerous different types of motors. The chief difference between the various types of single-phase motors lies in the method for starting them. For more details of the types of single-phase induction motors, see, for example, Ref. 5.

In this section, the equivalent circuit of a single-phase induction motor is developed. This equivalent circuit is valid for all types of single-phase motors.

8.5.1 Equivalent Circuit and Performance Analysis

A single-phase motor receives its power from a 60-Hz, single-phase source of electric power. This source causes a current to flow in the stator winding. In Chapter 5, the mmf of such a current was analyzed. It was determined that this mmf can be resolved into two revolving mmfs, rotating in opposite directions. The rotor is rotating at a speed of n_r. The component mmf rotating in the same direction as the rotor is called the forward-revolving field, and the oppositely rotating mmf is called the backward-rotating field. Each rotating mmf induces a voltage in the rotor winding. Therefore, two equivalent circuits are built: one for the forward component mmf and one for the backward-rotating component field. Then, the two component fields are combined, and the two equivalent circuits are interconnected.

The forward-rotating component field rotates at synchronous speed n_s in the same direction as the rotor. Therefore, the slip s^+ of the rotor with respect to the forward-rotating field may be expressed as

$$s^+ = s = \frac{n_s - n_r}{n_s} = 1 - \frac{n_r}{n_s} \qquad (8.42)$$

As shown in Chapter 5, the amplitude of the forward component mmf is one half of the stator mmf. Hence, one half of the stator current may be associated with the forward mmf. The equivalent circuit for this situation is similar to that of three-phase induction motor as shown in Fig. 8.1, with the modification that the core loss represented by R_c is omitted from the equivalent circuit of the single-phase machine. The core loss is treated separately and is usually lumped with the rotational losses. Thus, the equivalent circuit is as shown in Fig. 8.13. In the diagram, V_1^+ is the stator voltage corresponding to the forward component.

Next, the backward component field is considered. The stator current corresponding to this field is one half of the stator current, that is, $\frac{1}{2}I_1$. Because the rotor and the backward fields are rotating in opposite directions, the slip s^- of the rotor with respect to the backward-rotating field is expressed as

$$s^- = \frac{n_s - (-n_r)}{n_s} = 1 + \frac{n_r}{n_s} \qquad (8.43)$$

Substituting $s^+ = s$ from Eq. 8.42 into Eq. 8.43 yields

$$s^- = 2 - s^+ \qquad (8.44)$$

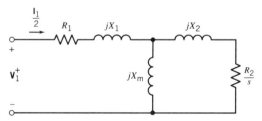

FIGURE 8.13   Equivalent circuit corresponding to forward mmf.

The equivalent circuit corresponding to the backward-revolving field is shown in Fig. 8.14.

The terminal voltage may be expressed in terms of its components:

$$\mathbf{V}_1 = \mathbf{V}_1^+ + \mathbf{V}_1^- \tag{8.45}$$

Since one half of the current $\mathbf{I}$ flowing in an impedance Z has the same performance effect as the current $\mathbf{I}$ flowing in the impedance $\frac{1}{2}Z$, the equivalent circuits shown in Figs. 8.13 and 8.14 may be combined to form the overall equivalent circuit shown in Fig. 8.15. In this equivalent circuit, the resistances $\frac{1}{2}R_1$ and $\frac{1}{2}R_1$ are combined into R_1, and $\frac{1}{2}X_1$ and $\frac{1}{2}X_1$ are likewise combined into X_1. This equivalent circuit can be used to analyze the performance of a single-phase induction motor.

EXAMPLE 8.10

A single-phase, $\frac{1}{4}$-hp, 110-V, four-pole, 60-Hz induction motor has the following parameters:

$$R_1 = 2.0\ \Omega \qquad R_2 = 4.0\ \Omega$$
$$X_1 = 2.5\ \Omega \qquad X_2 = 2.5\ \Omega \qquad X_m = 50.0\ \Omega$$

The core loss is 25 W, and the mechanical (friction and windage) losses are 15 W. The motor operates at rated voltage and rated frequency at a slip of 5%. Determine the following:

a. Motor speed

b. Input current

c. Output power

d. Efficiency of the motor

e. Output torque

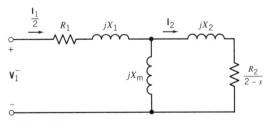

FIGURE 8.14   Equivalent circuit corresponding to backward mmf.

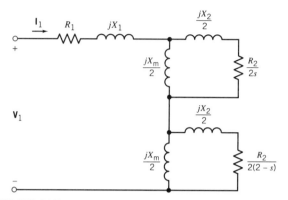

FIGURE 8.15 Single-phase induction motor equivalent circuit.

Solution

a. The motor speed is given by

$$n_r = (1-s)n_s = (1.0 - 0.05)1800 = 1710 \text{ rpm}$$
$$\omega_r = 2\pi(1710)/60 = 179.1 \text{ rad/s}$$

b. The equivalent circuit for the single-phase motor is shown in Fig. 8.16. The applied voltage V_1 is taken as reference phasor; thus,

$$V_1 = 110\underline{/0°}$$

The input impedance is found as follows:

$$Z_{in} = (2.0 + j2.5) + [j25\|(40 + j1.25)]$$
$$+ [j25\|(1.0256 + j1.25)]$$
$$= 13.85 + j21.56 = 25.62\underline{/57.3°}$$

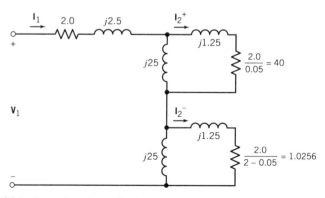

FIGURE 8.16 Equivalent circuit for the single-phase induction motor of Example 8.10.

Therefore, the input current is found as follows:

$$I_1 = \frac{110\,\underline{/0°}}{Z_{in}} = \frac{110\,\underline{/0°}}{25.62\,\underline{/57.3°}} = 4.29\,\underline{/-57.3°}$$

The components of the rotor current are computed as follows:

$$I_2^+ = I_1[j25/(j25 + 40 + j1.25)] = 2.24\,\underline{/-0.6°}$$

$$I_2^- = I_1[j25/(j25 + 1.0256 + j1.25)] = 4.08\,\underline{/-55.1°}$$

c. The input power factor is

$$PF = \cos 57.3° = 0.54 \text{ lagging}$$

Thus, the power input to the motor is given by

$$P_{in} = V_1 I_1 PF = (110)(4.29)(0.54) = 254.8 \text{ W}$$

The losses of the motor include the following:

$$\text{Stator copper loss} = R_1 I_1^2 = (2.0)(4.29)^2 = 36.8 \text{ W}$$
$$\text{Rotor copper loss} = (R_2/2)(I_2^+)^2 + (R_2/2)(I_2^-)^2$$
$$= (2.0)(2.24)^2 + (2.0)(4.08)^2 = 43.3 \text{ W}$$
$$\text{Core loss} = 25.0 \text{ W}$$
$$\text{Rotational losses} = 15.0 \text{ W}$$

The sum of the above losses amounts to

$$\Sigma(\text{losses}) = 36.8 + 43.3 + 25.0 + 15.0 = 120.1 \text{ W}$$

Therefore, the power output of the motor is given by

$$P_{out} = P_{in} - \Sigma(\text{losses}) = 254.8 - 120.1 = 134.7 \text{ W}$$

d. The efficiency of the motor is computed as

$$\eta = P_{out}/P_{in} = (134.7/254.8)100\% = 53\%$$

e. The output torque of the motor is given by

$$T_{out} = P_{out}/\omega_r = 134.7/179.1 = 0.752 \text{ N-m}$$

DRILL PROBLEMS

D8.16   A single-phase, 60-Hz induction motor has six poles and runs at 1020 rpm. Determine (a) the slip with respect to the forward-rotating field and (b) the slip with respect to the backward-rotating field.

D8.17   A single-phase, 110-V, four-pole, 60-Hz induction motor is operating at 1710 rpm. The parameters of the motor equivalent circuit are $R_1 = R_2 = 8\ \Omega$, $X_1 = X_2 = 15\ \Omega$, and $X_m = 120\ \Omega$. Determine the following:

   a. Line current

   b. Power factor

   c. Developed torque

8.5.2   Starting Single-Phase Induction Motors

Single-phase induction motors are classified according to the method used to produce their starting torque. These starting techniques differ in cost and in the amount of starting torque produced. There are three major starting techniques:

   a. Split-phase windings

   b. Capacitor type

   c. Shaded pole

Split-Phase Motors   A split-phase motor has two stator windings, a main stator winding (M) and an auxiliary starting winding (A), with their axes displaced 90 electrical degrees in space. The auxiliary winding is designed to be switched out of the circuit at some set speed by a centrifugal switch. This winding has a higher R/X ratio than the main winding, so its current leads the main winding current. The higher R/X ratio is accomplished by using smaller wire for the auxiliary winding. A schematic diagram of a split-phase winding induction motor is shown in Fig. 8.17.

   The current in the auxiliary winding always peaks before the current in the main winding, and therefore the magnetic field of the auxiliary winding peaks before the magnetic field from the main winding. The direction of rotation of the motor is determined by whether the space angle of the magnetic field from the auxiliary winding is 90° ahead or 90° behind the angle of the main winding. That angle can be changed from 90° ahead to 90° behind just by switching connections on the auxiliary winding while the main winding connection is unchanged; then the direction of rotation of the motor is reversed.

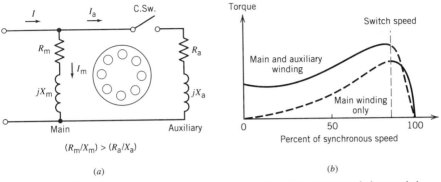

FIGURE 8.17 Split-phase induction motor: (a) connection; (b) torque-speed characteristic.

Split-phase motors have a moderate starting torque with a fairly low starting current. They are used for applications that do not require high starting torques such as fans, blowers, and centrifugal pumps. They are available for sizes in the fractional-hp range and are quite inexpensive.

Capacitor-Start Motors   A capacitor is placed in series with the auxiliary winding of the motor. By proper selection of capacitor size, the mmf of the starting current in the auxiliary winding can be adjusted to be equal to the mmf of the current in the main winding and the phase angle of the current in the auxiliary winding can be made to be 90° leading the current in the main winding. Since the two windings are physically separated, the 90° phase difference in current will yield a single rotating stator magnetic field, and the motor will behave as though it were starting from a three-phase supply. In this case, the starting torque of the motor can be more than 300% of its rated value. A schematic diagram of a capacitor-start induction motor is shown in Fig. 8.18.

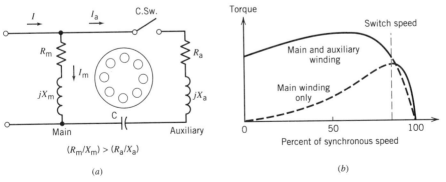

FIGURE 8.18 Capacitor-start induction motor: (a) connection; (b) torque-speed characteristic.

Capacitor-start motors are more expensive than split-phase motors. They are used in applications in which a high starting torque is absolutely required, such as in compressors, air conditioners, and pumps.

If the capacitor is left permanently in the motor circuit, the motor is called a permanent split-capacitor or capacitor-start-and-run motor. It is simpler because the starting switch is not needed. At normal loads, it is more efficient and has a higher power factor and a smoother torque characteristic. It has a lower starting torque because the capacitor is sized to balance the currents in the windings at normal load. A schematic diagram of a capacitor-start-and-run induction motor is shown in Fig. 8.19.

If the largest starting torque and the best running conditions are both needed, two capacitors are used with the auxiliary winding. This motor is called a capacitor-start–capacitor-run, or two-value capacitor, induction motor. The larger capacitor ensures that the currents in the two windings are balanced during starting, yielding very high torques. When the motor gets up to speed, the centrifugal switch opens and only the smaller capacitor, which is just large enough to balance currents at normal loads, remains connected. Thus, the motor operates at high power factor and high torque. The permanent capacitor is 10% to 20% of the starting capacitor. A capacitor-start–capacitor-run induction motor is shown in Fig. 8.20.

The direction of rotation of any capacitor-type single-phase induction motor may be reversed by switching the connections of its auxiliary winding.

Shaded-Pole Motors   A shaded-pole induction motor has only one (main) winding. It has salient poles; one portion of each pole is surrounded by a short-circuited coil called a shading coil. The varying flux in the poles produced by the main winding induces a voltage and a current (in the shading coil) that opposes the original change in flux. This opposition retards the flux changes

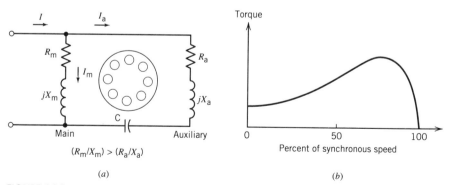

FIGURE 8.19   Capacitor-start-and-run induction motor: (a) connection; (b) torque-speed characteristic.

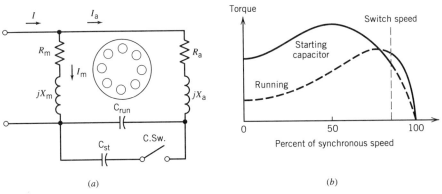

FIGURE 8.20 Capacitor-start–capacitor-run induction motor: (*a*) connection; (*b*) torque-speed characteristic.

under the shaded portion and produces an imbalance between the stator fields. The net torque and net rotation are in the direction from the unshaded to the shaded portion of the pole face.

Shaded-pole motors produce the least starting torque. They are less efficient and have much higher slip. They are also the cheapest design available. The direction of rotation cannot be reversed. A schematic diagram of a shaded-pole induction motor is shown in Fig. 8.21.

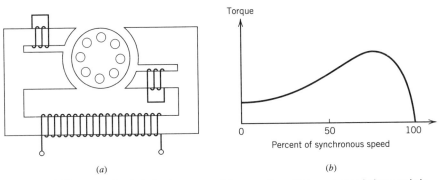

FIGURE 8.21 Shaded-pole induction motor: (*a*) connection; (*b*) torque-speed characteristic.

300 CHAPTER 8 INDUCTION MOTORS

REFERENCES

1. Bergseth, F. R., and S. S. Venkata. *Introduction to Electric Energy Devices.* Prentice Hall, Englewood Cliffs, N.J., 1987.

2. Chapman, Stephen J. *Electric Machinery Fundamentals.* 2nd ed. McGraw-Hill, New York, 1991.

3. Del Toro, Vincent. *Electric Machines and Power Systems.* Prentice-Hall, Englewood Cliffs, N.J., 1985.

4. El-Hawary, Mohamed E. *Principles of Electric Machines with Power Electronics Applications.* Prentice Hall, Englewood Cliffs, N.J., 1986.

5. Fitzgerald, A. E., Charles Kingsley, and Stephen Umans. *Electric Machinery.* 5th ed. McGraw-Hill, New York, 1990.

6. Kosow, Irving L. *Electric Machinery and Transformers.* Prentice Hall, Englewood Cliffs, N.J., 1991.

7. Macpherson, George, and Robert D. Laramore. *An Introduction to Electrical Machines and Transformers.* 2nd ed. Wiley, New York, 1990.

8. Nasar, Syed A. *Electric Energy Conversion and Transmission.* Macmillan, New York, 1985.

9. Ramshaw, Raymond, and R. G. van Heeswijk. *Energy Conversion Electric Motors and Generators.* Saunders College Publishing, Philadelphia, 1990.

10. Sen, P. C. *Principles of Electrical Machines and Power Electronics.* Wiley, New York, 1989.

11. Wildi, Theodore. *Electrical Machines, Drives, and Power Systems.* 2nd ed. Prentice Hall, Englewood Cliffs, N.J., 1991.

PROBLEMS

8.1 A three-phase, 100-hp, 480-V, 60-Hz, wound-rotor induction motor runs with no load on the shaft and is observed to run at 1194 rpm. Determine (a) the number of poles and (b) the frequency of the rotor currents at no load.

8.2 At full load, the motor of Problem 8.1 slows down to 1146 rpm. Find the frequency of the rotor currents.

8.3 A synchronous generator has four poles and is running at 1500 rpm. The generator supplies a six-pole induction motor, which is used to drive a load at a speed of 750 rpm. Determine the frequency of the rotor current of the motor.

8.4 A three-phase, six-pole induction motor is operating at 960 rpm from a 50-Hz, 230-V supply. The voltage induced in the rotor when the motor is blocked is 180 V. Determine

    a. Slip speed

    b. Rotor frequency

    c. Rotor voltage at 960 rpm

8.5 A three-phase, 100-hp induction motor operates at rated load and a speed of 855 rpm when it is connected to a 440-V, 60-Hz source. The slip at this load is 0.05. Determine

  a. Synchronous speed
  b. Number of stator poles
  c. Rotor frequency

8.6 A three-phase, 60-Hz, six-pole, wye-connected induction motor has a full-load speed of 1140 rpm. Calculate the (a) full-load slip and (b) slip rpm.

8.7 A three-phase, 15-hp, 208-V, 60-Hz, wye-connected induction motor runs at 1728 rpm when it delivers rated output power. Determine the following:

  a. Number of poles of the machine
  b. Slip at full load
  c. Frequency of the rotor currents
  d. Speed of the rotor field with respect to the stator
  e. Speed of the rotor field with respect to the stator rotating field

8.8 A three-phase, 480-V, four-pole, 50-Hz induction motor operates at a slip of 3.5%. Determine the following:

  a. Speed of the stator and rotor magnetic fields
  b. Speed of the rotor
  c. Slip speed of the rotor
  d. Rotor frequency

8.9 A three-phase induction motor is supplied with power from a 60-Hz source. At full load the motor speed is 1728 rpm, and at no load the speed is nearly 1800 rpm. At full-load conditions, determine the following:

  a. Number of poles
  b. Slip
  c. Frequency of the rotor voltages
  d. Speed of the rotor field with respect to the rotor
  e. Speed of the rotor field with respect to the stator
  f. Speed of the rotor field with respect to the stator field

8.10 A three-phase induction motor is connected to a three-phase 60-Hz source. The motor runs at almost 1200 rpm at no load and at 1140 rpm at full load. Determine the following:

  a. Number of poles
  b. Slip at full load
  c. Frequency of the rotor voltages

d. Speed of the rotor field with respect to the rotor

e. Speed of the rotor field with respect to the stator

8.11 A three-phase, 220-V, four-pole, 60-Hz induction motor has the following parameters referred to the stator.

$$R_1 = 0.3 \ \Omega \qquad R_2 = 0.2 \ \Omega$$
$$X_1 = 0.5 \ \Omega \qquad X_2 = 0.5 \ \Omega \qquad X_m = 15 \ \Omega$$

The total rotational and core loss is 500 W. For a slip of 5%, calculate:

a. The motor speed

b. The input current

c. The input power factor

d. The shaft torque

e. The efficiency of the motor

8.12 A three-phase, 440-V, four-pole, 60-Hz, wye-connected induction motor takes a stator current of 50 A at 0.8 power factor while operating at a slip of 5%. The stator copper loss is 2500 W, and the total core and rotational losses are 3200 W. Calculate the efficiency of the motor.

8.13 A three-phase, 440-V, four-pole, 60-Hz induction motor produces 100 hp at the shaft at 1728 rpm. Determine the efficiency of the motor if rotational losses are 3200 W and stator copper losses are 2700 W.

8.14 A three-phase, 440-V, six-pole, 60-Hz, wye-connected induction motor takes 30 kVA at 0.8 power factor, and it runs at a slip of 3.5%. The stator copper losses are 500 W, and the rotational losses are 350 W. Compute the following:

a. Rotor copper losses

b. Shaft output torque and hp

c. Efficiency of the motor

8.15 A three-phase, 60-Hz, six-pole, wye-connected induction motor is rated at 20 hp and 440 V. The motor operates at rated conditions and a slip of 5%. The friction and windage losses are 250 W, and the core losses are 225 W. Find the following:

a. Shaft speed

b. Output power

c. Load torque

d. Induced torque

8.16 A three-phase, 440-V, 60-Hz, six-pole induction motor supplies an output power of 36 kW when operating at 1158 rpm. The combined rotational loss is 1.5 kW, the stator copper loss is 1.2 kW, and the input power factor is 0.866. Determine the following:

a. Copper losses of the rotor circuit

b. Stator current

c. Total input power

8.17 The following test results are obtained for a three-phase, 100-hp, 440-V, eight-pole, wye-connected induction machine.

	No Load (at 60 Hz)	Blocked Rotor (at 15 Hz)	DC
Voltage (V)	440	80	8
Current (A)	38	110	50
Power (W)	3800	4800	

Determine the parameters of the equivalent circuit.

8.18 A three-phase, 20-hp, 440-V, 60-Hz, wye-connected induction motor, operating at rated conditions, draws a line current of 18 A. Data from blocked-rotor and no-load tests at rated frequency and from a DC test are as follows:

	No Load	Blocked Rotor	DC
Voltage (V)	440	62	25
Current (A)	9	24	34
Power (W)	3750	1800	

Determine R_1, R_2, X_1, X_2, X_m, and $P_{\text{rotational}}$.

8.19 A three-phase induction motor is subjected to no-load and blocked-rotor tests, and the following data are obtained.

	No-Load Test (at 60 Hz)	Blocked-Rotor Test (at 15 Hz)
Voltage (V)	440	150
Current (A)	10	40
Power (W)	4500	6100

The stator resistance between any two leads is 1.2 Ω. Determine the parameters of the equivalent circuit of the motor.

8.20 A three-phase, 15-hp, 440-V, 60-Hz, wye-connected induction motor draws a line current of 16 A when operating at rated load conditions. No-load and blocked-rotor tests at rated frequency and a DC test provide the following data:

	No Load	Blocked Rotor	DC
Voltage (V)	440	35	6
Current (A)	6	16	12
Power (W)	1200	750	

Determine the parameters of the equivalent circuit of the motor.

8.21 A three-phase, 100-hp, 2400-V, 6-pole, 60-Hz induction motor is tested, and the following data are obtained:

	No-Load Test (at 60 Hz)	Blocked-Rotor Test (at 15 Hz)
Voltage (V)	2400	420
Current (A)	6	35
Power (W)	4200	13,500

The stator resistance is measured at 1.75 ohms per phase.

a. Find the parameters of the equivalent circuit.

b. At a slip of 0.25, calculate the torque and power output and the efficiency of this motor.

8.22 A three-phase, 25-hp, 230-V, 60-Hz, four-pole, induction motor operates at rated load. The motor has a rotor copper loss of 350 W and friction and windage loss of 275 W. Determine

a. Mechanical power developed

b. Air-gap power

c. Shaft speed

d. Shaft torque

8.23 A three-phase, 60-Hz, six-pole, wye-connected induction motor is rated at 100 hp and 440 V. The parameters of the equivalent circuit are

$$R_1 = 0.12 \ \Omega \qquad R_2 = 0.10 \ \Omega$$
$$X_1 = 0.25 \ \Omega \qquad X_2 = 0.20 \ \Omega \qquad X_m = 15 \ \Omega$$

The friction and windage and core losses are 1.4 kW and 1.2 kW, respectively. For a slip of 3.5%, determine the following:

a. Line current and power factor

b. Air-gap power

c. Power converted from electrical to mechanical form

8.24 A three-phase, 25-hp, 440-V, 60-Hz, six-pole induction motor is operating at a slip of 4%. The core loss and the friction and windage losses at this load are 250 W and 125 W, respectively. The motor is wye connected, and the motor parameters in Ω/phase are

$$R_1 = 0.35 \qquad R_2 = 0.40$$
$$X_1 = 1.25 \qquad X_2 = 1.50 \qquad X_m = 25$$

Determine the following:

a. Line current and power factor

b. Real and reactive power input

c. Air-gap power

d. Mechanical power and torque developed

e. Shaft horsepower and torque

f. Efficiency of the motor

8.25 A six-pole, 60-Hz induction motor runs at 1020 rpm. The input to the rotor circuit is 4 kW. Calculate the rotor copper loss.

8.26 An induction motor delivers 50 kW to a load connected to its shaft. At this load condition, the efficiency of the motor is 88% and stator copper loss = rotor copper loss = core losses = friction and windage losses. Determine the slip.

8.27 A three-phase, 15-hp, 208-V, 60-Hz, six-pole, wye-connected induction motor has the following parameters per phase:

$$R_1 = 0.15 \ \Omega \qquad R_2 = 0.10 \ \Omega$$
$$X_1 = 0.50 \ \Omega \qquad X_2 = 0.50 \ \Omega \qquad X_m = 20 \ \Omega$$

The friction and windage losses and the hysteresis and eddy-current losses are 350 W and 200 W, respectively. For a slip of 3.5%, find the following:

a. Line current and power factor

b. Horsepower output

c. Starting torque

8.28 A three-phase, four-pole, 60-Hz, three-phase induction machine is rated at 10 hp, 208 V, and 1755 rpm. The parameters of the equivalent circuit of the motor are as follows:

$$R_1 = 0.15 \ \Omega \qquad R_2 = 0.15 \ \Omega$$
$$X_1 = 0.40 \ \Omega \qquad X_2 = 0.25 \ \Omega \qquad X_m = 30 \ \Omega$$

The combined rotational (friction and windage plus hysteresis and eddy-current) losses amount to 500 W. The motor operates at rated speed when connected to a 208-V and 60-Hz source. Calculate

a. Line current and power factor

b. Output torque

c. Efficiency of the motor

d. Starting current and torque

8.29 A three-phase, six-pole, 60-Hz, wye-connected induction motor delivers 20 kW at the shaft at a slip of 4.5%. The motor has total rotational losses of 1500 W. Calculate the following:

a. Rotor input

b. Output torque

c. Developed torque

8.30 A three-phase, 440-V, four-pole, 60-Hz induction motor takes 120 A of stator current at starting, and the motor draws 20 A while running at full load. The starting torque is 1.8 times the torque at full load at rated voltage. It is desired that the starting torque be equal to the full-load torque. Determine (a) the applied voltage and (b) the corresponding line current.

8.31 A three-phase, 440-V, wye-connected induction motor has a stator impedance of $1.0 + j1.6\ \Omega$ per phase. The rotor impedance referred to the stator is $0.8 + j1.4\ \Omega$ per phase. Determine the maximum electromagnetic power developed by the motor.

8.32 For the motor in Problem 8.23, determine the (a) slip at maximum (pullout) torque and (b) pullout torque.

8.33 A three-phase, 10-hp, 230-V, four-pole, 60-Hz, wye-connected, induction motor operates at rated voltage and rated frequency, and it develops full-load torque at a slip of 3.5%. The mechanical and core losses can be neglected. The impedance parameters of the motor in Ω/phase are as follows:

$$R_1 = 0.25 \qquad X_1 = 0.35 \qquad X_2 = 0.45 \qquad X_m = 40$$

Determine the (a) maximum torque and the slip at maximum torque and (b) starting torque.

8.34 A three-phase, 20-hp, 480-V, 60-Hz, six-pole, wound-rotor induction motor operates at rated conditions, and the motor runs at 1164 rpm with the rotor rheostat shorted. The motor parameters in Ω/phase are

$$R_1 = 0.30 \qquad R_2 = 0.40$$
$$X_1 = 1.50 \qquad X_2 = 2.00 \qquad X_m = 300$$

Determine the following:

a. Slip at which maximum torque occurs
b. Maximum torque
c. Resistance to be inserted in the rotor circuit to operate the motor at rated torque and a speed of 1074 rpm

8.35 A three-phase, 25-hp, 440-V, 60-Hz, 1750 rpm, wound-rotor induction motor has the following equivalent circuit parameters:

$$R_1 = 0.20\ \Omega \qquad X_1 = 1.0\ \Omega$$
$$R_2 = 0.15\ \Omega \qquad X_2 = 0.8\ \Omega \qquad X_m = 30\ \Omega$$

The motor is connected to a three-phase, 440-V, 60-Hz supply.

a. Calculate the starting torque.
b. Determine the resistance of the rheostat in the rotor circuit such that the maximum torque occurs at starting.

8.36 A single-phase, $\frac{1}{4}$-hp, 110-V, 1725-rpm, 60-Hz, four-pole, capacitor-start induction motor has the following equivalent circuit parameters for the main winding.

$$R_1 = 2.5 \; \Omega \qquad R_2 = 4.0 \; \Omega$$
$$X_1 = 3.5 \; \Omega \qquad X_2 = 3.5 \; \Omega \qquad X_m = 50 \; \Omega$$

The core loss at 110 V is 25 W and the friction and windage loss is 20 W. The motor is connected to a 110-V, 60-Hz supply, and it runs at a slip of 5%. Determine the following:

   a. Input current and power factor

   b. Input power

   c. Developed torque

   d. Efficiency of the motor

8.37 A single-phase, $\frac{1}{2}$-hp, 110-V, four-pole, 60-Hz induction motor has the following constants in ohms:

$$R_1 = 1.8 \qquad X_1 = 2.5$$
$$R_2 = 3.5 \qquad X_2 = 2.5 \qquad X_m = 60$$

The motor core loss is 35 W, and the friction and windage loss is 15 W. The motor operates at a slip of 0.05. Determine the (a) mechanical power output and (b) starting torque.

Nine

Transmission Lines

9.1 INTRODUCTION

An electric power system is a collection of equipment and devices whose common objectives are the production, conversion, transmission, distribution, and consumption of electric energy. The major components of an electric power system are generators, which convert mechanical energy into electricity; transformers, which change the voltage or current levels of an electric supply; power transmission lines, which are used to transfer power from one location to another; and a variety of auxiliary control and regulating equipment intended to vary the system characteristics.

Electric energy is produced in large quantities at various electric power plants by converting different forms of energy—fossil fuels, nuclear energy, water power, and so forth. Electric energy is transformed by the use of transformers to different voltage levels most suitable for transmission, distribution, and consumption.

Electric power is transmitted using overhead or cable lines to consumers at varied distances from its source. Electric energy is utilized by various conversion devices and mechanisms, such as electric motors, ovens, and electric lighting.

The need for power transmission lines arises from the fact that bulk electric power generation is done at large electric power plants remote from consumers. However, consumers require rather small amounts of energy, and they are scattered over wide areas. Thus, the transmission of electric energy over a distance offers a number of advantages, such as:

1. Use of remote energy sources
2. Reduction of the total power reserve of generators

308

3. Utilization of the time difference between various time zones when the peak demands are not coincident

4. Improved reliability of electric power supply

Power transmission lines are subdivided into overhead and cable lines. Overhead lines are made up of metal conductors suspended on insulators from a tower or post by suitable clamps. The tower or post may be made of metal, wood, or reinforced concrete depending on the purpose of the line, the operating voltage, economic considerations, and so on. Most modern overhead lines are constructed using *aluminum cable steel reinforced (ACSR)* conductors consisting of a central steel core on which aluminum wires are wound. The steel core increases the mechanical strength of the line, and the aluminum wires have good electrical properties. Multiwire conductors are preferred over single-wire conductors of equivalent cross-sectional area because of their improved flexibility and reduced skin effect.

The conductors of a transmission line may be configured in various ways. A few examples are shown in Fig. 9.1. The towers or poles carrying six transmission wires are called double-circuit lines. As described in Chapter 2, steel ground wires are sometimes placed atop the towers or poles for protection against direct lightning strikes.

In cable lines, the conductors are insulated from one another and are enclosed in protective sheaths. The practice is to lay the underground cables directly in the soil, or in a bed of sand, or within special cable ducts. At present, AC overhead transmission lines are favored because of their lower first costs and ease and simplicity of repair and maintenance. However, there is a tendency

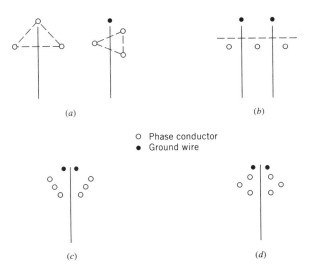

FIGURE 9.1 Transmission line configurations: (*a*) delta type; (*b*) horizontal; (*c*) reverse herring-bone; (*d*) barrel type.

toward more extensive utilization of cable lines, especially in new constructions in urban areas.

The maximum power that a transmission line can transmit, within some constraints, is called its carrying capacity. For overhead AC power transmission lines, the maximum power that can be transmitted may be approximated as being directly proportional to the square of the voltage and inversely proportional to the length of the transmission line. Similarly, the construction costs are also approximately proportional to voltage. Hence, to increase the power levels transmitted over long distances, the usual practice of electric utilities is to raise the transmission voltage.

In this chapter, the basic parameters of the conductors that are used in transmission lines are discussed. The various formulas for these parameters are presented, and the proper use of tables of conductor characteristics for obtaining values of the parameters is described. Models of transmission lines are then presented for use in power system studies.

9.2   RESISTANCE

The DC resistance of a conductor at a specified temperature T is given by

$$R_{DC,T} = \frac{\rho l}{A}$$

(9.1)

where

  l = length of the conductor

  A = cross-sectional area of the conductor

  ρ = resistivity of the conductor

Two sets of units are generally used: the English system and the Système International (SI).

In the English system, the length l is given in feet, the cross-sectional area A is expressed in *circular mils (CM)*, and the resistivity ρ is specified in ohm–circular mil per foot (Ω-CM/ft). Note that one inch is equivalent to 1000 mils, and the cross-sectional area measured in circular mils is obtained by squaring the diameter expressed in mils. In other words, one circular mil is the area of a circle having a diameter of 1 mil.

In the SI system of units, the length is measured in meters, the area is given in square meters, and resistivity is expressed in ohm-meter (Ω-m).

Resistivity depends on the conductor material. Annealed copper is the international standard for measuring resistivity. At a temperature of 20°C, the resistivity of annealed copper is given as 10.37 Ω-CM/ft or 1.72×10^{-8} Ω-m.

When the DC resistance of a conductor at a certain temperature is known, the DC resistance at any other temperature may be found as follows:

$$R_{T2} = \left(\frac{M + T_2}{M + T_1}\right) R_{T1} \qquad (9.2)$$

where

R_{T1} = DC resistance at temperature T_1 in °C

R_{T2} = DC resistance at temperature T_2 in °C

M = temperature constant in °C

The constant M is the temperature at which the extrapolated resistance of a particular conductor becomes zero. The temperature constants of certain materials, as well as their resistivities and percent conductivities, are given in Table 9.1.

The resistance of nonmagnetic conductors varies not only with temperature but also with frequency because of the *skin effect*. Electric current distribution in the conductor is not uniform. As frequency increases, current tends to flow nearer the outer surface of the conductor. This decreases the effective cross-sectional area of the conductor, thus increasing the resistance.

The resistances at 25, 50, and 60 Hz, as well as the DC resistance, are given in the tables of electrical characteristics of conductors in Appendix A. For other frequencies, the AC resistance is found as follows:

$$R_{AC} = K R_{DC} \ \Omega/\text{mi} \qquad (9.3)$$

where R_{DC} is the DC resistance in Ω/mi, and K is found from Table 5 in Appendix A as a function of the variable X, which is defined as follows:

$$X = 0.0636 \sqrt{\frac{\mu_r f}{R_{DC}}} \qquad (9.4)$$

where the relative permeability μ_r is 1.0 for nonmagnetic materials.

Table 9.1 Resistivity and Temperature Constants of Different Materials

Material	Conductivity (%)	Resistivity Ω-m	Resistivity Ω-CM/ft	Temperature Constant (°C)
Annealed copper	100.0	1.72×10^{-8}	10.37	234.5
Hard-drawn copper	97.3	1.77×10^{-8}	10.66	241.5
Aluminum	61.0	2.83×10^{-8}	17.00	228.1
Iron	17.2	10.00×10^{-8}	60.00	180.0
Silver	108.0	1.59×10^{-8}	9.60	243.0

EXAMPLE 9.1

Determine the 60-Hz AC resistance in Ω/mi of a 1-inch-diameter, 97.3% conductivity, hard-drawn copper conductor at 75°C.

Solution The resistivity of 100% conductivity copper at 20°C is

$$\rho_{100} = 10.37 \ \Omega\text{-CM/ft}$$

Thus, the resistivity of 97.3% conductivity copper at 20°C is

$$\rho_{97.3} = 10.37/0.973 = 10.66 \ \Omega\text{-CM/ft}$$

Hence, the resistance at 20°C is

$$R_{DC,20} = \frac{\rho l}{A} = \frac{(10.66)(5280)}{(1000)^2} = 0.0563 \ \Omega/\text{mi}$$

Therefore, the DC resistance at 75°C is given by

$$R_{DC,75} = \left(\frac{M + T_{75}}{M + T_{20}}\right) R_{DC,20} = \left(\frac{241.5 + 75}{241.5 + 20}\right) 0.0563 = 0.0681 \ \Omega/\text{mi}$$

The variable X is computed as follows:

$$X = 0.0636 \sqrt{\frac{\mu_r f}{R_{DC,75}}}$$

$$= 0.0636 \sqrt{\frac{(1.0)(60)}{0.0681}} = 1.8878 \cong 1.9$$

From Table 5 in Appendix A, for $X = 1.9$, read off the value of K as 1.0644. Therefore, the AC resistance is

$$R_{AC} = K R_{DC}$$
$$= (1.0644)(0.0681) = 0.0725 \ \Omega/\text{mi}$$

DRILL PROBLEMS

D9.1 One thousand circular mils is designated by the abbreviation MCM. Derive an equation to convert from area expressed in MCM to area expressed in m^2.

D9.2   Determine the cross-sectional area in m^2 of a 795-MCM conductor. Find the AC resistance in Ω/km of a 795-MCM ACSR conductor at 50°C.

9.3   INDUCTANCE AND INDUCTIVE REACTANCE

The series inductance of a transmission line consists of two components: its internal inductance, which is due to magnetic flux inside the conductor, and its external inductance, which is due to magnetic flux outside the conductor. The internal inductance is considered first.

9.3.1   Internal Inductance

Consider the conductor of radius r carrying the current I shown in Fig. 9.2. The magnetic field intensity at a distance x from the center of the conductor is denoted by H_x.

The mmf ($H_x l_{\text{cir}}$) around the closed circular path indicated by broken lines in Fig. 9.2 is equal to the current I_x enclosed by the path. The length of the path is $l_{\text{circle}} = 2\pi x$; thus,

$$H_x 2\pi x = I_x$$

$$= \left(\frac{\pi x^2}{\pi r^2}\right) I \qquad (9.5)$$

Therefore,

$$H_x = \frac{x}{2\pi r^2} I \qquad (9.6)$$

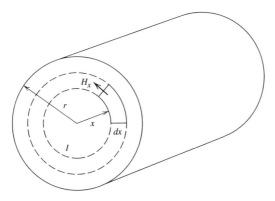

FIGURE 9.2   A current-carrying conductor.

The flux density B_x is given by

$$B_x = \mu H_x = \frac{\mu x}{2\pi r^2} I \qquad (9.7)$$

The differential magnetic flux is equal to the product of the magnetic flux density and the cross-sectional area $dA = (dx)$(axial length). Thus, the flux per unit length is

$$d\phi = \frac{\mu x}{2\pi r^2} I \, dx \qquad (9.8)$$

The flux linkages $d\lambda$ are equal to the flux times the fraction of current linked:

$$d\lambda = \left(\frac{\pi x^2}{\pi r^2}\right) d\phi = \frac{\mu x^3}{2\pi r^4} I \, dx \qquad (9.9)$$

Integrating Eq. 9.9 over the full radius of the conductor yields the total flux linkages inside the conductor.

$$\lambda_{\text{internal}} = \int_0^r \frac{\mu I}{2\pi r^4} x^3 \, dx = \frac{\mu I}{2\pi r^4}\left(\frac{x^4}{4}\bigg|_0^r\right) = \frac{\mu I}{8\pi} \qquad (9.10)$$

For a relative permeability $\mu_r = \mu/\mu_0 = 1.0$, the permeability μ is equal to the permeability of free space $\mu_0 = 4\pi \times 10^{-7}$. Thus,

$$\lambda_{\text{internal}} = \tfrac{1}{2} I \times 10^{-7} \qquad \text{Wb-t/m} \qquad (9.11)$$

Therefore, since $L = \lambda/I$, the internal inductance is found as

$$L = \frac{\mu_0 \mu_r}{8\pi}$$

$$L_{\text{internal}} = \tfrac{1}{2} \times 10^{-7} \qquad \text{H/m} \qquad (9.12)$$

9.3.2 External Inductance

Consider the external region surrounding the conductor carrying current I shown in Fig. 9.3. At the tubular element located at a distance x from the conductor center, the magnetic field intensity H_x is related to the current I as follows:

$$H_x 2\pi x = I \qquad (9.13)$$

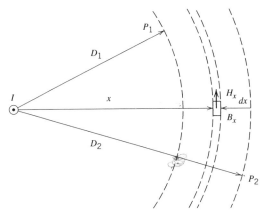

FIGURE 9.3 Surrounding region of a current-carrying conductor.

Solving for the magnetic field intensity from Eq. 9.13 yields

$$H_x = \frac{I}{2\pi x} \qquad (9.14)$$

The magnetic flux density B_x in the element is expressed as

$$B_x = \mu H_x = \frac{\mu I}{2\pi x} \qquad (9.15)$$

The magnetic flux per unit length of the tubular element of thickness dx is given by

$$d\phi = \frac{\mu I}{2\pi x} dx \qquad (9.16)$$

Since the flux links the full current carried by the conductor,

$$d\lambda = d\phi = \frac{\mu I}{2\pi x} dx \qquad (9.17)$$

Therefore, the total flux linkages between P_1 and P_2 are obtained by integrating Eq. 9.17 from $x = D_1$ to $x = D_2$.

$$\lambda_{12} = \int_{D_1}^{D_2} \frac{\mu I}{2\pi x} dx = \frac{\mu I}{2\pi} (\ln x |_{D_1}^{D_2} = \frac{\mu I}{2\pi} \ln \left(\frac{D_2}{D_1}\right) \qquad (9.18)$$

For a relative permeability of 1, the permeability μ is equal to the permeability of free space $\mu_0 = 4\pi \times 10^{-7}$. Thus,

$$\lambda_{12} = 2 \times 10^{-7} I \, \ln\left(\frac{D_2}{D_1}\right) \qquad \text{Wb-t/m} \qquad (9.19)$$

Therefore, the inductance due only to the flux between P_1 and P_2 is

$$L_{12} = \frac{\lambda_{12}}{I} = 2 \times 10^{-7} \, \ln\left(\frac{D_2}{D_1}\right) \qquad \text{H/m} \qquad (9.20)$$

9.3.3 Inductance of a Single-Phase Line

Consider the two-wire transmission line with a separation distance D between conductors shown in Fig. 9.4. A current I flows toward the plane of the paper in conductor 1 and returns through conductor 2.

A line of flux produced by current in conductor 1 at a distance equal to or greater than $(D + r_2)$ from the center of conductor 1 does not link the circuit and cannot induce a voltage in the circuit. Stated differently, such a line of flux links a net current of zero.

Therefore, the inductance of the two-wire circuit due to current flowing in conductor 1 is given by

$$L_1 = L_{1,\text{internal}} + L_{1,\text{external}} \qquad (9.21)$$

where

$$L_{1,\text{internal}} = \tfrac{1}{2} \times 10^{-7} \qquad \text{H/m} \qquad (9.22)$$

$$L_{1,\text{external}} = 2 \times 10^{-7} \, \ln\left(\frac{D}{r_1}\right) \qquad \text{H/m} \qquad (9.23)$$

Substituting Eqs. 9.22 and 9.23 into Eq. 9.21 and simplifying yields

$$L_1 = 2 \times 10^{-7} \, \ln\left(\frac{D}{r_1 e^{-1/4}}\right) \qquad \text{H/m} \qquad (9.24)$$

or

$$L_1 = 2 \times 10^{-7} \, \ln\left(\frac{D}{r_1'}\right) \qquad \text{H/m} \qquad (9.25)$$

where $r_1' = r_1 e^{-1/4} = 0.7788 r_1$ is the *equivalent radius* of conductor 1.

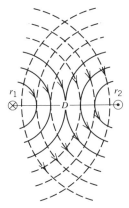

FIGURE 9.4   A two-wire transmission line.

Similarly, the inductance of the circuit due to current in conductor 2 is

$$L_2 = 2 \times 10^{-7} \ln\left(\frac{D}{r_2'}\right) \quad \text{H/m} \tag{9.26}$$

where $r_2' = r_2 e^{-1/4} = 0.7788 r_2$ is the equivalent radius of conductor 2. Thus, the total inductance for the complete circuit is given by

$$L = L_1 + L_2 = 2 \times 10^{-7} \left[\ln\left(\frac{D}{r_1'}\right) + \ln\left(\frac{D}{r_2'}\right)\right]$$

$$= 4 \times 10^{-7} \ln\left(\frac{D}{\sqrt{r_1' r_2'}}\right) \quad \text{H/m} \tag{9.27}$$

If $r_1' = r_2' = r'$, the total inductance reduces to

$$L = 4 \times 10^{-7} \ln\left(\frac{D}{r'}\right) \quad \text{H/m} \tag{9.28}$$

9.3.4   Inductance of a Three-Phase Circuit

Consider a three-phase transmission line whose conductors are not equilaterally spaced. This condition would result in different flux linkages and inductances in each phase. Since the unbalances are small, they may be neglected. Thus, the transmission line is assumed to be transposed, and each conductor occupies the original positions of the other conductors over equal distances. Figure 9.5 shows one transposition cycle of a completely transposed three-phase circuit.

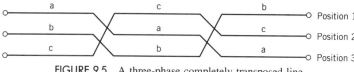

FIGURE 9.5   A three-phase completely transposed line.

For a completely transposed transmission line, the average inductance of each conductor over a complete cycle is the same. To find this average, a point F that is remote from the line is chosen as reference as shown in Fig. 9.6.

The flux linking phase a between conductor a and point F due only to the current in phase a is given by

$$\lambda_{aF,a} = 2 \times 10^{-7} I_a \ln\left(\frac{D_{aF}}{r'_a}\right) \qquad (9.29)$$

The flux linking phase a between conductor a and point F due only to the current in phase b is

$$\lambda_{aF,b} = 2 \times 10^{-7} I_b \ln\left(\frac{D_{bF}}{D_{12}}\right) \qquad (9.30)$$

The flux linking phase a between conductor a and point F due only to the current in phase c is

$$\lambda_{aF,c} = 2 \times 10^{-7} I_c \ln\left(\frac{D_{cF}}{D_{31}}\right) \qquad (9.31)$$

The total flux linking phase a between conductor a out to point F is the sum of the flux linkages due to the three phase currents. Thus, by adding Eqs. 9.29, 9.30, and 9.31 and grouping similar terms, Eq. 9.32 is obtained.

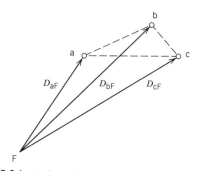

FIGURE 9.6   A three-phase unsymmetrically spaced line.

$$\lambda_{aF} = 2 \times 10^{-7}\left[I_a \ln\left(\frac{1}{r'_a}\right) + I_b \ln\left(\frac{1}{D_{12}}\right) + I_c \ln\left(\frac{1}{D_{31}}\right)\right]$$
$$+ 2 \times 10^{-7}[I_a \ln D_{aF} + I_b \ln D_{bF} + I_c \ln D_{cF}] \quad (9.32)$$

If the point F is moved farther and farther away from the line, in the limit as F goes to infinity the three distances D_{aF}, D_{bF}, and D_{cF} become equal to each other:

$$D_{aF} = D_{bF} = D_{cF} = D \quad (9.33)$$

Thus, Eq. 9.32 can be written as

$$\lambda_{aF} = 2 \times 10^{-7}\left[I_a \ln\left(\frac{1}{r'_a}\right) + I_b \ln\left(\frac{1}{D_{12}}\right) + I_c \ln\left(\frac{1}{D_{31}}\right)\right]$$
$$+ 2 \times 10^{-7}(I_a + I_b + I_c) \ln D \quad (9.34)$$

If I_a, I_b, and I_c form a balanced set of three-phase currents, then their sum is equal to zero:

$$I_a + I_b + I_c = 0 \quad (9.35)$$

Therefore, the total flux linkage for phase a in the first transposition position is

$$\lambda_{a1} = 2 \times 10^{-7}\left[I_a \ln\left(\frac{1}{r'_a}\right) + I_b \ln\left(\frac{1}{D_{12}}\right) + I_c \ln\left(\frac{1}{D_{31}}\right)\right] \quad (9.36)$$

Similarly, the total flux linkage for phase a in the second transposition position is given by

$$\lambda_{a2} = 2 \times 10^{-7}\left[I_a \ln\left(\frac{1}{r'_a}\right) + I_b \ln\left(\frac{1}{D_{23}}\right) + I_c \ln\left(\frac{1}{D_{12}}\right)\right] \quad (9.37)$$

In the same manner, the total flux linkage for phase a in the third transposition position is given by

$$\lambda_{a3} = 2 \times 10^{-7}\left[I_a \ln\left(\frac{1}{r'_a}\right) + I_b \ln\left(\frac{1}{D_{31}}\right) + I_c \ln\left(\frac{1}{D_{23}}\right)\right] \quad (9.38)$$

To find the flux linkage for phase a over one complete transposition cycle, Eqs. 9.36, 9.37, and 9.38 are added together and divided by three; upon simplification, this yields

$$\lambda_a = \frac{\lambda_{a1} + \lambda_{a2} + \lambda_{a3}}{3}$$

$$= \frac{2 \times 10^{-7}}{3} \left[3I_a \ln\left(\frac{1}{r'_a}\right) + (I_b + I_c) \ln\left(\frac{1}{D_{12}D_{23}D_{31}}\right) \right] \qquad (9.39)$$

According to Eq. 9.35, the sum of the phase currents is equal to zero, or $(I_b + I_c) = -I_a$; hence,

$$\lambda_a = 2 \times 10^{-7} I_a \left[\ln\left(\frac{1}{r'_a}\right) - \ln\left(\frac{1}{D_{12}D_{23}D_{31}}\right)^{1/3} \right] \qquad (9.40)$$

Therefore, the inductance for phase a of the completely transposed three-phase line is given by

$$L_a = 2 \times 10^{-7} \ln \frac{\sqrt[3]{D_{12}D_{23}D_{31}}}{r'_a} \qquad \mathrm{H/m} \qquad (9.41)$$

Calculations for phases b and c yield similar results. Thus,

$$L_b = 2 \times 10^{-7} \ln \frac{\sqrt[3]{D_{12}D_{23}D_{31}}}{r'_b} \qquad \mathrm{H/m} \qquad (9.42)$$

$$L_c = 2 \times 10^{-7} \ln \frac{\sqrt[3]{D_{12}D_{23}D_{31}}}{r'_c} \qquad \mathrm{H/m} \qquad (9.43)$$

If all three phase conductors have the same equivalent radius $r' = r e^{-1/4}$, then the inductance per phase is given by

$$L_a = L_b = L_c = 2 \times 10^{-7} \ln \frac{\sqrt[3]{D_{12}D_{23}D_{31}}}{r'}$$

$$= 2 \times 10^{-7} \ln \frac{\mathrm{GMD}}{r'} \qquad \mathrm{H/m} \qquad (9.44)$$

where $\mathrm{GMD} = (D_{12}D_{23}D_{31})^{1/3}$ is the *geometric mean distance*.

For a three-phase, equilaterally spaced transmission line, the phase conductors have equal separation distances between them; that is, $D_{12} = D_{23} = D_{31} = D$. Therefore, the inductance per phase is given by

$$L_a = 2 \times 10^{-7} \ln\left(\frac{D}{r'}\right) \qquad \mathrm{H/m} \qquad (9.45)$$

9.3.5   Stranded Conductors and Bundled Conductors

Stranded conductors are composed of two or more elements or strands of wire that are electrically in parallel. It is assumed that all the strands are identical and that they share the total current equally. For an n-strand conductor, the equivalent radius of the stranded conductor is referred to as its *geometric mean radius (GMR)*, and it is derived from the equivalent radius of a strand and the distances from the center of a strand to the centers of the rest of the strands. Thus,

$$\text{GMR} = \sqrt[N]{(r'D_{12}D_{13}\cdots D_{1n})(D_{21}r'D_{23}\cdots D_{2n})\cdots(D_{n1}D_{n2}\cdots r')} \quad (9.46)$$

where
$$N = n^2$$
$$r' = re^{-1/4} = 0.7788r = \text{equivalent radius of each strand}$$
$$D_{km} = \text{distance from the center of strand } k \text{ to the center of strand } m$$

Therefore, in terms of the GMR of the phase conductors, the inductance per phase given by Eq. 9.44 reduces to

$$L_a = L_b = L_c = 2 \times 10^{-7} \ln \frac{\text{GMD}}{\text{GMR}} \quad \text{H/m} \quad (9.47)$$

Bundled conductors consist of two or more conductors that belong to the same phase and are close together in comparison with the separation distances between the phases. The bundle consists of two, three, or four conductors. The three-conductor bundle usually has the conductors at the vertices of an equilateral triangle, and the four-conductor bundle usually has its conductors at the corners of a square.

Bundling of the conductors of a transmission line has several advantages. The increased number of conductors in each phase reduces the effects of corona for the very high transmission voltages. Bundling also results in reduced resistance and reduced reactance. The GMR for a bundled conductor is calculated in the same way as that for a stranded conductor. For this case, each conductor of the bundle is treated as a strand in the multistrand conductor.

The equivalent GMR_b of the bundled conductor is related to the conductor GMR_c as follows:

For a two-conductor bundle

$$\text{GMR}_b = \sqrt[4]{d^2\,\text{GMR}_c^2} = \sqrt{d\,\text{GMR}_c} \quad (9.48)$$

For a three-conductor bundle

$$\text{GMR}_b = \sqrt[9]{d^3 d^3\,\text{GMR}_c^3} = \sqrt[3]{d^2\,\text{GMR}_c} \quad (9.49)$$

For a four-conductor bundle

$$\text{GMR}_b = \sqrt[16]{d^4 d^4 (\sqrt{2}d)^4 \text{GMR}_c^4} = 1.09 \sqrt[4]{d^3 \text{GMR}_c} \qquad (9.50)$$

9.3.6   Inductive Reactance of a Single-Phase Line

Inductive reactance, rather than inductance, is usually desired. For a single-phase, two-conductor distribution line with a separation distance D, the inductive reactance of one conductor is derived from Eq. 9.25 as follows:

$$X_L = 2\pi f L = 2\pi f \times 2 \times 10^{-7} \ln\left(\frac{D}{r'}\right)$$

$$= 4\pi f \times 10^{-7} \ln\left(\frac{D}{r'}\right) \qquad \Omega/\text{m} \qquad (9.51)$$

$$X_L = 2.022 \times 10^{-3} f \ln\left(\frac{D}{r'}\right)$$

$$= 0.2794 \left(\frac{f}{60}\right) \log\left(\frac{D}{r'}\right) \qquad \Omega/\text{mi} \qquad (9.52)$$

$$X_L = 1.256 \times 10^{-3} f \ln\left(\frac{D}{r'}\right)$$

$$= 0.1736 \left(\frac{f}{60}\right) \log\left(\frac{D}{r'}\right) \qquad \Omega/\text{km} \qquad (9.53)$$

EXAMPLE 9.2

Determine the 60-Hz impedance in Ω/mi of the single-phase distribution line employing the 1-inch-diameter copper conductor of Example 9.1 with a separation distance of 4 feet.

Solution   From Example 9.1:

$$R_a = 0.0725 \ \Omega/\text{mi}$$

$$r' = re^{-1/4} = (0.5/12)e^{-1/4} = 0.0324 \text{ ft}$$

By Eq. 9.52, the inductive reactance of one conductor is

$$X_L = 2.022 \times 10^{-3} f \ \ln(D/r')$$
$$= (2.022 \times 10^{-3})(60) \ \ln(4/0.0324) = 0.584 \ \Omega/\text{mi}$$

Therefore, the impedance of the single-phase line is

$$Z = 2(R_a + jX_L)$$
$$= 2(0.0725 + j0.584) = 0.145 + j1.168 = 1.177 \underline{/82.9°} \ \Omega/\text{mi}$$

DRILL PROBLEMS

D9.3 A single-phase, 60-Hz transmission line consists of two 1/0 copper conductors, hard drawn, 97.3% conductivity whose centers are 5 m apart. Find the inductive reactance in Ω/mi.

D9.4 A single-phase, 60-Hz transmission line consists of two 636,000 CM (or 636 MCM), 26/7, ACSR conductors. The total inductive reactance per phase is 1.65 Ω/mi. Find the distance between the two conductor centers.

D9.5 A single-phase, 60-Hz transmission line consists of two bundles, each arranged vertically. One bundle contains two conductors, each with a radius of 2.5 cm, that are 10 cm apart. The other bundle consists of three conductors, each of radius 1.5 cm, with a separation distance between adjacent conductors of 5 cm. The horizontal distance between the bundles is 8 m.

a. Find the geometric mean radius for each bundle.

b. Compute the geometric mean distance.

c. Determine the total inductance of the transmission line in mH/km and in mH/mi.

d. Find the total inductive reactance of the transmission line in Ω/mi.

9.3.7   Inductive Reactance of a Three-Phase Circuit

Consider the three-phase circuit consisting of three identical conductors. If the three conductors are symmetrically spaced, no transposition is necessary to maintain balanced conditions. The inductance was derived in Section 9.3.4, and the inductance formula is given by Eq. 9.45 for a separation distance D between conductors. Therefore, the inductive reactance per phase is found as follows:

$$X_L = 2\pi f L_a = 2\pi f, \times 2 \times 10^{-7} \ \ln\left(\frac{D}{\text{GMR}}\right)$$

$$= 4\pi f \times 10^{-7} \ \ln\left(\frac{D}{\text{GMR}}\right) \qquad \Omega/\text{m} \qquad (9.54)$$

$$X_L = 2.022 \times 10^{-3} f \, \ln\left(\frac{D}{\text{GMR}}\right)$$

$$= 0.2794\left(\frac{f}{60}\right) \log\left(\frac{D}{\text{GMR}}\right) \quad \Omega/\text{mi} \quad (9.55)$$

$$X_L = 1.256 \times 10^{-3} f \, \ln\left(\frac{D}{\text{GMR}}\right)$$

$$= 0.1736\left(\frac{f}{60}\right) \log\left(\frac{D}{\text{GMR}}\right) \quad \Omega/\text{km} \quad (9.56)$$

When the conductors are unsymmetrically spaced, the voltage drop for each conductor is different. For this case, the conductors are transposed; that is, the positions of the conductors are exchanged at regular intervals along the line so that each conductor occupies the original position of every other conductor over an equal distance. The average inductance of a transposed line was derived in Section 9.3.4, and the inductance formula is given by Eq. 9.44. Therefore, the inductive reactance per phase for the transposed three-phase circuit is found as follows:

$$X_L = 2\pi f L_a = 2\pi f \times 2 \times 10^{-7} \, \ln\left(\frac{\text{GMD}}{\text{GMR}}\right)$$

$$= 4\pi f \times 10^{-7} \, \ln\left(\frac{\text{GMD}}{\text{GMR}}\right) \quad \Omega/\text{m} \quad (9.57)$$

$$X_L = 2.022 \times 10^{-3} f \, \ln\left(\frac{\text{GMD}}{\text{GMR}}\right)$$

$$= 0.2794\left(\frac{f}{60}\right) \log\left(\frac{\text{GMD}}{\text{GMR}}\right) \quad \Omega/\text{mi} \quad (9.58)$$

$$X_L = 1.256 \times 10^{-3} f \, \ln\left(\frac{\text{GMD}}{\text{GMR}}\right)$$

$$= 0.1736\left(\frac{f}{60}\right) \log\left(\frac{\text{GMD}}{\text{GMR}}\right) \quad \Omega/\text{km} \quad (9.59)$$

9.3.8 The Use of Tables to Find Inductive Reactance

The inductive reactance of one conductor of a single-phase distribution line is given by Eq. 9.52, which may be rewritten as

$$X_L = 0.2794 \left(\frac{f}{60}\right) \log\left(\frac{1}{r'}\right)$$

$$+ 0.2794 \left(\frac{f}{60}\right) \log D \qquad \Omega/\text{mi} \qquad (9.60)$$

If both r' and D are given in feet, the first term on the right-hand side is denoted as x_a and is called *inductive reactance at 1-ft spacing* because it is equal to the inductive reactance of one conductor of a two-conductor line 1 foot apart. The second term on the right-hand side is called the *inductive reactance spacing factor* x_d. Thus, for the single-phase circuit, the total inductive reactance per conductor is given by

$$X_L = x_a + x_d \qquad \Omega/\text{mi} \qquad (9.61)$$

The values of x_a at 25, 50, and 60 Hz and the values of GMR at 60 Hz are given in the tables of conductor characteristics (Tables 1–4) in Appendix A. The values of x_d for various spacings are given in Table 6, also in Appendix A.

The inductive reactance of a three-phase transmission line with an equivalent spacing factor GMD is given by Eq. 9.58, which may be rewritten as

$$X_L = 0.2794 \left(\frac{f}{60}\right) \log\left(\frac{1}{\text{GMR}}\right)$$

$$+ 0.2794 \left(\frac{f}{60}\right) \log \text{GMD} \qquad \Omega/\text{mi} \qquad (9.62)$$

For the case of an equilaterally spaced, three-phase circuit, Eq. 9.62 equally applies with the GMD represented by the distance between conductors, that is, GMD = D.

If both GMD and GMR are expressed in feet, Eq. 9.62 can also be expressed as follows:

$$X_L = x_a + x_d \qquad \Omega/\text{mi} \qquad (9.63)$$

where x_a is the per-phase inductive reactance at 1-foot spacing and x_d is the per-phase inductive reactance spacing factor. As mentioned previously, the values of x_a and x_d may be found from the tables in Appendix A.

A simple procedure that can be used to find the inductive reactance of a three-phase circuit is as follows:

1. Read x_a from the tables of conductor characteristics.
2. (a) Evaluate GMD and read x_d from Table 6 using the GMD value as spacing factor, or (b) find the average of the three inductive reactance spacing factors corresponding to the distances between phase conductors.

EXAMPLE 9.3

Find the impedance of 2.5 miles of a single-phase ungrounded circuit consisting of 3 No. 5, 40% conductivity copperweld (CW) conductors that are spaced 6 feet apart and are expected to operate at 75% of current-carrying capacity.

Solution From Table 4-B of Appendix A, for 3 No. 5, 40% conductivity, copperweld (CW) conductors:

$$r_a = 1.772 \ \Omega/mi$$
$$x_a = 0.617 \ \Omega/mi$$

From Table 6, at 6-feet spacing:

$$x_d = 0.217 \ \Omega/mi$$

Therefore, the impedance per unit length is

$$z = r_a + j(x_a + x_d)$$
$$= 1.772 + j(0.617 + 0.217) = 1.772 + j0.834 \ \Omega/mi$$

Hence, the total impedance for the two 2.5-mi conductors is

$$Z = (2)(2.5)(1.772 + j0.834) = 8.860 + j4.172 \ \Omega$$

EXAMPLE 9.4

Determine the impedance of 80 miles of a 220-kV, three-phase circuit consisting of 795,000 CM, 26/7 ACSR conductors that are arranged in flat spacing with 20 feet between conductors.

Solution From Table 2-A in Appendix A, for 795 MCM ACSR,

$$r_a = 0.1288 \ \Omega/mi$$
$$x_a = 0.3990 \ \Omega/mi$$

From Table 6,

$$x_{d,20} = 0.3635 \ \Omega/mi$$
$$x_{d,20} = 0.3635 \ \Omega/mi$$
$$x_{d,40} = 0.4476 \ \Omega/mi$$

$$x_d = \tfrac{1}{3}(0.3635 + 0.3635 + 0.4476) = 0.3915 \ \Omega/mi$$

Alternatively,

$$\text{GMD} = \sqrt[3]{(20)(20)(40)} = 25.2 \text{ ft}$$

From Table 6,

$$x_{d,26} = 0.3953 \ \Omega/\text{mi}$$
$$x_{d,25} = 0.3906 \ \Omega/\text{mi}$$

Interpolating,

$$x_d = 0.3906 + (0.2)(0.3953 - 0.3906) = 0.3915 \ \Omega/\text{mi}$$

Thus,

$$z = 0.1288 + j(0.3990 + 0.3915) = 0.1288 + j0.7905 \ \Omega/\text{mi}$$

Therefore, the total impedance for 80 miles of line is given by

$$Z = 80z = 80(0.1288 + j0.7905) = 10.30 + j63.24 \ \Omega$$

DRILL PROBLEMS

D9.6   A three-phase, 60-Hz transmission line has solid cylindrical copper conductors arranged in the form of an equilateral triangle with 2 m spacing between conductors. Each conductor has a diameter of 2.5 cm. Calculate the inductance in H/m and the inductive reactance in Ω/km.

D9.7   A three-phase, 60-Hz transmission line has its conductors arranged in a triangular formation so that two of the distances between conductors are 8 m and the third distance is 10 m. The conductors are 795-MCM 26/7 ACSR. Determine the inductance and inductive reactance per phase per mile.

D9.8   A three-phase transmission line consists of bundled conductors that are placed at the corners of an equilateral triangle with a separation distance of 10 m between the centers of each bundle. Each bundle contains three conductors, each with a radius of 1 cm, with an equilateral spacing between conductors of 25 cm. Determine the (a) inductance per phase in mH/km and in mH/mi and (b) inductive reactance per phase in Ω/km and in Ω/mi.

9.4 CAPACITANCE AND CAPACITIVE REACTANCE

When a voltage is applied to a pair of conducting plates (or cylinders) separated by a nonconducting dielectric medium, charge of equal magnitude but opposite sign accumulates on the plates (or cylinders). The charge deposited is proportional to the applied voltage, the proportionality constant being the capacitance C.

Similarly, for a transmission line the potential difference between the conductors causes them to be charged just like a capacitor. The effective capacitance of the transmission line is dependent on the size and separation distance of the conductors.

In AC power systems, the transmission line is energized by a time-varying voltage. This alternating voltage also causes the charges on the conductors to vary with time. The alternately increasing and decreasing charges on the conductors give rise to the charging current. The charging current affects the power transmitted and the operating power factor, as well as the voltage drop along the transmission line.

Consider a long, straight, cylindrical conductor having a uniform charge throughout its length. According to *Gauss's law,* the electric flux density at a point P is equal to the flux leaving the conductor per unit length divided by the area of the surface in a unit length:

$$D = \frac{q}{2\pi x} \tag{9.64}$$

where

D = electric flux density, coulombs/m^2

q = charge on the conductor, coulombs/m

x = distance in meters from conductor to point P

The electric field intensity at point P is equal to the electric flux density divided by the permittivity ϵ; thus,

$$E = \frac{D}{\epsilon} = \frac{q}{2\pi\epsilon x} \tag{9.65}$$

The permittivity may be expressed in terms of the relative permittivity and the permittivity of free space; that is,

$$\epsilon = \epsilon_r\epsilon_0 \tag{9.66}$$

where

ϵ_r = relative permittivity (dielectric constant) = 1 for air

ϵ_0 = permittivity of free space

= $(1/36\pi) \times 10^{-9} = 8.854 \times 10^{-12}$ F/m

Next, consider a point P_1 located at a distance D_1 from the center of a conductor and another point P_2 at a distance D_2 from the center. The instantaneous potential difference, or voltage drop, from point P_1 to P_2 is found by integrating the electric field intensity over a radial path between the two equipotential surfaces passing through the two points as shown in Fig. 9.7.

Thus, the voltage drop V_{12} is found as follows:

$$V_{12} = \int_{D_1}^{D_2} E \, dx$$

$$= \int_{D_1}^{D_2} \frac{q}{2\pi\epsilon x} \, dx = \frac{q}{2\pi\epsilon} \ln\left(\frac{D_2}{D_1}\right) \tag{9.67}$$

9.4.1 Capacitance of a Single-Phase Line

The voltage V_{ab} between the two conductors of the transmission line shown in Fig. 9.4, which has a separation distance D, is found as follows.

The voltage drop due to the charge q_a on conductor a is given by

$$V_{ab,a} = \frac{q_a}{2\pi\epsilon} \ln\left(\frac{D}{r_a}\right) \tag{9.68}$$

Similarly, the voltage drop due to the charge q_b on conductor b is given by

$$V_{ba,b} = \frac{q_b}{2\pi\epsilon} \ln\left(\frac{D}{r_b}\right) \tag{9.69}$$

or

$$V_{ab,b} = -\frac{q_b}{2\pi\epsilon} \ln\left(\frac{D}{r_b}\right) \tag{9.70}$$

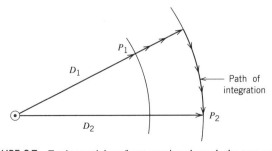

FIGURE 9.7 Equipotential surfaces passing through the two points.

By the principle of superposition,

$$V_{ab} = V_{ab,a} + V_{ab,b}$$

$$= \frac{q_a}{2\pi\epsilon} \ln\left(\frac{D}{r_a}\right) - \frac{q_b}{2\pi\epsilon} \ln\left(\frac{D}{r_b}\right) \qquad (9.71)$$

Since $q_a = -q_b$,

$$V_{ab} = \frac{q_a}{2\pi\epsilon} \left[\ln\left(\frac{D}{r_a}\right) + \ln\left(\frac{D}{r_b}\right)\right]$$

$$= \frac{q_a}{2\pi\epsilon} \ln\left(\frac{D^2}{r_a r_b}\right) \qquad (9.72)$$

If $r_a = r_b = r$, then

$$V_{ab} = \frac{q_a}{\pi\epsilon} \ln\left(\frac{D}{r}\right) \qquad (9.73)$$

Therefore, the capacitance between the conductors is given by

$$C_{ab} = \frac{q_a}{V_{ab}} = \frac{\pi\epsilon}{\ln(D/r)} \qquad \text{F/m} \qquad (9.74)$$

The potential difference between each conductor and ground is half of the potential difference between the two conductors. Thus, the capacitance to ground, or capacitance to neutral, is twice the capacitance from line to line. Therefore,

$$C_n = C_{an} = C_{bn} = \frac{2\pi\epsilon}{\ln(D/r)} \qquad \text{F/m} \qquad (9.75)$$

9.4.2 Capacitance of a Three-Phase Circuit

If the three-phase circuit is unsymmetrically spaced, it is assumed that it is completely transposed as shown in Fig. 9.5. Thus, the voltage to neutral in any transposition cycle will be the same. The voltage to neutral is derived in a manner similar to the derivation of Eq. 9.75. For the first transposition section,

$$V_{ab1} = \frac{1}{2\pi\epsilon} \left[q_a \ln\left(\frac{D_{12}}{r}\right) + q_b \ln\left(\frac{r}{D_{12}}\right) + q_c \ln\left(\frac{D_{23}}{D_{31}}\right)\right] \qquad (9.76)$$

For the second transposition section,

$$V_{ab2} = \frac{1}{2\pi\epsilon}\left[q_a \ln\left(\frac{D_{23}}{r}\right) + q_b \ln\left(\frac{r}{D_{23}}\right) + q_c \ln\left(\frac{D_{31}}{D_{12}}\right)\right] \qquad (9.77)$$

For the third transposition section,

$$V_{ab3} = \frac{1}{2\pi\epsilon}\left[q_a \ln\left(\frac{D_{31}}{r}\right) + q_b \ln\left(\frac{r}{D_{31}}\right) + q_c \ln\left(\frac{D_{12}}{D_{23}}\right)\right] \qquad (9.78)$$

Thus, the average voltage across conductors a and b is

$$V_{ab} = \frac{1}{3}\frac{1}{2\pi\epsilon}\left[q_a \ln\left(\frac{D_{12}D_{23}D_{31}}{r^3}\right) + q_b \ln\left(\frac{r^3}{D_{12}D_{23}D_{31}}\right)\right.$$
$$\left. + q_c \ln\left(\frac{D_{12}D_{23}D_{31}}{D_{12}D_{23}D_{31}}\right)\right] \qquad (9.79)$$

Simplifying,

$$V_{ab} = \frac{1}{2\pi\epsilon}\left[q_a \ln\left(\frac{\sqrt[3]{D_{12}D_{23}D_{31}}}{r}\right) - q_b \ln\left(\frac{\sqrt[3]{D_{12}D_{23}D_{31}}}{r}\right)\right]$$
$$= \frac{1}{2\pi\epsilon}\left[q_a \ln\left(\frac{\text{GMD}}{r}\right) - q_b \ln\left(\frac{\text{GMD}}{r}\right)\right] \qquad (9.80)$$

Similarly, the average value of the voltage across conductors a and c is given by

$$V_{ac} = \frac{1}{2\pi\epsilon}\left[q_a \ln\left(\frac{\text{GMD}}{r}\right) - q_c \ln\left(\frac{\text{GMD}}{r}\right)\right] \qquad (9.81)$$

For a balanced wye system, assuming a positive, or abc, phase sequence, the phase voltage may be calculated from the average values of the line-to-line voltages as follows:

$$V_{an} = \frac{V_{ab} + V_{ac}}{3} \qquad (9.82)$$

Substituting Eqs. 9.80 and 9.81 into Eq. 9.82 yields

$$V_{an} = \frac{1}{3}\frac{1}{2\pi\epsilon}\left[2q_a \ln\left(\frac{\text{GMD}}{r}\right) - q_b \ln\left(\frac{\text{GMD}}{r}\right) - q_c \ln\left(\frac{\text{GMD}}{r}\right)\right] \qquad (9.83)$$

For a balanced system,

$$q_a = -(q_b + q_c) \tag{9.84}$$

Hence,

$$V_{an} = \frac{1}{2\pi\epsilon} q_a \ln\left(\frac{GMD}{r}\right) \tag{9.85}$$

Therefore, the capacitance per phase is

$$C_n = \frac{q_a}{V_{an}} = \frac{2\pi\epsilon}{\ln(GMD/r)} \quad F/m \tag{9.86}$$

9.4.3 Capacitive Reactance

For a single-phase line, the capacitance to neutral C_n is given by Eq. 9.75. When ϵ_0 is substituted for the permittivity ϵ, the capacitive reactance from one conductor to neutral is given by the following:

$$X_C = \frac{1}{2\pi f C_n} = \frac{2.862}{f} \times 10^9 \ln\left(\frac{D}{r}\right) \quad \Omega\text{-m} \tag{9.87}$$

$$X_C = \frac{1.779}{f} \times 10^6 \ln\left(\frac{D}{r}\right) \quad \Omega\text{-mi} \tag{9.88}$$

It should be noted that the capacitance to neutral C_n given by Eq. 9.75 is in units of farads per unit length. Thus, the total capacitance to neutral is obtained by multiplying C_n by the total length of the line. Therefore, to find the total capacitive reactance to neutral in ohms for the entire length of the line, the expressions given by Eqs. 9.87 and 9.88 must be divided by the length of the line.

For a three-phase, completely transposed transmission line with equivalent spacing factor GMD, the capacitance to neutral is given by Eq. 9.86. The capacitive reactance from one conductor to neutral, with ϵ_0 substituted for permittivity ϵ, is given by the following:

$$X_C = \frac{1}{2\pi f C_n} = \frac{2.862}{f} \times 10^9 \ln\left(\frac{GMD}{r}\right) \quad \Omega\text{-m} \tag{9.89}$$

$$X_C = \frac{1.779}{f} \times 10^6 \ln\left(\frac{GMD}{r}\right) \quad \Omega\text{-mi} \tag{9.90}$$

Again, it may be noted that the capacitive reactance X_C is given in units of Ω-m, or Ω-mi. Therefore, to obtain the total capacitive reactance in Ω, the expressions given by Eqs. 9.89 and 9.90 must be divided by the total length of the line.

9.4.4 The Use of Tables to Find Capacitive Reactance

The capacitive reactance from one conductor to neutral of a single-phase distribution line is given by Eq. 9.88 in Ω-mi. This may be rewritten as follows:

$$X_C = \frac{1.779}{f} \times 10^6 \ln\left(\frac{1}{r}\right) + \frac{1.779}{f} \times 10^6 \ln D \qquad \Omega\text{-mi} \qquad (9.91)$$

If both D and r are expressed in feet, the first term on the right-hand side is called *capacitive reactance at 1-ft spacing* x_a' and the second term is called *capacitive reactance spacing factor* x_d'. Therefore, the capacitive reactance is given by

$$X_C = x_a' + x_d' \qquad (9.92)$$

The values of x_a' may be found from Tables 1–4 in Appendix A. The values of x_d' are given in Table 8 as a function of spacing.

For a three-phase transmission line with an equivalent spacing factor GMD, the capacitive reactance from one conductor to neutral is given by Eq. 9.90 in Ω-mi. This may be rewritten as follows:

$$X_C = \frac{1.779}{f} \times 10^6 \ln\left(\frac{1}{r}\right) + \frac{1.779}{f} \times 10^6 \ln \text{ GMD} \qquad (9.93)$$

If both GMD and r are given in feet, Eq. 9.93 can also be expressed as follows:

$$X_C = x_a' + x_d' \qquad (9.94)$$

where x_a' is the per-phase shunt capacitive reactance at 1-foot spacing and x_d' is the per-phase shunt capacitive reactance spacing factor. The values of x_a' may be found from Tables 1–4 and the values of x_d' from Table 8 in Appendix A.

EXAMPLE 9.5

Determine the capacitive reactance of the three-phase circuit of Example 9.4. The transmission line is 80 miles long, and the conductors are of 795-MCM, 26/7 ACSR that are arranged in flat spacing with 20 feet between conductors.

Solution From Table 2-A in Appendix A:

$$x'_a = 0.0912 \text{ M}\Omega\text{-mi}$$

From Table 8:

$$x'_{d,20} = 0.0889 \text{ M}\Omega\text{-mi}$$
$$x'_{d,40} = 0.1094 \text{ M}\Omega\text{-mi}$$
$$x'_d = \tfrac{1}{3}(0.0889 + 0.0889 + 0.1094) = 0.0957 \text{ M}\Omega\text{-mi}$$

Therefore, the capacitive reactance is

$$x_C = x'_a + x'_d = 0.0912 + 0.0957 = 0.1869 \text{ M}\Omega\text{-mi}$$

For the 80-mile long transmission line,

$$X_C = x_C/80 = 186,900/80 = 2336 \ \Omega$$

DRILL PROBLEMS

D9.9 A single-phase, 60-Hz transmission line consists of identical conductors each having a radius of 2.5 cm. The separation distance between the conductors is 5 m. Determine (a) the capacitance of the line in μF/km and in μF/mi and (b) the capacitive reactance in Ω-mi.

D9.10 Calculate the capacitance to neutral in F/m and the admittance per phase in S/km for the three-phase transmission line described in Drill Problem D9.6.

D9.11 A three-phase, 60-Hz, transmission line has a total length of 120 miles. The conductors are 715.5-MCM 26/7 ACSR and are arranged in a triangular formation so that two of the distances between conductors are 20 ft and the third distance is 30 ft. Determine (a) the capacitance to neutral in μF/mi and the capacitive reactance to neutral in Ω-mi and (b) the total capacitance to neutral and the total capacitive reactance.

D9.12 A three-phase, 60-Hz, bundled transmission line consists of three 900-MCM ACSR conductors per bundle arranged with equilateral spacing of 40 cm between conductors. The spacing between bundle centers is 8, 8, and 12 m. Calculate the capacitive reactance to neutral in Ω-km.

9.5 TRANSMISSION LINE MODELS

Transmission lines are classified as short lines, medium-length lines, or long lines. A transmission line is considered a short line if it is less than 50 miles (or 80 km). Medium-length lines are those with lengths roughly between 50 miles and 150 miles (or 240 km). Transmission lines extending more than 150 miles are considered long lines and are usually represented in terms of their distributed parameters. For some applications, however, lines as long as 200 miles (or 320 km) can still be approximated using equivalent lumped parameters. In the following sections, various models are described for the transmission line corresponding to the three classifications.

9.5.1 The Short Transmission Line

The equivalent circuit of a short transmission line consists of a series combination of a resistance and an inductive reactance. It is shown in Fig. 9.8. This equivalent circuit is applicable for transmission lines up to 50 miles in length.

Since there are no shunt branches, the current will be the same at the sending and receiving ends:

$$\mathbf{I_S} = \mathbf{I_R} \tag{9.95}$$

The voltage at the sending end is given by

$$\mathbf{V_S} = \mathbf{V_R} + \mathbf{I_R}Z \tag{9.96}$$

The voltage regulation of a transmission line is defined as the increase in receiving-end voltage as the load is reduced from full load to no load with the sending-end voltage held constant. It is expressed in percent of the full-load voltage; thus,

$$\text{Voltage regulation} = \frac{V_{R,nl} - V_{R,fl}}{V_{R,fl}} 100\% \tag{9.97}$$

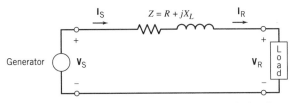

FIGURE 9.8 Equivalent circuit for a short transmission line.

EXAMPLE 9.6

A 60-Hz, three-phase transmission line is 40 miles long. It has a total series impedance of $35 + j140\ \Omega$ and negligible shunt admittance. It delivers 40 MW at 220 kV and 0.9 power factor lagging.

a. Find the voltage, current, and power factor at the sending end of the line.

b. Compute the voltage regulation and efficiency of the line.

Solution For a load of 40 MW, at 220 kV and 0.9 power factor lagging, the receiving-end current is given by

$$\mathbf{I_R} = \frac{40,000}{\sqrt{3}(220)(0.9)}\ \underline{/-\cos^{-1} 0.9} = 116.6\ \underline{/-25.8^\circ}$$

The sending-end voltage is found as follows:

$$\mathbf{V_S} = \mathbf{V_R} + Z\mathbf{I_R}$$
$$= (220/\sqrt{3}) \times 10^3\ \underline{/0^\circ} + (35 + j140)(116.6\ \underline{/-25.8^\circ})$$
$$= 127,000\ \underline{/0^\circ} + (144.3\ \underline{/76^\circ})(116.6\ \underline{/-25.8^\circ})$$
$$= 138.4\ \underline{/5.4^\circ}\ \text{kV} \quad \text{(line-to-neutral)}$$
$$= 239.7\ \underline{/35.4^\circ}\ \text{kV} \quad \text{(line-to-line)}$$

Since there is no shunt branch, the sending-end current is the same as the receiving-end current. Thus,

$$\mathbf{I_S} = \mathbf{I_R} = 116.6\ \underline{/-25.8^\circ}$$

The sending-end power factor is

$$\text{PF}_S = \cos[5.4^\circ - (-25.8^\circ)] = 0.86\ \text{lagging}$$

The percent voltage regulation is computed as

$$\text{Voltage regulation} = \frac{V_{R,nl} - V_{R,fl}}{V_{R,fl}}100\%$$
$$= \frac{239.7 - 220}{220}100\% = 8.95\%$$

The sending-end real power is

$$P_S = 3V_SI_S\text{PF}_S = 3(138.4 \times 10^3)(116.6)(0.86)$$
$$= 41.6 \times 10^6 \text{ W} = 41.6 \text{ MW}$$

Therefore, the efficiency of the line is found as follows:

$$\text{Efficiency} = (P_R/P_S)100\% = (40/41.6)100\% = 96.2\%$$

9.5.2   The Medium-Length Line

The shunt admittance, often a pure capacitance, is included in the equivalent circuit of a medium-length line. The total shunt capacitive admittance of the line is divided into two equal parts and placed at the sending and receiving ends of the line forming a *nominal π circuit*. The equivalent circuit is shown in Fig. 9.9 and is used to represent transmission lines that are between 50 to 150 miles long.

The sending-end voltage and current may be expressed in terms of the receiving-end voltage and current by using the ABCD transmission parameters as follows:

$$\mathbf{V}_S = A\mathbf{V}_R + B\mathbf{I}_R \tag{9.98}$$
$$\mathbf{I}_S = C\mathbf{V}_R + D\mathbf{I}_R \tag{9.99}$$

where
$A = D = ZY/2 + 1$
$B = Z$
$C = Y(ZY/4 + 1)$
$Z = $ total series impedance of the line
$Y = $ total shunt admittance of the line

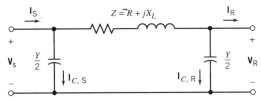

FIGURE 9.9   Equivalent circuit for a medium-length line.

At no load, the receiving-end voltage is $1/A$ times the sending-end voltage; thus, the voltage regulation is expressed as follows:

$$\text{Voltage regulation} = \frac{V_S/A - V_{R,fl}}{V_{R,fl}} 100\% \qquad (9.100)$$

EXAMPLE 9.7

A 60-Hz, three-phase transmission line is 125 miles long. It has a total series impedance of $35 + j140\ \Omega$ and a shunt admittance of $930 \times 10^{-6}\ \underline{/90°}$ S. It delivers 40 MW at 220 kV and 0.9 power factor lagging.

a. Determine the voltage, current, and power factor at the sending end of the line.

b. Find the voltage regulation and efficiency of the line.

Solution For a load of 40 MW, at 220 kV and 0.9 power factor lagging, the receiving-end current is given by

$$\mathbf{I_R} = \frac{40,000}{\sqrt{3}(220)(0.9)} \underline{/-\cos^{-1} 0.9} = 116.6 \underline{/-25.8°}\ A$$

For the nominal π equivalent circuit, the ABCD parameters are computed as follows:

$$
\begin{aligned}
A = D &= ZY/2 + 1 \\
&= (35 + j140)(930 \times 10^{-6}\ \underline{/90°})/2 + 1 \\
&= 0.935 + j0.0163 = 0.935 \underline{/1°} \\
B = Z &= 35 + j140 = 144.3 \underline{/76°}\ \Omega \\
C &= Y(ZY/4 + 1) \\
&= (930 \times 10^{-6}\ \underline{/90°})[(35 + j140)(930 \times 10^{-6}\ \underline{/90°})/4 + 1] \\
&= (-7.57 + j899.7) \times 10^{-6} = 900 \times 10^{-6}\ \underline{/90.5°}\ S
\end{aligned}
$$

Thus, the sending-end voltage and current are given by

$$
\begin{aligned}
\mathbf{V_S} &= A\mathbf{V_R} + B\mathbf{I_R} \\
&= (0.935 \underline{/1°})(127,000 \underline{/0°}) + (144.3 \underline{/76°})(116.6 \underline{/-25.8°}) \\
&= 130.4 \underline{/6.6°}\ kV \quad \text{(line-to-neutral)} \\
&= 225.8 \underline{/36.6°}\ kV \quad \text{(line-to-line)}
\end{aligned}
$$

$$I_S = CV_R + DI_R$$
$$= (900 \times 10^{-6} \underline{/90.5°})(127,000 \underline{/0°}) + (0.935 \underline{/1°})(116.6 \underline{/-25.8°})$$
$$= 97.97 + j68.57 = 119.6 \underline{/35°} \text{ A}$$

The sending-end power factor is

$$PF_S = \cos(6.6° - 35.0°) = 0.88 \text{ leading}$$

At no load, the receiving-end voltage is $1/A$ times the sending-end voltage; thus, the voltage regulation is computed as follows:

$$\text{Voltage regulation} = \frac{V_S/A - V_{R,fl}}{V_{R,fl}} 100\%$$

$$= \frac{225.8/0.935 - 220}{220} 100\% = 9.77\%$$

The sending-end real power is

$$P_S = 3V_S I_S PF_S = 3(130.4 \times 10^3)(119.6)(0.88)$$
$$= 41.17 \times 10^6 \text{ W} = 41.17 \text{ MW}$$

Therefore, the efficiency of the line is found as follows:

$$\text{Efficiency} = (P_R/P_S)100\% = (40/41.17)100\% = 97.2\%$$

EXAMPLE 9.8

Solve Example 9.7 using per unit representation. Choose a power base of 50 MVA and a voltage base of 220 kV.

Solution

a. For the chosen $S_{base} = 50$ MVA and $V_{base} = 220$ kV, the current base is computed as follows:

$$I_{base} = \frac{S_{base}}{\sqrt{3}V_{base}} = \frac{50,000}{220\sqrt{3}} = 131.2 \text{ A}$$

The impedance base is calculated as

$$Z_{base} = \frac{(V_{base})^2}{S_{base}} = \frac{(220)^2}{50} = 968 \text{ }\Omega$$

Correspondingly, the admittance base is given by

$$Y_{base} = 1/Z_{base} = 1/968 = 1033 \times 10^{-6} \text{ S}$$

The series impedance is converted to per unit as follows.

$$Z = (35 + j140)/968 = 0.036 + j0.145 \text{ pu} = 0.149 \underline{/76°} \text{ pu}$$

The shunt admittance in per unit is found as follows:

$$Y = (j930 \times 10^{-6})/(1033 \times 10^{-6}) = j0.90 \text{ pu}$$

The ABCD parameters are computed as follows:

$$A = D = ZY/2 + 1$$
$$= (0.149 \underline{/76°})(j0.90)/2 + 1 = 0.935 \underline{/1°}$$
$$B = Z = 0.149 \underline{/76°} \text{ pu}$$
$$C = Y(ZY/4 + 1)$$
$$= j0.90[(0.149 \underline{/76°})(j0.90)/4 + 1] = 0.871 \underline{/90.5°} \text{ pu}$$

The receiving-end voltage is taken as reference phasor. Thus,

$$\mathbf{V}_R = 1.0 \underline{/0°} \text{ pu}$$

The line delivers a per-unit power of

$$\mathbf{S}_R = \frac{40}{(0.9)(50)} \underline{/\cos^{-1} 0.9} = 0.889 \underline{/25.8°} \text{ pu}$$

Thus, the receiving-end current is given by

$$\mathbf{I}_R = \frac{(0.889 \underline{/25.8°})^*}{(1.0 \underline{/0°})^*} = 0.889 \underline{/-25.8°} \text{ pu}$$

The sending-end voltage and current are computed as follows:

$$\mathbf{V}_S = A\mathbf{V}_R + B\mathbf{I}_R$$
$$= (0.935 \underline{/1°})(1.0 \underline{/0°}) + (0.149 \underline{/76°})(0.889 \underline{/-25.8°})$$
$$= 1.0264 \underline{/6.6°} \text{ pu} = 225.8 \text{ kV}$$

$$\mathbf{I}_S = C\mathbf{V}_R + D\mathbf{I}_R$$
$$= (0.871\ \underline{/1°})(1.0\ \underline{/0°}) + (0.935\ \underline{/1°})(0.889\ \underline{/-25.8°})$$
$$= 0.9114\ \underline{/35°}\ \text{pu} = 119.6\ \text{A}$$

The sending end power factor is

$$\text{PF}_S = \cos(6.6° - 35°) = 0.88\ \text{leading}$$

b. At no load, the receiving-end voltage is $1/A$ times the sending-end voltage; thus, the voltage regulation is given by

$$\text{Voltage regulation} = \frac{V_S/A - V_R}{V_R}100\%$$
$$= \frac{1.0264/0.935 - 1.0}{1.0}100\% = 9.77\%$$

The receiving-end real power is

$$P_R = 40/50 = 0.80\ \text{pu}$$

The sending-end real power is found as follows:

$$P_S = V_S I_S \text{PF}_S = (1.0264)(0.9114)(0.88) = 0.823\ \text{pu} = 41.16\ \text{MW}$$

The efficiency of the transmission line is given by

$$\text{Efficiency} = (P_R/P_S)100\% = (0.80/0.823)100\% = 97.2\%$$

9.5.3   The Long Transmission Line

In the analysis of long transmission lines of more than 150 miles, the lumped-parameter models that were used for short and medium-length lines may no longer provide the desired accuracy. It is then necessary to consider that the parameters are distributed uniformly throughout the length of the line.

Consider a point on the transmission line at a distance x from the receiving end of the line. The line-to-neutral voltage and the current flowing toward the receiving end are expressed, in terms of the distributed impedance and admittance parameters of the line, by differential equations of the form

$$\frac{dV}{dx} = zI \tag{9.101}$$

$$\frac{dI}{dx} = yV \tag{9.102}$$

where

z = series impedance per unit length

y = shunt admittance per unit length

If Eq. 9.101 is differentiated and the expression for dI/dx is substituted into Eq. 9.102, the following is obtained.

$$\frac{d^2V}{dx^2} = yzV \tag{9.103}$$

Similarly, Eq. 9.102 can be differentiated and the expression for dV/dx substituted into Eq. 9.101 to obtain

$$\frac{d^2I}{dx^2} = yzI \tag{9.104}$$

The solutions of Eqs. 9.103 and 9.104 have been derived in several of the references listed at the end of the chapter; for example, see Ref. 11. Thus, the solutions are presented here without derivation. At a distance x from the receiving end of the line, the voltage and current are given by

$$\mathbf{V}_x = \tfrac{1}{2}(\mathbf{V}_R + \mathbf{I}_R Z_0)e^{\gamma x} + \tfrac{1}{2}(\mathbf{V}_R - \mathbf{I}_R Z_0)e^{-\gamma x} \tag{9.105}$$

$$\mathbf{I}_x = \tfrac{1}{2}\left(\frac{\mathbf{V}_R}{Z_0} + \mathbf{I}_R\right)e^{\gamma x} - \tfrac{1}{2}\left(\frac{\mathbf{V}_R}{Z_0} - \mathbf{I}_R\right)e^{-\gamma x} \tag{9.106}$$

respectively, where

$Z_0 = \sqrt{z/y}$ = characteristic or surge impedance

$\gamma = \sqrt{zy}$ = propagation constant

To obtain the equations relating the voltage and current at the sending end to the voltage and current at the receiving end, the distance is set equal to the length of the line; that is, $x = l$. Therefore,

$$\mathbf{V}_S = \frac{1}{2}(e^{\gamma l} + e^{-\gamma l})\mathbf{V}_R + \frac{1}{2}Z_0(e^{\gamma l} - e^{-\gamma l})\mathbf{I}_R$$

$$= (\cosh \gamma l)\mathbf{V}_R + (Z_0 \sinh \gamma l)\mathbf{I}_R \tag{9.107}$$

$$\mathbf{I}_S = \frac{1}{2}\frac{1}{Z_0}(e^{\gamma l} - e^{-\gamma l})\mathbf{V}_R + \frac{1}{2}(e^{\gamma l} + e^{-\gamma l})\mathbf{I}_R$$

$$= \frac{1}{Z_0}(\sinh \gamma l)\mathbf{V}_R + (\cosh \gamma l)\mathbf{I}_R \tag{9.108}$$

The transmission parameters of the long line are

$$A = D = \cosh \gamma l = \frac{1}{2}(e^{\gamma l} + e^{\gamma l}) \qquad (9.109)$$

$$B = Z_0 \sinh \gamma l = \frac{1}{2}Z_0(e^{\gamma l} - e^{\gamma l}) \qquad (9.110)$$

$$C = \frac{1}{Z_0} \sinh \gamma l = \frac{1}{2}\frac{1}{Z_0}(e^{\gamma l} - e^{\gamma l}) \qquad (9.111)$$

The equivalent circuit of the long transmission line is found by comparing its voltage and current equations with those for a nominal π circuit. Thus,

$$Z' = Z_0 \sinh \gamma l = Z\frac{\sinh \gamma l}{\gamma l} \qquad (9.112)$$

$$\frac{Y'}{2} = \frac{1}{Z_0}\frac{\cosh \gamma l - 1}{\sinh \gamma l}$$

$$= \frac{Y}{2}\frac{\tanh(\gamma l/2)}{\gamma l/2} \qquad (9.113)$$

where

$Z = zl$ = total series impedance
$Y = yl$ = total shunt admittance

The equivalent circuit is shown in Fig. 9.10.

EXAMPLE 9.9

A 60-Hz, three-phase transmission line is 175 miles long. It has a total series impedance of $35 + j140\ \Omega$ and a shunt admittance of $930 \times 10^{-6}\ \underline{/90^\circ}$ S. It delivers 40 MW at 220 kV and 0.9 power factor lagging.

    a. Find the voltage, current, and power factor at the sending end of the line.

    b. Determine the voltage regulation and efficiency of the line.

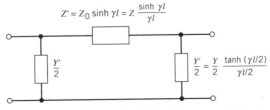

FIGURE 9.10 Equivalent circuit for a long transmission line.

Solution   For a load of 40 MW, at 220 kV and 0.9 power factor lagging, the receiving-end current is given by

$$\mathbf{I}_R = \frac{40{,}000}{\sqrt{3}(220)(0.9)} \angle{-\cos^{-1} 0.9} = 116.6 \angle{-25.8°}$$

For the long-line model, the transmission parameters are computed as follows:

$$Z_0 = \sqrt{\frac{35 + j140}{930 \times 10^{-6} \angle{90°}}} = 393.9 \angle{-7°}$$

$$\gamma l = \sqrt{(35 + j140)(930 \times 10^{-6} \angle{90°})}$$
$$= 0.3663 \angle{83°} = 0.0446 + j0.3636$$

$$e^{\gamma l} = e^{(0.0446 + j0.3636)} = 1.0456 \angle{20.8°}$$

$$e^{-\gamma l} = e^{(-0.0446 - j0.3636)} = 0.9563 \angle{-20.8°}$$

$$A = D = \cosh \gamma l = \tfrac{1}{2}(e^{\gamma l} + e^{-\gamma l})$$
$$= \tfrac{1}{2}(1.0456 \angle{20.8°} + 0.9563 \angle{-20.8°}) = 0.936 \angle{1°}$$

$$B = Z_0 \sinh \gamma l = \tfrac{1}{2}Z_0(e^{\gamma l} - e^{-\gamma l})$$
$$= \tfrac{1}{2}(393.9 \angle{-7°})(1.0456 \angle{20.8°} - 0.9563 \angle{-20.8°}) = 141.2 \angle{76°}$$

$$C = (1/Z_0) \sinh \gamma l = \tfrac{1}{2}(1/Z_0)(e^{\gamma l} - e^{-\gamma l})$$
$$= \tfrac{1}{2}(1/393.9 \angle{-7°})(1.0456 \angle{20.8°} - 0.9563 \angle{20.8°})$$
$$= 910 \times 10^{-6} \angle{90°}$$

Thus, the sending-end voltage and current are given by

$$\mathbf{V}_s = (\cosh \gamma l)\mathbf{V}_R + (Z_0 \sinh \gamma l)\mathbf{I}_R$$
$$= (0.936 \angle{1°})(127{,}000 \angle{0°})$$
$$+ (141.2 \angle{76°})(116.6 \angle{-25.8°})$$
$$= 130.1 \angle{7°} \text{ kV}\quad \text{(line-to-neutral)}$$
$$= 225.4 \angle{37°} \text{ kV}\quad \text{(line-to-line)}$$

$$\mathbf{I}_s = (1/Z_0)(\sinh \gamma l)\mathbf{V}_R + (\cosh \gamma l)\mathbf{I}_R$$
$$= (910 \times 10^{-6} \angle{90°})(127{,}000 \angle{0°})$$
$$+ (0.936 \angle{1°})(116.6 \angle{-25.8°})$$
$$= 120.6 \angle{35.3°} \text{ A}$$

The sending-end power factor is

$$PF_S = \cos(7.0° - 35.3°) = 0.88 \text{ leading}$$

The voltage regulation is computed as follows:

$$\text{Voltage regulation} = \frac{V_S/A - V_{R,fl}}{V_{R,fl}} 100\%$$

$$= \frac{225.4/0.936 - 220}{220} 100\% = 9.46\%$$

The sending-end real power is

$$P_S = 3V_S I_S PF_S = 3(130.1 \times 10^3)(120.6)(0.88)$$
$$= 41.42 \times 10^6 \text{ W} = 41.42 \text{ MW}$$

Therefore, the efficiency of the line is found as follows:

$$\text{Efficiency} = (P_R/P_S)100\% = (40/41.42)100\% = 96.6\%$$

DRILL PROBLEMS

D9.13   Solve Example 9.6 using per unit representation. Choose a power base of 100 MVA and a voltage base of 220 kV.

D9.14   A three-phase transmission line is 45 mi long, and it delivers 50 MVA at 0.866 power factor lagging to a load connected to its receiving end at 115 kV. The transmission line is composed of 795-MCM, 26/7 ACSR conductors with flat horizontal spacing of 4 m between adjacent conductors. Determine the sending-end voltage, current, and power.

D9.15   A three-phase, 60-Hz transmission line delivers 25 MW at 69 kV and 0.8 power factor lagging to a load center substation at its receiving end. The transmission line consists of 336-MCM, 26/7 ACSR conductors arranged horizontally with a spacing of 5 m between adjacent conductors. The line is 75 km long. Determine the voltage, current, power, and power factor at the sending end of the line.

D9.16   A three-phase, 60-Hz, completely transposed transmission line has a length of 100 km and has a series impedance per phase of $0.25 + j0.85$ Ω/mi and a shunt admittance of 5.0×10^{-5} S/mi. The transmission line delivers 150 MW at 0.85 lagging power factor to a load connected to its receiving end. The line-to-line voltage at the receiving end is 138 kV. Determine the following:

a. Sending-end voltage and current

b. Sending-end real and reactive power

c. Total power dissipated in the transmission line

D9.17 A three-phase, 60-Hz transmission line has the following parameters: $R = 0.25 \, \Omega/\text{mi}$, $X = 1.0 \, \Omega/\text{mi}$, and $Y = 5.0 \, \mu\text{S/mi}$. The line is 200 mi long, and it delivers 150 MW at 220 kV and unity power factor to a substation at its receiving end.

a. Calculate the voltage, current, power factor, and real power at the sending end.

b. Calculate the voltage regulation of the line.

REFERENCES

1. Del Toro, Vincent. *Electric Power Systems*. Prentice Hall, Englewood Cliffs, N.J., 1992.

2. *Electrical Transmission and Distribution Reference Book*. 4th ed. ABB Power T&D Co., Raleigh, N.C., 1964.

3. El-Hawary, Mohamed E. *Electric Power Systems: Design and Analysis*. Reston Publishing, Reston, Va., 1983.

4. Glover, J. Duncan, and Mulukutla Sarma. *Power System Analysis and Design*. PWS Publishers, Boston, 1987.

5. Gönen, Turan. *Electric Power Transmission System Engineering Analysis and Design*. Wiley, New York, 1988.

6. ———. *Modern Power System Analysis*. Wiley, New York, 1988.

7. Gross, Charles A. *Power System Analysis*. 2nd ed. Wiley, New York, 1986.

8. Gungor, Behic R. *Power Systems*. Harcourt Brace Jovanovich, New York, 1988.

9. Hayt, William H., Jr. *Engineering Electromagnetics*. 4th ed. McGraw-Hill, New York, 1981.

10. Kraus, John D. *Electromagnetics*. 3rd ed. McGraw-Hill, New York, 1984.

11. Stevenson, William D., Jr. *Elements of Power System Analysis*. 4th ed. McGraw-Hill, New York, 1982.

12. Venikov, V. A., and E. V. Putyatin. *Introduction to Energy Technology*. Mir, Moscow, 1981.

PROBLEMS

9.1 Find the AC resistance in Ω/km of a $\frac{3}{4}$-inch-diameter conductor made of 97.3% conductivity hard-drawn copper at 60 Hz and 75°C.

9.2 Find the AC resistance in Ω/mi of a 500-MCM conductor made of 97.3% conductivity hard-drawn copper at 50 Hz and 60°C.

9.3 A single-phase, 60-Hz transmission line has solid cylindrical copper conductors 1.5 cm in diameter. The conductors are arranged in a horizontal configuration with 1.25 m spacing between adjacent conductors. Calculate the following:

    a. Inductance in mH/km due to internal flux linkages only

    b. Inductance in mH/km due to both internal and external flux linkages

    c. The total inductance of the line in mH/km.

9.4 A single-phase, 60-Hz transmission line consists of solid round aluminum wires each having a diameter of 1.5 cm. The conductor spacing is 3 m. Determine the inductance of the line in mH/mi.

9.5 Find the inductive reactance in Ω/km of a single-phase, 60-Hz transmission line consisting of 500-MCM ACSR conductors with a separation distance of 4 m.

9.6 A single-phase, 60-Hz transmission line consists of two bundles each containing two conductors arranged horizontally. Each conductor of a bundle has a radius of 2.0 cm, and they are 8 cm apart. The distance between the nearest conductors of each bundle is 6 m.

    a. Find the geometric mean radius of each bundle.

    b. Calculate the geometric mean distance.

    c. Determine the total inductive reactance in Ω/km and in Ω/mi of the transmission line.

9.7 A single-phase, 60-Hz transmission line consists of two bundled conductors. Each bundle consists of three 500-MCM ACSR conductors placed at the corners of an equilateral triangle standing on one of its sides with 5 cm separation between conductor centers. The distance between the nearest conductors of each bundle is 8 m.

    a. Compute the total inductance of the transmission line in mH/km and mH/mi.

    b. Find the total inductive reactance in Ω/km and in Ω/mi.

9.8 A three-phase line has equilateral spacing between phase conductors with a separation distance of 4 m. Each conductor has a radius of 1.5 cm. Determine (a) the inductance per phase in mH/mi and (b) the inductive reactance per phase in Ω/km and in Ω/mi.

9.9 A three-phase transmission line has its conductors arranged horizontally and spaced such that $D_{13} = 2D_{12} = 2D_{23}$. The conductors are completely transposed. Determine the spacing between adjacent conductors in order to obtain an equivalent spacing of 5 m.

9.10 A three-phase, 230-kV, 60-Hz, completely transposed transmission line has one ACSR 1113-MCM conductor per phase and flat horizontal spacing with 5 m between adjacent conductors. Find the inductance in H/m and the inductive reactance in Ω/km.

9.11 A three-phase, 60-Hz transmission line has flat horizontal spacing. The conductors are ACSR, each having a GMR of 0.0278 ft, with a separation distance of 35 ft between adjacent conductors. Determine the inductive reactance per phase in ohms per mile. What is the name of this conductor?

9.12   A three-phase, 60-Hz, bundled transmission line consists of three 900-MCM ACSR conductors per bundle arranged with an equilateral spacing of 40 cm between conductors. The spacing between bundle centers is 8, 8, and 12 m. Calculate the inductive reactance in ohms per kilometer.

9.13   A three-phase, bundled, 345-kV, 60-Hz, completely transposed transmission line has flat horizontal spacing with 10 m between centers of adjacent bundles. Each bundle consists of three 954-MCM ACSR conductors per bundle, with equilateral spacing of 40 cm between conductors in the bundle. Calculate the inductive reactance in Ω/km.

9.14   A three-phase, 60-Hz transmission line has a flat horizontal spacing of 10 m between adjacent conductors. Each phase conductor is a 1113-MCM ACSR. Compare the inductive reactance in Ω/mi/phase of this transmission line with that of another transmission line using a bundle of two ACSR 26/7 conductors in each phase having the same total cross-sectional area of aluminum as the single-conductor line and 10 m spacing between bundle centers. The spacing between conductors in the bundle is 25 cm.

9.15   A single-phase, 60-Hz power line is supported on a horizontal crossarm, and its conductors are 2.5 m apart and carrying a current of 100 A. A telephone line is also supported on a horizontal crossarm 1.5 m directly below the power line, and the spacing between the conductors is 1.0 m. Determine (a) the mutual inductance between the power and telephone circuits and (b) the induced voltage per mile in the telephone line.

9.16   The single-phase power line and the telephone line described in Problem 9.15 are placed in the same horizontal plane. The distance between the nearest conductors of the two lines is 10 m. Determine (a) the mutual inductance between the power and telephone circuits and (b) the induced voltage per mile in the telephone line.

9.17   A three-phase, 60-Hz, power line is supported on a horizontal crossarm, and the separation distance between adjacent conductors is 2 m. The current flowing in the conductors is 100 A. A telephone line is supported on a horizontal crossarm that is placed 1.5 m below the power line, and the telephone wires are 1 m apart. Find the expression for the mutual inductance between the power line and the telephone line. Calculate the induced voltage per kilometer in the telephone line.

9.18   A three-phase power line and a telephone line are supported on the same tower. The three-phase line is arranged horizontally with a separation distance of 2.5 m between adjacent conductors. The telephone conductors are arranged vertically 0.5 m apart, and the top conductor is located 1 m directly below the middle conductor of the three-phase line. Determine the expression for the mutual inductance between the power line and the telephone line.

9.19   A single-phase transmission line consists of two conductors each having a diameter $\frac{1}{4}$ in. The conductors are 10 ft apart and are placed 40 ft above ground. Calculate the capacitance to neutral in farads per meter.

9.20   Calculate the capacitance to neutral in F/m and the admittance to neutral in S/km for the single-phase transmission line described in Problem 9.3.

9.21 A single-phase, 60-Hz transmission line consists of two bundles. Each bundle contains two identical conductors of 1 cm radius and placed 10 cm apart. The separation distance between the bundle centers is 5 m. Compute the capacitance and the capacitive reactance of 50 mi of this transmission line.

9.22 A three-phase, 60-Hz transmission line is 100 mi long and has flat horizontal spacing. Each phase conductor has an outside diameter of 3 cm, and the separation distance between adjacent conductors is 10 m. Calculate the capacitive reactance to neutral in Ω-m and the total capacitive reactance of the line in ohms.

9.23 Calculate the capacitance per phase in F/m and the admittance per phase in S/km for the three-phase transmission line described in Problem 9.10. Also calculate the total charging current taken by the transmission line capacitance when it is 125 km long and is operated at 345 kV.

9.24 A three-phase, 60-Hz transmission line is completely transposed. The separation distances between conductors are 7, 7, and 10 m. Each conductor has a diameter of 5 cm. The line-to-line voltage is 138 kV, and the length of the transmission line is 100 km. Determine the following:

  a. Capacitance per phase

  b. Total capacitive reactance per phase

  c. Charging current taken by the total capacitance of the transmission line

9.25 A three-phase, 60-Hz transmission line has an equivalent equilateral spacing of 5 ft, and its capacitive reactance to neutral is given as 196.1 kΩ-mi. What would be the value of the capacitive reactance to neutral in kΩ-mi at 1-ft spacing and 25 Hz?

9.26 Calculate the capacitance per phase in F/m and the admittance per phase in S/km for the three-phase transmission line described in Problem 9.13. Also calculate the total reactive power in MVAR/km supplied by the line capacitance when it is operated at 345 kV.

9.27 A three-phase, 60-Hz transmission line has a flat horizontal spacing of 10 m between adjacent conductors. Each phase conductor is a 1113-MCM ACSR. Compare the capacitive reactance in Ω-mi/phase of this transmission line with that of another transmission line using a bundle of two ACSR 26/7 conductors in each phase having the same total cross-sectional area of aluminum as the single-conductor line and 10 m spacing between bundle centers. The spacing between conductors in the bundle is 25 cm.

9.28 A three-phase, 60-Hz transmission line is composed of 300-MCM 26/7 ACSR conductors that are equilaterally spaced with 1.5 m between conductor centers. The line is 50 km long, and it delivers 2.5 MW at 13.8 kV to a load connected to its receiving end. Find the sending-end voltage, current, real power, and reactive power for the following conditions.

  a. 80% power factor lagging

  b. Unity power factor

  c. 90% power factor leading

9.29 A three-phase, 34.5-kV, 60-Hz, 40-km transmission line has a series impedance $z = 0.20 + j0.50 \ \Omega/\text{km}$. The load at the receiving end absorbs 10 MVA at 33 kV. Calculate the following:

    a. $ABCD$ parameters

    b. Sending-end voltage at a power factor of 0.9 lagging

    c. Sending-end voltage at a power factor of 0.9 leading

9.30 A three-phase, 60-Hz, completely transposed transmission line has the following total parameters:

$$Z_{\text{series}} = 10 + j50 \ \Omega$$
$$Y_{\text{shunt}} = j30 \times 10^{-5} \ \text{S}$$

The transmission line is 80 mi long, and the line-to-line voltage at the receiving end is 230 kV. The load connected to the receiving end of the line may be represented by the load impedance $Z_L = 150 \ \underline{/36.9^\circ} \ \Omega$.

    a. Determine the current and the line-to-line voltage at the sending end.

    b. Find the voltage regulation.

    c. Calculate the real and reactive power at the sending end.

    d. Find the efficiency of the transmission line.

9.31 A three-phase, 230-kV, 60-Hz, 200-km transmission line has a series impedance $z = 0.10 + j0.35 \ \Omega/\text{km}$ and a shunt admittance $y = j5.0 \times 10^{-6} \ \text{S/km}$. The line delivers 250 MW at 220 kV and 0.95 power factor lagging to a substation connected to its receiving end. Using the nominal π circuit, calculate:

    a. The ABCD parameters

    b. The sending-end voltage and current

    c. The percent voltage regulation

9.32 A three-phase transmission line is 120 mi long. It has a total series impedance of $25 + j110 \ \Omega$ and a total shunt admittance of $j5 \times 10^{-4} \ \text{S}$. It delivers 175 MW at 220 kV and 0.9 power factor lagging to a load connected to its receiving end. Use per-unit representation and find

    a. The voltage, current, power factor, and real power at the sending end

    b. The voltage regulation

9.33 A three-phase, 60-Hz, 350-km transmission line has the following parameters: $R = 0.10 \ \Omega/\text{km}$, $X_L = 0.4 \ \Omega/\text{km}$, and $X_C = 350 \ \text{k}\Omega\text{-km}$. The line delivers 150 MW at 230 kV and unity power factor to its receiving end. Determine the sending-end voltage, current, real power, and power factor. Use per-unit representation.

9.34 A three-phase, 60-Hz, 300-mi transmission line has a total series impedance $Z = 40 + j175 \ \Omega$ and a total shunt admittance $Y = j10^{-3} \ \text{S}$. The line delivers 300 MW at 220 kV and 0.9 power factor lagging.

    a. Determine the sending-end voltage, current, power factor, and real power.

b. Repeat part (a) when the transmission line delivers 300 MW at 220 kV and unity power factor.

9.35 A three-phase, 500-kV, 60-Hz, 300-km transmission line has a series impedance $z = 0.04 + j0.50$ Ω/km and a shunt admittance $y = j3.5 \times 10^{-6}$ S/km. The line delivers 1000 MW at 480 kV and unity power factor to its receiving end. Calculate the following:

    a. Characteristic impedance Z_0

    b. Exact ABCD parameters

    c. Sending-end voltage and current

    d. Percent voltage regulation

9.36 Determine the equivalent π circuit for the transmission line of Problem 9.35 and compare it with the nominal π circuit.

Ten

Power Flow Solutions

10.1   INTRODUCTION

The planning, design, and operation of electric power systems require continuing and comprehensive analysis in order to determine system performance and to evaluate alternative system expansion plans. Because of the increasing cost of system additions and modifications, as well as soaring fuel costs, it is imperative that utilities consider a range of design options. In-depth analysis is necessary to determine the effectiveness of each alternative in alleviating operating problems and supplying load demands during normal and abnormal operating conditions, for peak and off-peak loadings, and for both present and future power systems.

As the size of the network grows, the determination of such system variables as voltage levels and power flows in the transmission lines, transformers, and generators becomes more and more difficult. A large volume of network data must be collected and handled accurately. This is aggravated by the dynamic nature of the system as it is reconfigured to meet changing system conditions. Thus, as the electric system grows in size and the number of its interconnections increases, planning for future expansion becomes increasingly complex.

The availability of fast and large computers has somewhat eased the work load of the power system engineer. Routine calculations can now be accomplished more efficiently and more extensively. Advances in device and system modeling, as well as developments in computational techniques, have greatly enhanced the analysis and planning tasks.

The high costs of system operations and system expansions and modifications have paved the way for the use of computers in power systems. To assist the engineer in his or her role as system operator and planner, the engineer is provided with digital computers and highly developed computer programs.

352

10.1.1   Scope of Power System Analysis

The performance of the power system is routinely analyzed under normal steady-state operating conditions, as well as under abnormal conditions in the presence of sudden, large disturbances. The principal power system problems usually considered include power flow analysis, fault analysis, transient stability analysis, and system protection.

Power flow calculations provide the power flows and voltages for a particular steady-state operating condition of the system. Power flow analysis is performed for small or large power systems, for high-voltage and low-voltage systems, and for existing and future systems. The power flow problem formulation and the popularly accepted solution methods such as Gauss–Seidel, Newton–Raphson, and fast-decoupled power flow methods are presented in this chapter.

The analysis of symmetrical and unsymmetrical faults, the description and applications of protection systems, and transient stability analysis are discussed in Chapter 11.

It is customary in power system analysis to use per-unit representation. The circuit parameters are expressed in per unit, and all calculations are performed in per unit. The power base is usually chosen to be the rating of one of the major pieces of equipment or is set as a company policy. The power base is the same for all parts of the power system. The voltage bases are normally chosen as the nominal voltages in the various parts of the system or are selected to be the voltage ratings of the transformer windings in order to maintain a unity per-unit turns ratio of the transformer. The current and impedance bases are computed on the basis of the previously selected values of power base and voltage bases.

10.1.2   One-Line Diagrams

A typical power system involves the interconnection between many different components, including generators, transformers, transmission lines, loads, circuit breakers, and switches. These interconnections are commonly referred to as *nodes* or *buses*.

A power system that operates under balanced conditions may be studied using per-phase analysis. Thus, it is sufficient to represent the three-phase power system by a single-phase power system in a simplified diagram commonly referred to as a *one-line, or single-line, diagram*. A one-line diagram of a sample power system is shown in Fig. 10.1.

The complexity of representation in the one-line diagram depends mainly on the type of study to be undertaken. For example, information about circuit breakers and relays may be important and have to be present in a system protection study; however, they may be omitted completely in a power flow analysis. The various real and reactive power demands at the different buses are

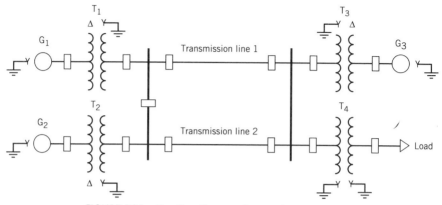

FIGURE 10.1   One-line diagram of a sample power system.

needed in power flow studies but may sometimes be neglected in short-circuit analysis.

The one-line diagram is used to build the equivalent circuit or impedance diagram of the power system. The models to be used for representing the generators, electrical loads, transmission lines, and transformers are selected and are included in the impedance diagram. The models are chosen depending on the analysis to be performed.

10.1.3   Power System Modeling

The equivalent circuit for the synchronous generator was derived in Chapter 7. The circuit model for the cylindrical-rotor machine consists of the per-phase generated voltage E_a and the per-phase armature resistance R_a and synchronous reactance X_s in series. Thus, the synchronous generator is represented by the equivalent circuit shown in Fig. 10.2.

The equivalent circuit of a transmission line was derived in Chapter 9 and is shown in Fig. 10.3. A nominal π circuit is chosen over a nominal T circuit in order not to introduce an additional node at the midpoint of the line. The series branch represents the series impedance per phase, and the shunt branch

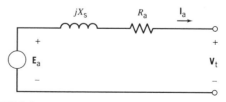

FIGURE 10.2   Equivalent circuit of a synchronous generator.

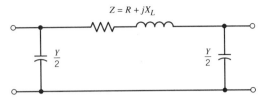

FIGURE 10.3   Equivalent circuit of a transmission line.

represents one half of the total capacitance of the line. For short transmission lines, the shunt branches are neglected.

The equivalent circuit of the three-phase transformer was derived in Chapter 4. The series impedance is given in terms of the per-phase equivalent resistance and leakage reactance of the transformer windings. The shunt magnetizing branch is generally neglected except when the transformer efficiency is being considered; then the core loss is included. Thus, the equivalent circuit is as shown in Fig. 10.4.

For the sample power system whose one-line diagram is shown in Fig. 10.1, the impedance diagram is derived by representing the various system components by their respective equivalent circuits. When all the resistances are neglected, the impedance diagram reduces to the reactance diagram shown in Fig. 10.5. From the figure, it may be seen that all shunt elements have also been assumed to be negligible. Thus, this model may be useful in short-circuit analysis, which will be discussed in Section 11.2.

The models and equivalent circuits described in this section are valid for studying balanced system conditions. Thus, they are useful for power flow analysis, as well as for balanced three-phase faults and transient stability analysis. They are also needed for unbalanced or unsymmetrical fault analysis, together with other additional models which are described later in Section 11.2.3.

10.2   POWER FLOW ANALYSIS

Power flow analysis is the solution for the static operating condition of a power system. It is the most frequently performed of routine power network calculations in digital computers. Power flow analysis is used in power system

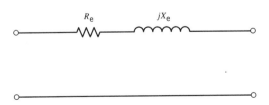

FIGURE 10.4   Equivalent circuit of a three-phase transformer.

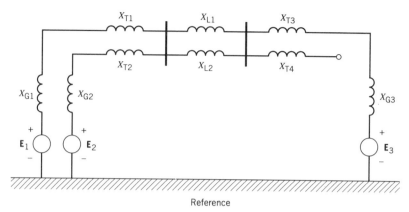

FIGURE 10.5 Impedance diagram for power system of Fig. 10.1.

planning, operational planning, and operations/control. It is also employed in multiple assessments, stability analysis, and system optimization.

Power flow programs compute the voltage magnitudes and phase angles at each bus of the network under steady-state operating conditions. These programs also compute the power flows and power losses for all equipment, including transformers and transmission lines. Thus, overloaded transformers and transmission lines are identified, and remedial measures can be designed and implemented.

Present-day computers have sufficiently large storage and high speeds. These enable the power system engineer to simulate and analyze the many different base cases, summer and winter peak conditions, and various contingency cases encountered in system operation and planning.

10.2.1   Power Flow Concepts

A power system comprising three buses is shown in Fig. 10.6. A generator is connected to each of the first two buses, and an electrical load is connected to

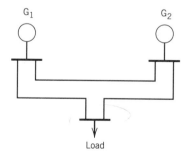

FIGURE 10.6 A three-bus power system.

the third bus. The real power and reactive power demands are known for the load bus, bus 3, and the generator voltages at buses 1 and 2 are also specified. The three transmission lines interconnecting the buses contain both resistance and reactance; thus, electric current flowing through these lines results in electrical losses.

The two generators must jointly supply the total load requirements and the power losses in the transmission lines. While doing so, the generators are constrained to operate within their power generation capabilities, and the power they deliver has to be at the desired voltage at the load. In addition, there should be no overloading of equipment, including transmission lines and transformers. Furthermore, there should be no bus voltages either above or below specified operating limits. In case of an equipment overload or voltage-limit violation, the generation schedules have to be adjusted, power flows in the transmission lines have to be rerouted, or capacitor banks have to be switched in order to bring the system conditions back to normal.

The requirements of power system operations described above are satisfied to a certain extent by undertaking power flow studies to determine appropriate measures and proper procedures. These analytical studies are also incorporated in power system planning and engineering design. These ensure that future expansions and modifications in the power system will redound to the ultimate objectives of reliability, economy, and quality of service.

10.2.2   Node-Voltage Equations

A three-bus power system consisting of two generators and one load interconnected by transmission lines is shown in Fig. 10.6. A per-phase impedance diagram that can be used in a power flow study for this three-bus system is shown in Fig. 10.7. This impedance diagram is basically an electrical network containing four nodes. The phasor voltages at nodes 1, 2, and 3 may be expressed with respect to the fourth node, which is taken as reference or ground.

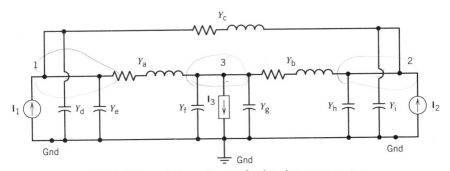

FIGURE 10.7   Impedance diagram for three-bus power system.

Kirchhoff's current law is applied to each node of the network of Fig. 10.7 successively, with the sum of currents flowing into the node set equal to the sum of currents flowing away from the node. Thus, the node-voltage equations for the three-bus system of Fig. 10.6 are obtained as follows:

$$\mathbf{I}_1 = (\mathbf{V}_1 - \mathbf{V}_2)Y_c + (\mathbf{V}_1 - \mathbf{V}_3)Y_a + \mathbf{V}_1(Y_d + Y_e) \qquad (10.1)$$

$$\mathbf{I}_2 = (\mathbf{V}_2 - \mathbf{V}_3)Y_b + (\mathbf{V}_2 - \mathbf{V}_1)Y_c + \mathbf{V}_2(Y_h + Y_i) \qquad (10.2)$$

$$-\mathbf{I}_3 = (\mathbf{V}_3 - \mathbf{V}_1)Y_a + (\mathbf{V}_3 - \mathbf{V}_2)Y_b + \mathbf{V}_3(Y_f + Y_g) \qquad (10.3)$$

Rearranging these equations yields

$$\mathbf{I}_1 = (Y_a + Y_c + Y_d + Y_e)\mathbf{V}_1 - Y_c\mathbf{V}_2 - Y_a\mathbf{V}_3 \qquad (10.4)$$

$$\mathbf{I}_2 = -Y_c\mathbf{V}_1 + (Y_b + Y_c + Y_h + Y_i)\mathbf{V}_2 - Y_b\mathbf{V}_3 \qquad (10.5)$$

$$-\mathbf{I}_3 = -Y_a\mathbf{V}_1 - Y_b\mathbf{V}_2 + (Y_a + Y_b + Y_f + Y_g)\mathbf{V}_3 \qquad (10.6)$$

These three equations may be written in standard matrix form as

$$\begin{bmatrix} \mathbf{I}_1 \\ \mathbf{I}_2 \\ -\mathbf{I}_3 \end{bmatrix} = \begin{bmatrix} Y_{11} & Y_{12} & Y_{13} \\ Y_{21} & Y_{22} & Y_{23} \\ Y_{31} & Y_{32} & Y_{33} \end{bmatrix} \begin{bmatrix} \mathbf{V}_1 \\ \mathbf{V}_2 \\ \mathbf{V}_3 \end{bmatrix} \qquad (10.7)$$

The vector on the left-hand side is called the bus injection-current vector $\mathbf{I}_{bus}$. On the right-hand side are the bus voltage vector $\mathbf{V}_{bus}$ and the *bus admittance matrix* $\mathbf{Y}_{bus}$, which relates the bus voltages to the injection currents.

The bus admittance matrix is square, sparse, and symmetrical when there are no phase shifters and no mutual coupling. A diagonal element Y_{kk} is called the self-admittance of bus k and is found by summing the primitive admittances of all lines and transformers connected to bus k, plus the admittances of shunt connections from bus k to reference. The off-diagonal element Y_{km} is the mutual admittance between bus k and bus m, and it is equal to the negative of the admittance of the line or transformer between buses k and m; thus, it is zero if there is no connection from bus k to bus m.

EXAMPLE 10.1

The one-line diagram of a four-bus power system is shown in Fig. 10.8. The line impedances are given in per unit. Find the bus admittance matrix $\mathbf{Y}_{bus}$.

Solution   The elements of the bus admittance matrix are computed as follows:

$$Y_{12} = Y_{21} = -1.0/(0.05 + j0.15) = -2 + j6$$

$$Y_{13} = Y_{31} = -1.0/(0.10 + j0.30) = -1 + j3$$

$$Y_{14} = Y_{41} = 0 + j0$$

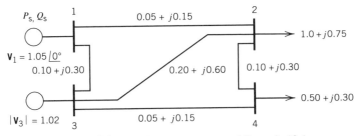

FIGURE 10.8 Four-bus power system of Example 10.1.

$$Y_{11} = (2 - j6) + (1 - j3) = 3 - j9$$
$$Y_{23} = Y_{32} = -1.0/(0.20 + j0.60) = -0.5 + j1.5$$
$$Y_{24} = Y_{42} = -1.0/(0.10 + j0.30) = -1 + j3$$
$$Y_{22} = (2 - j6) + (0.5 - j1.5) + (1 - j3) = 3.5 - j10.5$$
$$Y_{34} = Y_{43} = -1.0/(0.05 + j0.15) = -2 + j6$$
$$Y_{33} = (1 - j3) + (0.5 - j1.5) + (2 - j6) = 3.5 - j10.5$$
$$Y_{44} = (1 - j3) + (2 - j6) = 3 - j9$$

Hence, the bus admittance matrix is assembled as follows:

$$\mathbf{Y}_{bus} = \begin{bmatrix} 3.0 - j9.0 & -2.0 + j6.0 & -1.0 + j3.0 & 0.0 + j0.0 \\ -2.0 + j6.0 & 3.5 - j10.5 & -0.5 + j1.5 & -1.0 + j3.0 \\ -1.0 + j3.0 & -0.5 + j1.5 & 3.5 - j10.5 & -2.0 + j6.0 \\ 0.0 + j0.0 & -1.0 + j3.0 & -2.0 + j6.0 & 3.0 - j9.0 \end{bmatrix}$$

DRILL PROBLEMS

D10.1 The power system shown in Fig. 10.1 is operating at steady state. The reactance of each transmission line is $X = 20\ \Omega$. The generators and transformers are rated as follows:

> G1: 20 MVA, 12 kV, $X = 1.20$ pu
> G2: 60 MVA, 13.8 kV, $X = 1.40$ pu
> G3: 50 MVA, 13.2 kV, $X = 1.40$ pu
> T1: 25 MVA, 12/69 kV, $X = 0.08$ pu
> T2: 75 MVA, 13.8/69 kV, $X = 0.16$ pu
> T3: 60 MVA, 69/13.2 kV, $X = 0.14$ pu
> T4: 75 MVA, 69/13.8 kV, $X = 0.16$ pu

Choose a power base of 100 MVA and a voltage base of 12 kV in the circuit of generator 1. Form the bus admittance matrix $\mathbf{Y}_{bus}$.

D10.2 In the one-line diagram of the four-bus power system shown in Fig. 10.8, the per-unit line impedances are given as follows:

$$
\begin{aligned}
\text{Line } 1\text{–}2: \quad & Z_{12} = 0.20 + j0.60 \\
\text{Line } 1\text{–}3: \quad & Z_{13} = 0.10 + j0.30 \\
\text{Line } 2\text{–}3: \quad & Z_{23} = 0.30 + j0.90 \\
\text{Line } 2\text{–}4: \quad & Z_{24} = 0.10 + j0.30 \\
\text{Line } 3\text{–}4: \quad & Z_{34} = 0.20 + j0.60
\end{aligned}
$$

Find the bus admittance matrix $\mathbf{Y}_{\text{bus}}$.

10.2.3 Classification of Buses

The power system model used in power flow analysis shows the interconnections between generating plants and substations, substations and transmission lines, distribution lines and loads, and so forth. Three major types of nodes or buses are identified in the power network. These are listed ahead, together with their corresponding descriptive equations:

1. *Load bus, or P-Q bus k.* At a load bus, the net real and reactive power demands are specified, or scheduled; that is,

$$
\begin{aligned}
S_k^{\text{sp}} &= P_k^{\text{sp}} + jQ_k^{\text{sp}} \\
&= (P_{Gk} - P_{Dk}) + j(Q_{Gk} - Q_{Dk})
\end{aligned} \tag{10.8}
$$

   where

$$
\begin{aligned}
P_k^{\text{sp}}, \ Q_k^{\text{sp}} &= \text{specified real and reactive power at bus } k \\
P_{Gk}, \ Q_{Gk} &= \text{real and reactive power generation at bus } k \\
P_{Dk}, \ Q_{Dk} &= \text{real and reactive power demand at bus } k
\end{aligned}
$$

   A physical load may or may not be present at bus k; in the latter case, P_{Dk} and Q_{Dk} are simply set equal to zero. When a generator is present at a P-Q bus, its real and reactive power generations are given and are fixed.

2. *Generator bus, or P-V bus m.* The generators are connected to P-V buses. At these buses, the real power generation and the voltage magnitude are specified; thus,

$$
P_m^{\text{sp}} = P_{Gm} - P_{Dm} \tag{10.9}
$$

$$
|V_m| = V_m^{\text{sp}} \tag{10.10}
$$

where V_m^{sp} is the specified voltage magnitude at bus m. The voltage magnitude at a P-V bus is fixed at its specified, or controlled, value provided the reactive power generation is able to support the voltage; that is, the reactive power generated is within operating limits. Otherwise, the voltage is allowed to seek its proper level, and the bus is converted into a P-Q bus by specifying the reactive power at this bus.

3. *System slack, or swing bus s*. Because the system losses are not known precisely before the power flow solution, it is not possible to specify the real power injected at every bus. Hence, the real power of one of the generator buses is allowed to swing, and it supplies the slack between the scheduled real power generations and the sum of all loads and system losses. Therefore, the swing bus voltage magnitude is specified and its voltage phase angle is usually chosen as the system reference and set equal to zero. Thus,

$$|\mathbf{V}_s| = V_s^{sp} \qquad (10.11)$$

$$\delta_s = 0° \qquad (10.12)$$

where V_s and δ_s are the swing bus voltage magnitude and angle, respectively.

10.2.4   The Gauss–Seidel Method

In Section 10.2.2, the power system is represented as an electrical network that is characterized by node equations. Thus at bus k, the phasor current $\mathbf{I}_k$ injected into the bus is given by

$$Y_{k1}\mathbf{V}_1 + Y_{k2}\mathbf{V}_2 + Y_{k3}\mathbf{V}_3 + \cdots + Y_{kN}\mathbf{V}_N = \mathbf{I}_k \qquad (10.13)$$

where

$\mathbf{V}_m$ = phasor voltage at bus m

Y_{km} = element of the bus admittance matrix

N = total number of buses in the power system

The specified, or scheduled, complex power injection at bus k may be expressed in terms of the current injected into the bus and the bus voltage phasor as follows:

$$S_k^{sp} = P_k^{sp} + jQ_k^{sp} = \mathbf{V}_k\mathbf{I}_k^* \qquad (10.14)$$

where

$(\ )^*$ = the complex conjugate

P_k^{sp} = specified real power injected into bus k

Q_k^{sp} = specified reactive power injected into bus k

Solving Eq. 10.14 for $\mathbf{I}_k$ yields

$$\mathbf{I}_k = \frac{P_k^{\text{sp}} - jQ_k^{\text{sp}}}{\mathbf{V}_k^*} \tag{10.15}$$

Substituting Eq. 10.15 into Eq. 10.13,

$$Y_{k1}\mathbf{V}_1 + Y_{k2}\mathbf{V}_2 + \cdots + Y_{km}\mathbf{V}_m + \cdots + Y_{kN}\mathbf{V}_N = \frac{P_k^{\text{sp}} - jQ_k^{\text{sp}}}{\mathbf{V}_k^*} \tag{10.16}$$

Transposing all terms, except for the term $\mathbf{V}_k Y_{kk}$, from the left-hand side of Eq. 10.16, and dividing by Y_{kk} yields the expression for the phasor voltage $\mathbf{V}_k$ at bus k.

$$\mathbf{V}_k = \frac{1}{Y_{kk}}\left[\frac{P_k^{\text{sp}} - jQ_k^{\text{sp}}}{\mathbf{V}_k^*} - (Y_{k1}\mathbf{V}_1 + Y_{k2}\mathbf{V}_2\right.$$
$$\left. + \cdots Y_{km}\mathbf{V}_m + \cdots + Y_{kN}\mathbf{V}_N)\right] \quad \text{for } m \neq k \tag{10.17}$$

The swing bus voltage is taken as the reference phasor, and its voltage magnitude is specified and phase angle set equal to zero. Equation 10.17 is used to write a system of $(N-1)$ simultaneous algebraic equations relating the phasor voltage at the individual buses with the corresponding power injections at the bus and phasor voltages at all the buses. These equations are coupled through the elements of the bus admittance matrix. This system of equations may be solved by an iterative technique such as the Gauss–Seidel method.

The *Gauss–Seidel method* is an iterative technique for solving a system of nonlinear algebraic equations such as that given in Eq. 10.17 for the unknown bus phasor voltages. With this procedure, the phasor voltage at a bus is found by using the latest computed values of the phasor voltages at the other buses. Thus, Eq. (10.17) may be rewritten as follows to demonstrate the procedure.

$$\mathbf{V}_k^{(i+1)} = \frac{1}{Y_{kk}}\left[\frac{P_k^{\text{sp}} - jQ_k^{\text{sp}}}{\mathbf{V}_k^{(i)*}} - \sum_{\substack{m=1 \\ m \neq k}}^{N} Y_{km}\mathbf{V}_m^{(\beta)}\right]$$
$$\beta = i \quad \text{for } m > k$$
$$\beta = i + 1 \quad \text{for } m < k \tag{10.18}$$

where i represents the iteration count.

In the solution of Eq. 10.18, the most recently computed values of $\mathbf{V}_k$ are used. The solution procedure bypasses the swing bus because the phasor voltage of the swing bus is already known. For a P-Q (or load) bus, the procedure is applied directly to obtain an improved estimate of the phasor voltage inasmuch as P_k^{sp} and Q_k^{sp} are both given.

In a P-V (or generator) bus, the bus voltage $\mathbf{V}_k$ is specified; hence, Q_k^{sp} is not directly available. Therefore, an estimate Q_k^{calc} is computed by using the current estimates of the phasor voltages as follows:

$$Q_k^{\text{calc}} = -\text{Im} \, [\mathbf{V}_k^*(Y_{k1}\mathbf{V}_1 + Y_{k2}\mathbf{V}_2 + \cdots + Y_{km}\mathbf{V}_m + \cdots + Y_{kN}\mathbf{V}_N)] \qquad (10.19)$$

In Eq. 10.19, Im[ ] indicates the imaginary part of the expression in the brackets. The computed value of the reactive power Q_k^{calc} is used to replace Q_k^{sp} in Eq. 10.18, and V_k is recalculated. Then the magnitude of V_k is reset to its specified value V_k^{sp}, but the new value of its phase angle is retained.

The Gauss–Seidel method has the attractions of simplicity, comparatively good performance, and nonstorage of previous values. It has a very reliable convergence characteristic, but the rate of convergence is quite slow. The rate of convergence is improved somewhat with the use of an accelerating factor α. Therefore, at the $(i + 1)$th iteration, the phasor voltage is modified as follows:

$$[\mathbf{V}_k^{(i+1)}]_{\text{mod}} = \mathbf{V}_k^{(i)} + \alpha \, \Delta\mathbf{V}_k^{(i+1)} \qquad (10.20)$$

In the second term on the right-hand side of Eq. 10.20, $\Delta\mathbf{V}_k^{(i+1)}$ is defined as

$$\Delta\mathbf{V}_k^{(i+1)} = \mathbf{V}_k^{(i+1)} - \mathbf{V}_k^{(i)} \qquad (10.21)$$

When α is left at its default value of 1.0, the phasor voltage is not modified. The best value of acceleration factor to use is usually different for different systems. A value of 1.6 for α is generally considered a good choice.

The iterative process is said to have converged when the process no longer yields any improvement on the solution. At this point, the phasor voltages at all buses have been found and may be used to derive other information about the steady-state operating characteristics of the power system. The real and reactive power generation of the slack bus and the reactive power generations of the other generators may be found by using Eqs. 10.16 and 10.8. The real and reactive powers flowing through any transformer or transmission line connected between buses k and m may be found as follows:

$$\mathbf{S}_{km} = P_{km} + jQ_{km} = \mathbf{V}_k \left(\frac{\mathbf{V}_k - \mathbf{V}_m}{Z_{km}} + \frac{Y_{km}}{2}\mathbf{V}_k \right)^* \qquad (10.22)$$

The real power and reactive power losses of any transformer or transmission line are derived from the line power flows computed by using Eq. 10.22.

$$\mathbf{S}_{km}^{\text{loss}} = \mathbf{S}_{km} + \mathbf{S}_{mk}$$
$$P_{km}^{\text{loss}} + jQ_{km}^{\text{loss}} = (P_{km} + P_{mk}) + j(Q_{km} + Q_{mk}) \qquad (10.23)$$

By adding the losses in all transformers and transmission lines, the total system loss is determined.

EXAMPLE 10.2

With the bus data given for the power system shown in Fig. 10.8, determine the value of the voltage V_2 at bus 2 that is produced by the first iteration of the Gauss–Seidel method. Bus number 1 is taken as swing bus.

Solution   The real and reactive powers at bus 2 are

$$P_2 = -1.00 \text{ pu}$$
$$Q_2 = -0.75 \text{ pu}$$

The initial values of the bus voltages are

$$V_1 = 1.05 \underline{/0°} \text{ pu} \qquad V_2 = 1.00 \underline{/0°} \text{ pu}$$
$$V_3 = 1.02 \underline{/0°} \text{ pu} \qquad V_4 = 1.00 \underline{/0°} \text{ pu}$$

The voltage at bus 2, for the first iteration, is calculated as follows:

$$V_2 = \frac{1}{Y_{22}} \left[\frac{P_2 - jQ_2}{(V_2)*} - Y_{21}V_1 - Y_{23}V_3 - Y_{24}V_4 \right]$$

Substituting the initial values of the bus voltages and the elements of the admittance matrix,

$$V_2 = \frac{1}{3.5 - j10.5} \left[\frac{-1.00 + j0.75}{(1.00 \underline{/0°})*} - (-2.0 + j6.0)(1.05 \underline{/0°}) \right.$$
$$\left. - (-0.5 + j1.5)(1.02 \underline{/0°}) - (-1.0 + j3.0)(1.00 \underline{/0°}) \right]$$

Simplifying,

$$V_2 = 0.94 \underline{/-3.9°} \text{ pu}$$

DRILL PROBLEMS

D10.3   A two-bus power system has a bus admittance matrix given by

$$Y_{bus} = \begin{bmatrix} 2.0 \underline{/-75°} & 1.5 \underline{/105°} \\ 1.5 \underline{/105°} & 2.0 \underline{/-75°} \end{bmatrix} \text{ pu}$$

The power demand on bus 1 is $S_{D1} = (1.2 + j0.9)$ pu and the power demand on bus 2 is $S_{D2} = (0.4 - j0.3)$ pu. Bus 1 is selected as swing bus and its voltage is specified as $V_1 = 1.05 \underline{/0°}$. Use the Gauss–Seidel iterative technique to find the per-unit voltage at bus 2.

D10.4 Calculate the per-unit complex power supplied by the swing generator of the power system described in Problem D10.3.

D10.5 A two-bus power system has a generator connected to bus 1. Bus 2 is connected to bus 1 through a transmission line with a series impedance $Z = (0.02 + j0.10)$ pu and a shunt admittance $Y = j0.20$ pu. The load demands on the buses are

$$S_{L1} = (0.4 + j0.1) \text{ pu}$$
$$S_{L2} = (0.8 + j0.2) \text{ pu}$$

Bus 1 is chosen as the slack or swing bus, and its phasor voltage is set to $V_1 = 1.0 \underline{/0°}$.

a. Form the bus admittance matrix Y_{bus} with the elements expressed in rectangular form.

b. Reformulate Y_{bus} with the matrix elements in polar form.

c. Write the expressions for P_1 and Q_1 and P_2 and Q_2 in terms of the bus voltage magnitudes and angles.

---

*10.2.5   The Newton–Raphson Method

The specified, or scheduled, complex power injection into a bus k of the power system may be expressed as

$$P_k^{sp} + jQ_k^{sp} = V_k I_k^* \tag{10.24}$$

Substituting the expression for the bus current injection given by Eq. 10.13 into Eq. 10.24 yields

$$P_k^{sp} + jQ_k^{sp} = V_k \sum_{m=1}^{N} Y_{km}^* V_m^* = \sum_{m=1}^{N} V_k Y_{km}^* V_m^* \tag{10.25}$$

The bus voltages and the elements of the bus admittance matrix may be expressed in terms of their magnitudes and phase angles as

$$V_k = V_k \underline{/\delta_k}$$
$$V_m = V_m \underline{/\delta_m} \tag{10.26}$$
$$Y_{km} = Y_{km} \underline{/\theta_{km}}$$

Substituting the relations of Eq. 10.26 into Eq. 10.25 yields

$$P_k^{\text{sp}} + jQ_k^{\text{sp}} = \sum_{m=1}^{N} V_k V_m Y_{km} \underline{/\delta_k - \delta_m - \theta_{km}} \qquad (10.27)$$

The real and imaginary parts of Eq. 10.27 are the specified real power and specified reactive power, respectively. Thus,

$$P_k^{\text{sp}} = P_{Gk} - P_{Dk} = \sum_{m=1}^{N} V_k V_m Y_{km} \cos(\delta_k - \delta_m - \theta_{km}) \qquad (10.28)$$

$$Q_k^{\text{sp}} = Q_{Gk} - Q_{Dk} = \sum_{m=1}^{N} V_k V_m Y_{km} \sin(\delta_k - \delta_m - \theta_{km}) \qquad (10.29)$$

Equations 10.28 and 10.29 are used to write a set of nonlinear algebraic equations for the various buses of the power system, and the resulting equations are referred to as *power flow equations*.

It is sometimes more convenient to express the bus admittance elements in terms of their conductance and susceptance components, that is,

$$Y_{km} = G_{km} + jB_{km} \qquad (10.30)$$

Thus, Eqs. 10.28 and 10.29 may be expressed alternatively as follows:

$$P_k^{\text{sp}} = V_k \sum_{m=1}^{N} V_m [G_{km} \cos(\delta_k - \delta_m) + B_{km} \sin(\delta_k - \delta_m)] \qquad (10.31)$$

$$Q_k^{\text{sp}} = V_k \sum_{m=1}^{N} V_m [G_{km} \sin(\delta_k - \delta_m) - B_{km} \cos(\delta_k - \delta_m)] \qquad (10.32)$$

To find the solution of the power flow problem, Eqs. 10.31 and 10.32 are used to write as many equations as necessary to compute the unknown bus voltage magnitudes and phase angles. For each bus k, except the swing bus, an equation for the bus real power (10.31) is written, because the bus real power has been specified or scheduled. For these buses, the voltage angles are unknown except for the swing bus angle, which is used as reference and set equal to zero. An equation for the reactive power is written for each P-Q, or load, bus, because the bus reactive power has been specified and the voltage magnitude at this bus is to be computed. No reactive power equations are written for P-V, or generator, buses because the reactive power generations are not known; however, the bus voltage magnitudes are already known and are controlled (maintained) at their specified values as far as physically possible.

The resulting system of power flow equations to be solved consists of non-linear algebraic equations involving the products of voltage magnitudes and trigonometric functions of the phase angles. These equations are linearized using a Taylor series expansion, and an iterative procedure is introduced. At every iteration, the system of linearized equations is solved to find improvements to the current estimate of the solution. An example of such a procedure is the Newton–Raphson method, also called Newton's method.

The generalized *Newton–Raphson method* is an iterative technique for solving a set of simultaneous nonlinear equations involving the same number of unknown variables. Consider the vector equation **F** of the vector of unknown variables **X**.

$$\mathbf{F(X)} = \mathbf{0} \tag{10.33}$$

At every iteration, Eq. 10.33 is approximated by the first two terms of a Taylor Series expansion. The linearized problem is used to solve for corrections, or improvements, to the current estimate of the vector of unknowns. The elements of the *Jacobian matrix J* are the first-order partial derivatives, which are evaluated at the current estimate of the solution. The Jacobian matrix is employed in the Newton–Raphson method as shown in the following.

$$\mathbf{J}\,\Delta\mathbf{X} = \left[\frac{\partial\mathbf{F}}{\partial\mathbf{X}}\right]\Delta\mathbf{X} = -\mathbf{F(X)} \tag{10.34}$$

Equation 10.34 is solved for the correction vector $\Delta\mathbf{X}$ by using any known method of solving a system of linear equations. The state vector is then updated as follows.

$$\mathbf{X}^{(\text{new})} = \mathbf{X}^{(\text{old})} + \Delta\mathbf{X} \tag{10.35}$$

The process is repeated until the solution converges to within a specified tolerance.

In power flow calculations, the set of nonlinear power flow equations given in Eqs. 10.31 and 10.32 may be rewritten as Eqs. 10.36 and 10.37, respectively, which represent the real power and reactive power mismatches at bus k. These mismatches give the difference between the injected power at the bus and the sum of the powers flowing through the various lines connected to the other buses.

$$\Delta P_k = V_k \sum_{m=1}^{N} V_m [G_{km}\,\cos(\delta_k - \delta_m)$$
$$+ B_{km}\,\sin(\delta_k - \delta_m)] - P_k^{\text{sp}} = 0 \tag{10.36}$$

$$\Delta Q_k = V_k \sum_{m=1}^{N} V_m [G_{km}\,\sin(\delta_k - \delta_m)$$
$$- B_{km}\,\cos(\delta_k - \delta_m)] - Q_k^{\text{sp}} = 0 \tag{10.37}$$

Equations 10.36 and 10.37 are linearized, and the resulting equations to be solved at each iteration are given ahead. In these equations, the real power mismatches and reactive power mismatches at the buses are expressed as linear functions of the incremental voltage angles and incremental voltage magnitudes.

$$\begin{bmatrix} \mathbf{H} & \mathbf{N} \\ \mathbf{M} & \mathbf{L} \end{bmatrix} \begin{bmatrix} \Delta\boldsymbol{\delta} \\ \Delta\mathbf{V} \end{bmatrix} = - \begin{bmatrix} \Delta\mathbf{P} \\ \Delta\mathbf{Q} \end{bmatrix} \tag{10.38}$$

The submatrices $\mathbf{H}$, $\mathbf{N}$, $\mathbf{M}$, and $\mathbf{L}$ are found by taking the derivatives of Eqs. 10.36 and 10.37 with respect to the voltage angles and magnitudes. These expressions are given in Eqs. 10.39 to 10.42.

$$H_{km} = \frac{\partial P_k}{\partial \delta_m} = V_k V_m [G_{km} \sin(\delta_k - \delta_m) - B_{km} \cos(\delta_k - \delta_m)]$$

$$H_{kk} = \frac{\partial P_k}{\partial \delta_k} = -Q_k - B_{kk} V_k^2 \tag{10.39}$$

$$N_{km} = \frac{\partial P_k}{\partial V_m} = V_k [G_{km} \cos(\delta_k - \delta_m) + B_{km} \sin(\delta_k - \delta_m)]$$

$$N_{kk} = \frac{\partial P_k}{\partial V_k} = \frac{1}{V_k}(P_k + G_{kk} V_k^2) \tag{10.40}$$

$$M_{km} = \frac{\partial Q_k}{\partial \delta_m} = V_k V_m [-G_{km} \cos(\delta_k - \delta_m) - B_{km} \sin(\delta_k - \delta_m)]$$

$$M_{kk} = \frac{\partial Q_k}{\partial \delta_k} = P_k - G_{kk} V_k^2 \tag{10.41}$$

$$L_{km} = \frac{\partial Q_k}{\partial V_m} = V_k [G_{km} \sin(\delta_k - \delta_m) - B_{km} \cos(\delta_k - \delta_m)]$$

$$L_{kk} = \frac{\partial Q_k}{\partial V_k} = \frac{1}{V_k}(Q_k - B_{kk} V_k^2) \tag{10.42}$$

Because the swing bus voltage is taken as the reference phasor (i.e., $V_s = 1.0 \underline{/0°}$), real and reactive power mismatches are not included in Eq. 10.38. For P-V buses, reactive power mismatches and incremental voltage magnitudes are not included because the voltage magnitudes are set to the specified values. Also, if bus k is not directly connected to bus m, then the kmth element of submatrices $\mathbf{H}$, $\mathbf{N}$, $\mathbf{M}$, and $\mathbf{L}$ are zero. The Jacobian matrix $\mathbf{J}$ has sparsity characteristics similar to those of the bus admittance matrix $\mathbf{Y}_{\text{bus}}$.

Equation 10.38 is solved directly for the incremental voltage angles and incremental voltage magnitudes that are used as corrections to the current estimate of the solution. Any known solution technique for a system of linear algebraic equations may be used to solve Eq. 10.38. These computed corrections are used to determine new estimates of the voltage phase angles and voltage magnitudes as follows:

$$\delta^{new} = \delta^{old} + \Delta\delta^{new} \tag{10.43}$$

$$\mathbf{V}^{new} = \mathbf{V}^{old} + \Delta\mathbf{V}^{new} \tag{10.44}$$

The iterative process is said to have converged when the bus real power mismatch vector $\Delta\mathbf{P}$ and reactive power mismatch vector $\Delta\mathbf{Q}$, which are computed by using Eqs. 10.36 and 10.37, have been reduced to within a specified tolerance. At this point, the voltage magnitudes and phase angles at all buses have been found, and they may be used to calculate the real and reactive powers flowing through any transmission line or transformer, the real and reactive power generation of the slack bus, and the reactive power generations of the other generators. In addition, the real and reactive power losses of any transmission line or transformer, as well as of the total system, may be found by using the computed line power flows.

The Newton–Raphson method converges most rapidly of any of the power flow solution techniques, especially when the solution point is close. It has a quadratic convergence characteristic. It is reliable and has minimal sensitivity to factors that cause poor convergence, such as the choice of swing bus and the presence of series capacitors in the network, which could cause problems in the Gauss–Seidel method.

EXAMPLE 10.3

For the power system shown in Fig. 10.9, the bus admittance matrix $\mathbf{Y}_{bus}$ is given by

$$\mathbf{Y}_{bus} = \begin{bmatrix} 3.0 - j9.0 & -2.0 + j6.0 & -1.0 + j3.0 \\ -2.0 + j6.0 & 2.5 - j7.5 & -0.5 + j1.5 \\ -1.0 + j3.0 & -0.5 + j1.5 & 1.5 - j4.5 \end{bmatrix}$$

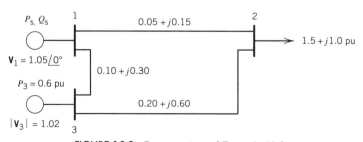

FIGURE 10.9   Power system of Example 10.3.

The per-unit bus voltages and power injections are also given in the figure.

a. For each bus k, specify the bus type, and determine which of the variables V_k, δ_k, P_k, and Q_k are input data and which are unknowns.

b. Assume an initial estimate of $\mathbf{V}_2 = 1.0\underline{/0°}$, and $\delta_3 = 0°$, and calculate the bus real and reactive power mismatches to be used in the first iteration of the Newton–Raphson power flow method.

c. Set up the linearized system of equations that are solved at each iteration of the Newton-Raphson power flow method.

Solution

a. The bus types and the classification of variables at each bus are shown in Table 10.1.

b. Using the initial estimates $\mathbf{V}_2 = 1.00\underline{/0°}$ and $\delta_3 = 0.0°$, the real and reactive bus power mismatches are calculated using Eqs. 10.36 and 10.37 with $\delta_{km} = (\delta_k - \delta_m)$ as follows:

$$\begin{aligned}
\Delta P_2 &= V_2[V_1(G_{21}\cos\delta_{21} + B_{21}\sin\delta_{21}) + V_2 G_{22} \\
&\quad + V_3(G_{23}\cos\delta_{23} + B_{23}\sin\delta_{23})] - P_2^{\text{sp}} \\
&= 1.00\,[1.05(-2.0\cos 0 + 6.0\sin 0) + 1.00(2.5) \\
&\quad + 1.02(-0.5\cos 0 + 1.5\sin 0)] - (-1.5) \\
&= 1.39 \text{ pu}
\end{aligned}$$

$$\begin{aligned}
\Delta P_3 &= V_3[V_1(G_{31}\cos\delta_{31} + B_{31}\sin\delta_{31}) \\
&\quad + V_2(G_{32}\cos\delta_{32} + B_{32}\sin\delta_{32}) + V_3 G_{33}] - P_3^{\text{sp}} \\
&= 1.02\,[1.05(-1.0\cos 0 + 3.0\sin 0) \\
&\quad + 1.00(-0.5\cos 0 + 1.5\sin 0) + 1.02(1.5)] - 0.6 \\
&= -0.62 \text{ pu}
\end{aligned}$$

$$\begin{aligned}
\Delta Q_2 &= V_2[V_1(G_{21}\sin\delta_{21} - B_{21}\cos\delta_{21}) - V_2 B_{22} \\
&\quad + V_3(G_{23}\sin\delta_{23} - B_{23}\cos\delta_{23})] - Q_2^{\text{sp}} \\
&= 1.00\,[1.05(-2.0\sin 0 - 6.0\cos 0) - 1.00(-7.5) \\
&\quad + 1.02(-0.5\sin 0 - 1.5\cos 0)] - (-1.00) \\
&= 0.67 \text{ pu}
\end{aligned}$$

Table 10.1 Bus Type and Classification of Variables for Power System of Example 10.3

Bus Number	Bus Type	Input Data	Unknown
1	Swing	V_1, δ_1	P_1, Q_1
2	Load	P_2, Q_2	V_2, δ_2
3	Generator	P_3, V_3	δ_3, Q_3

c. The real and reactive power mismatches calculated above are used in the linearized system of equations to find corrections, or improvements, in the values of the voltage magnitudes and angles as shown in the following equation.

$$
\begin{bmatrix}
\dfrac{\partial P_2}{\partial \delta_2} & \dfrac{\partial P_2}{\partial \delta_3} & \dfrac{\partial P_2}{\partial V_2} \\
\dfrac{\partial P_3}{\partial \delta_2} & \dfrac{\partial P_3}{\partial \delta_3} & \dfrac{\partial P_3}{\partial V_2} \\
\dfrac{\partial Q_2}{\partial \delta_2} & \dfrac{\partial Q_2}{\partial \delta_3} & \dfrac{\partial Q_2}{\partial V_2}
\end{bmatrix}
\begin{bmatrix}
\Delta\delta_2 \\ \Delta\delta_3 \\ \Delta V_2
\end{bmatrix}
= -
\begin{bmatrix}
\Delta P_2 \\ \Delta P_3 \\ \Delta Q_3
\end{bmatrix}
$$

* 10.2.6 The Fast-Decoupled Power Flow Method

In the Newton–Raphson method, described in the previous section, the Jacobian matrix is formed and evaluated at the current estimate of the solution at every iteration. This large matrix is used to solve for the corrections or improvements to the current estimate of the solution. Thus at each iteration, the Newton–Raphson method performs either a matrix inversion or a Gaussian elimination or a triangular factorization of the large Jacobian matrix. Therefore, the Newton–Raphson method is more complicated and requires more computations per iteration and more storage space than the Gauss–Seidel technique. To alleviate some of the shortcomings of the Newton–Raphson method, the fast-decoupled method has been developed and is discussed in this section.

Any practical transmission system operating in steady state exhibits the characteristic of strong interdependence between real powers and bus voltage angles and between reactive powers and voltage magnitudes. Correspondingly, the coupling between P and V, as well as between Q and δ, is relatively weak. Based on the physical property of the problem, the *fast-decoupled method* solves the power flow problem by "decoupling," that is, solving separately, the P-δ and Q-V subproblems.

The first step in the fast-decoupled approach is to neglect the submatrices **M** and **N** in Eq. 10.38, giving two separate matrix equations:

$$\mathbf{H}\Delta\boldsymbol{\delta} = -\Delta\mathbf{P} \tag{10.45}$$

$$\mathbf{L}\Delta\mathbf{V} = -\Delta\mathbf{Q} \tag{10.46}$$

where the elements of the matrices **H** and **L** are given by Eqs. 10.39 and 10.42 and are rewritten here as follows:

$$
H_{km} = V_k V_m [G_{km}\sin(\delta_k - \delta_m) - B_{km}\cos(\delta_k - \delta_m)]
$$
$$
H_{kk} = -Q_k - B_{kk}V_k^2 \tag{10.47}
$$

$$L_{km} = V_k[G_{km}\sin(\delta_k - \delta_m) - B_{km}\cos(\delta_k - \delta_m)]$$

$$L_{kk} = \frac{1}{V_k}(Q_k - B_{kk}V_k^2) \tag{10.48}$$

For practical power systems, the phase angles of the bus voltages are seen to be generally small and close to the values of neighboring buses. In addition, the transformer and line reactances are normally greater than the corresponding resistances. Hence, the following assumptions are further made.

$$\cos(\delta_k - \delta_m) \cong 1 \qquad G_{km} \ll B_{km}$$

$$\sin(\delta_k - \delta_m) \cong 0 \qquad Q_k \ll B_{kk}V_k^2 \tag{10.49}$$

Thus, good approximations to Eqs. 10.45 and 10.46 are given by

$$\sum_{m=1}^{N} V_k(-B'_{km})V_m\,\Delta\delta_m = -\Delta P_k \tag{10.50}$$

$$\sum_{m=1}^{N} V_k(-B''_{km})\,\Delta V_m = -\Delta Q_k \tag{10.51}$$

Equations 10.50 and 10.51 may be simplified by moving the V_k terms to the right-hand side and setting the V_m terms in Eq. 10.50 to unity. The final fast-decoupled power flow equations may thus be expressed as two sets of linear algebraic equations with constant coefficients as follows:

$$\mathbf{B'\Delta\delta} = \left[\frac{\Delta P_k}{V_k}\right] \tag{10.52}$$

$$\mathbf{B''\Delta V} = \left[\frac{\Delta Q_k}{V_k}\right] \tag{10.53}$$

Both $\mathbf{B'}$ and $\mathbf{B''}$ are real, sparse, and constant matrices. $\mathbf{B''}$ is symmetrical, and $\mathbf{B'}$ is also symmetrical if the effects of phase shifters are neglected.

The elements of the matrix $\mathbf{B'}$ may be simplified by neglecting all the resistances and omitting the elements that affect predominantly the Q-V problem, such as shunt susceptances and transformer off-nominal taps. When the resistances are neglected, $\mathbf{B'}$ is formed from the reactance network by using the same procedure used in forming the bus admittance matrix $\mathbf{Y}_{\text{bus}}$. The matrix $\mathbf{B''}$ is the imaginary part of $\mathbf{Y}_{\text{bus}}$. The elements of $\mathbf{B''}$ may be simplified by omitting phase shifters.

In the fast-decoupled power flow method, the bus real power mismatches $\Delta\mathbf{P}$ are evaluated using Eq. 10.36 at the current estimate of the voltage magnitudes and phase angles. Equation 10.52 is solved for the improvements $\Delta\boldsymbol{\delta}$. The improvements $\Delta\boldsymbol{\delta}$ are then used to update the voltage phase angles using Eq. 10.43.

Using the newly computed values of the phase angles and the current estimates of the voltage magnitudes, the reactive power mismatches $\Delta \mathbf{Q}$ are calculated by using Eq. 10.37. Equation 10.53 is solved for the improvements $\Delta \mathbf{V}$. The improvements $\Delta \mathbf{V}$ are then used to update the voltage magnitudes using Eq. 10.44.

It is seen that the phase angles and voltage magnitudes are alternately updated while using the latest solution estimates in the calculation of the power mismatches. Convergence is achieved when both the real and reactive power mismatches are within specified tolerances. It may be noted that the mismatches are computed using the exact expressions, without approximations or decoupling; thus, the method converges to the true and exact solution.

The fast-decoupled power flow Eqs. 10.52 and 10.53 may be solved by any known method of solving a system of linear equations. The matrices $\mathbf{B}'$ and $\mathbf{B}''$ in these equations are constant matrices, which have the same size as the Jacobian submatrices $\mathbf{H}$ and $\mathbf{L}$, respectively, and have the same sparsity characteristics as the bus admittance matrix $\mathbf{Y}_{\text{bus}}$. If matrix inversion is used, both matrices $\mathbf{B}'$ and $\mathbf{B}''$ need to be inverted only once at the beginning of the solution process. If triangular factorization is employed, the triangular factors are calculated only once at the beginning of the iterative process.

The fast-decoupled power flow method converges reliably. The speed per iteration is approximately five times faster than in the Newton–Raphson method, and the storage requirements for both $\mathbf{B}'$ and $\mathbf{B}''$ matrices are also less.

EXAMPLE 10.4

For the four-bus power system shown in Fig. 10.8, form the $\mathbf{B}'$ and $\mathbf{B}''$ matrices that are used in the fast-decoupled power flow solution method.

Solution The $\mathbf{B}'$ matrix is formed using the same procedure as in building the bus admittance matrix with the resistances of the lines neglected. The reactance network, derived from Fig. 10.8, is shown in Fig. 10.10, and it is used to obtain the $\mathbf{B}'$ matrix.

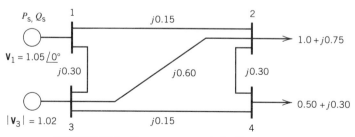

FIGURE 10.10 Reactance network for Example 10.4.

The elements of $\mathbf{B}'$ are found as follows:

$$jB_{12} = jB_{21} = -1.0/j0.15 = j6.667$$
$$jB_{13} = jB_{31} = -1.0/j0.30 = j3.333$$
$$jB_{14} = jB_{41} = 0$$
$$jB_{11} = -j6.667 - j3.333 = -j10$$
$$jB_{23} = jB_{32} = -1.0/j0.60 = j1.667$$
$$jB_{24} = jB_{42} = -1.0/j0.30 = j3.333$$
$$jB_{22} = -j6.667 - j1.667 - j3.333 = -j11.667$$
$$jB_{34} = jB_{43} = -1.0/j0.15 = j6.667$$
$$jB_{33} = -j3.333 - j1.667 - j6.667 = -j11.667$$
$$jB_{44} = -j3.333 - j6.667 = -j10$$

Thus, the $\mathbf{B}'$ matrix is given by

$$\mathbf{B}' = \begin{bmatrix} -10.000 & 6.667 & 3.333 & 0.000 \\ 6.667 & -11.667 & 1.667 & 3.333 \\ 3.333 & 1.667 & -11.667 & 6.667 \\ 0.000 & 3.333 & 6.667 & -10.000 \end{bmatrix}$$

The bus admittance matrix $\mathbf{Y}_{bus}$ was found in Example 10.1 and is repeated as follows:

$$\mathbf{Y}_{bus} = \begin{bmatrix} 3.0 - j9.0 & -2.0 + j6.0 & -1.0 + j3.0 & 0.0 + j0.0 \\ -2.0 + j6.0 & 3.5 - j10.5 & -0.5 + j1.5 & -1.0 + j3.0 \\ -1.0 + j3.0 & -0.5 + j1.5 & 3.5 - j10.5 & -2.0 + j6.0 \\ 0.0 + j0.0 & -1.0 + j3.0 & -2.0 + j6.0 & 3.0 - j9.0 \end{bmatrix}$$

The $\mathbf{B}''$ matrix is the imaginary component of bus admittance matrix $\mathbf{Y}_{bus}$. Thus,

$$\mathbf{B}'' = \begin{bmatrix} -9.0 & 6.0 & 3.0 & 0.0 \\ 6.0 & -10.5 & 1.5 & 3.0 \\ 3.0 & 1.5 & -10.5 & 6.0 \\ 0.0 & 3.0 & 6.0 & -9.0 \end{bmatrix}$$

10.3   AN APPLICATION OF POWER FLOW

To demonstrate the application of power flow for analyzing the performance of a power system, a 15-bus test system is considered. The one-line diagram of the test system is shown in Fig. 10.11.

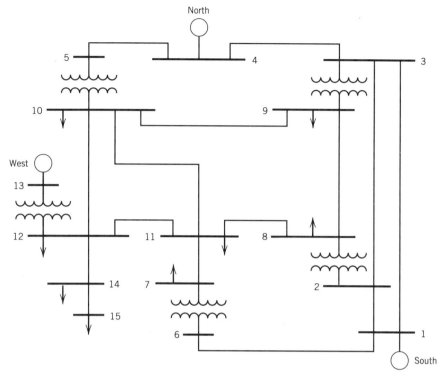

FIGURE 10.11   One-line diagram of test system.

The system power base used is 100 MVA, and the voltage bases are selected to match the nominal voltages in various parts of the system. The transformer data are provided in Table 10.2. Each of the transformers is fixed-tap, which is set at 100%, and the percent resistances and reactances given are based on the device ratings.

The system load data are given in Table 10.3, and the system generator data are shown in Table 10.4. The generators at buses 1 and 4 represent in-feeds from the neighboring power system that is charged with the bulk-power generation and transmission in the region. The South generator is connected to bus 1, which is chosen as slack bus.

Table 10.2   Transformer Data

Bus$_{From}$	Bus$_{To}$	Name	S_{Rated} (MVA)	V_{Rated} (kV)	% R	% X
2	8	Janice	200	230/115	0.5	6
3	9	Excel	200	230/115	0.5	6
5	10	Sharp	200	230/115	0.5	6
6	7	Kate	100	230/115	0.2	5
12	13	Center	150	115/13.8	0.5	5

Table 10.3 System Load Data

Bus	Name	P_{Load} (MW)	Q_{Load} (MVAR)
1	South	0	0
2	East	0	0
3	Main	0	0
4	North	0	0
5	Rich	0	0
6	Market	0	0
7	Ash	25	10
8	Cedar	40	20
9	Maple	18	10
10	Walnut	130	40
11	Oak	160	50
12	Center	100	30
13	West	0	0
14	Woods	50	20
15	Green	24	10

The transmission line parameters are calculated and converted to per unit based on the chosen bases. The transformer impedances are also converted to per unit on system bases. The load data and generator data are, likewise, converted to system bases.

The per-unit transformer data and transmission line data are presented in Tables 10.5 and 10.6, respectively. The system input bus data are given in Table 10.7. A "flat voltage" start is assumed; that is, the voltage phase angle at each bus is set equal to zero, while the voltage magnitude at each load bus is set equal to 1.0 per unit. The voltage at each generator bus is held at the specified value provided the generator reactive power stays within specified minimum and maximum limits.

The test system was analyzed using a noncommercial educational software package. This package includes programs employing either of the following power flow solution methods: Gauss–Seidel, Newton–Raphson, fast-decoupled; combined Gauss–Seidel/Newton–Raphson, or combined Gauss–Seidel/fast-decoupled. All programs are written in FORTRAN. Each solution method converges to the same solution.

Table 10.4 Generator Data

Bus	Name	S_{Rated} (MVA)	V_{Rated} (kV)	P_G^{spec} (MW)	Q_G^{min} (MVAR)	Q_G^{max} (MVAR)
1	South	400	230	—	—	—
4	North	200	230	170	−50	110
13	West	150	13.8	120	0	90

Table 10.5 Transformer Input Data in Per Unit

Bus$_{From}$	Bus$_{To}$	Name	S_{Rated}	R (pu)	X (pu)
2	8	Janice	200	0.0025	0.0300
3	9	Excel	200	0.0025	0.0300
5	10	Sharp	200	0.0025	0.0300
6	7	Kate	100	0.0020	0.0500
12	13	Center	150	0.0033	0.0333

The Gauss–Seidel power flow method was selected, and the corresponding program was run. A converged solution was reached in 15 iterations. The bus data from the converged solution are given in Table 10.8.

Results of Simulation The program flagged the West generator as having violated its reactive generation limit. As can be seen from Table 10.8, the reactive power generated by the West generator is 1.18 per unit, exceeding its upper limit of 0.9 per unit. This means that the specified voltage at bus 13 is too high, and the generator cannot maintain it at this level. To find the correct operating voltage of the generator, the P-V bus 13 is converted to a P-Q bus and the reactive power demand at this bus is set equal to the negative of the violated limit Q_G^{max}, or -0.90 per unit. The program was run again to solve for the voltage magnitude at this bus. It was found to be 1.00 per unit.

The program also automatically checks all bus voltages and flags those that are less than 0.95 per unit or higher than 1.05 per unit. The voltages at load buses 14 and 15 are 0.935 and 0.922 per unit, respectively, which are both

Table 10.6 Transmission Line Input Data in Per Unit

Bus$_{From}$	Bus$_{To}$	R (pu)	X (pu)	B (pu)	S_{Rated}
1	2	0.0100	0.0528	0.1119	3.585
1	3	0.0233	0.1232	0.2611	3.585
1	6	0.0083	0.0440	0.0933	3.585
2	3	0.0133	0.0704	0.1492	3.585
3	4	0.0100	0.0528	0.1119	3.585
4	5	0.0066	0.0352	0.0746	3.585
7	11	0.0277	0.1518	0.0271	2.012
8	9	0.0332	0.1822	0.0325	2.012
8	11	0.0222	0.1214	0.0217	2.012
9	10	0.0277	0.1518	0.0271	2.012
10	11	0.0277	0.1518	0.0271	2.012
10	12	0.0222	0.1214	0.0217	2.012
11	12	0.0166	0.0911	0.0163	2.012
12	14	0.0222	0.1214	0.0217	2.012
14	15	0.0166	0.0911	0.0163	2.012

Table 10.7   Bus Input Data in Per Unit

Bus	Name	Code	V	δ	P_G	P_L	Q_L
1	South	1	1.025	0	—	0	0
2	East	3	1.000	0	0	0	0
3	Main	3	1.000	0	0	0	0
4	North	2	1.025	0	1.7	0	0
5	Rich	3	1.000	0	0	0	0
6	Market	3	1.000	0	0	0	0
7	Ash	3	1.000	0	0	0.25	0.10
8	Cedar	3	1.000	0	0	0.40	0.20
9	Maple	3	1.000	0	0	0.18	0.10
10	Walnut	3	1.000	0	0	1.30	0.40
11	Oak	3	1.000	0	0	1.60	0.50
12	Center	3	1.000	0	0	1.00	0.30
13	West	2	1.035	0	1.2	0	0
14	Woods	3	1.000	0	0	0.50	0.20
15	Green	3	1.000	0	0	0.24	0.10
	Totals					5.47	1.90

Table 10.8   Bus Results

Bus	Name	Code	V	δ	P_G	Q_G
1	South	1	1.025	0	2.68	0.076
2	East	3	1.009	−3.43	0	0
3	Main	3	1.017	−3.65	0	0
4	North	2	1.025	−3.29	1.70	0.443
5	Rich	3	1.002	−6.19	0	0
6	Market	3	1.008	−2.22	0	0
7	Ash	3	0.996	−4.90	0	0
8	Cedar	3	0.998	−5.36	0	0
9	Maple	3	1.009	−4.82	0	0
10	Walnut	3	0.988	−8.81	0	0
11	Oak	3	0.968	−11.0	0	0
12	Center	3	0.994	−12.1	0	0
13	West	2	1.035	−10.0	1.20	1.18
14	Woods	3	0.935	−17.2	0	0
15	Green	3	0.922	−18.6	0	0
	Totals				5.58	1.70

Table 10.9 Final Bus Results

Bus	Name	Code	V	δ	P_G	Q_G
1	South	1	1.025	0	2.68	0.176
2	East	3	1.007	−3.42	0	0
3	Main	3	1.016	−3.64	0	0
4	North	2	1.025	−3.29	1.70	0.571
5	Rich	3	0.998	−6.16	0	0
6	Market	3	1.007	−2.21	0	0
7	Ash	3	0.993	−4.89	0	0
8	Cedar	3	0.994	−5.35	0	0
9	Maple	3	1.007	−4.81	0	0
10	Walnut	3	0.981	−8.79	0	0
11	Oak	3	0.958	−11.0	0	0
12	Center	3	0.975	−12.0	0	0
13	West	2	1.001	−9.76	1.20	0.682
14	Woods	3	0.950	−17.5	0	0
15	Green	3	0.950	−19.0	0	0
	Totals				5.58	1.42

less than the minimum allowable level of 0.95 per unit. These P-Q buses are converted to P-V buses, with their power generations set equal to zero. The voltage magnitudes are set equal to the violated limit V^{min}, or 0.95 per unit. The program is run again to find the generator reactive powers at buses 14 and 15. These are equal to the per-unit MVAR of the capacitors that have to be connected at the two buses to bring the voltages back within limits. These are 0.116 per unit and 0.139 per unit for buses 14 and 15, respectively.

Finally, with the generator voltage at bus 13 set to 1.00 per unit and capacitors of 0.116 and 0.139 per unit MVAR connected to buses 14 and 15, respectively, the program was run again. The final bus results are presented in Table 10.9. It may be noted that all previous constraint violations have been resolved.

REFERENCES

1. Anderson, Paul M. *Analysis of Faulted Power Systems*. Iowa State Press, Ames, Iowa, 1973.

2. Arrillaga, J., C. P. Arnold, and B. J. Harker. *Computer Modelling of Electrical Power Systems*. Wiley, New York, 1986.

3. Del Toro, Vincent. *Electric Power Systems*. Prentice Hall, Englewood Cliffs, N.J., 1992.

4. *Electrical Transmission and Distribution Reference Book*. 4th ed. ABB Power T&D Co., Raleigh, NC, 1964.

5. El-Hawary, Mohamed E. *Electric Power Systems: Design and Analysis*. Reston Publishing, Reston, Va., 1983.

6. Glover, J. Duncan, and Mulukutla Sarma. *Power System Analysis and Design*. PWS Publishers, Boston, 1987.

7. Gönen, Turan. *Electric Power Transmission System Engineering Analysis and Design*. Wiley, New York, 1988.

8. ———. *Modern Power System Analysis*. Wiley, New York, 1988.

9. Gross, Charles A. *Power System Analysis*. 2nd ed. Wiley, New York, 1986.

10. Gungor, Behic R. *Power Systems*. Harcourt Brace Jovanovich, New York, 1988.

11. Heydt, G. T. *Computer Analysis Methods for Power Systems*. Macmillan, New York, 1986.

12. Neuenswander, John R. *Modern Power Systems*. International Textbook, Scranton, Pa., 1971.

13. Shipley, Randall B. *Introduction to Matrices and Power Systems*. Wiley, New York, 1976.

14. Stagg, Glenn A., and A. H. El-Abiad. *Computer Methods in Power System Analysis*. McGraw-Hill, New York, 1968.

15. Stevenson, William D., Jr. *Elements of Power System Analysis*. 4th ed. McGraw-Hill, New York, 1982.

16. Stott, B., and O. Alsac. "Fast Decoupled Load Flow," *IEEE Trans. PAS-93*, 1974, pp. 859–869.

17. Tinney, W. F., and C. E. Hart. "Power Flow Solutions by Newton's Method," *IEEE Trans. PAS-86*, 1967, p. 1449.

18. Tinney, W. F., and J. W. Walker. "Direct Solution of Sparse Network Equations by Optimally Ordered Triangular Factorization," *IEEE Proc.*, Vol. 55, 1967, pp. 1801–1809.

19. Wallach, Y. *Calculations and Programs for Power System Networks*. Prentice-Hall, Englewood Cliffs, N.J., 1986.

PROBLEMS

10.1 The one-line diagram of an unloaded power system is shown in Fig. 10.12. The reactance of the transmission line is $X = 30\ \Omega$. The generators and transformers are

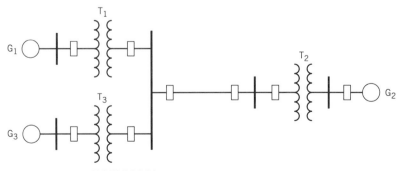

FIGURE 10.12 Power system of Problem 10.1.

rated as follows:

G$_1$:   20 MVA, 12 kV, X = 1.20 pu
G$_2$:   60 MVA, 13.8 kV, X = 1.40 pu
G$_3$:   20 MVA, 13.2 kV, X = 1.20 pu
T$_1$:   25 MVA, 12/69 kV, X = 0.08 pu
T$_2$:   60 MVA, 69/13.8 kV, X = 0.14 pu
T$_3$:   75 MVA, 13.2/69 kV, X = 0.16 pu

Choose a power base of 50 MVA and a voltage base of 12 kV in the circuit of generator 1. Form the bus admittance matrix.

10.2 The power system shown in Fig. 10.13 has the following generator and transformer data.

G$_1$:   50 MVA, 13.2 kV, X = 1.20 pu
G$_2$:   20 MVA, 13.2 kV, X = 1.10 pu
T$_1$:   60 MVA, 13.2/115 kV, X = 0.16 pu
T$_2$:   25 MVA, 34.5/13.2 kV, X = 0.10 pu
T$_3$:   60 MVA, 115/34.5 kV, X = 0.16 pu
T$_4$:   10 MVA, 4.16/34.5 kV, X = 0.05 pu

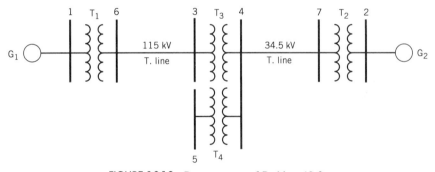

FIGURE 10.13   Power system of Problem 10.2.

The 115-kV transmission line is 50 mi long, and it has a reactance of 0.8 Ω/mi. The 34.5-kV line is 10 mi long, and it has a reactance of 1.2 Ω/mi. Choose a power base of 50 MVA and a voltage base of 115 kV for the high-voltage transmission line.

a. Choose the voltage bases in the other parts of the system in order that per-unit turns ratios of all transformers are 1:1.

b. Draw the one-line impedance diagram showing all impedance values in per unit.

c. Form the bus admittance matrix.

10.3 For the power system shown in Fig. 10.14, bus 1 is selected as the slack or swing bus and its voltage is set to V_1 = 1.0 $\underline{/0°}$ pu. The chosen power base is 100 MVA. Generator 2 delivers a real power of 0.75 pu at a voltage of 1.02 pu. The loads on buses 3 and 4 are S_{D3} = (0.40 + j0.30) pu and S_{D4} = (0.80 + j0.60) pu,

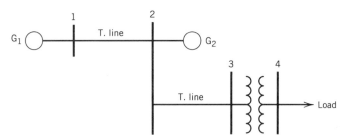

FIGURE 10.14   Power system of Problem 10.3.

respectively. The impedance parameters for the transmission lines referred to the given power base and a voltage base of 115 kV are

$$\text{Line 1–2:} \quad Z = (0.01 + j0.05) \text{ pu}; Y = j0.30 \text{ pu}$$
$$\text{Line 2–3:} \quad Z = (0.03 + j0.15) \text{ pu}; Y = j0.90 \text{ pu}$$

The transformer is connected between buses 3 and 4, and its reactance is 0.10 pu.

From a flat start, perform one iteration of the Gauss–Seidel iterative technique to find the voltages at buses 2, 3, and 4.

10.4 For the three-bus power system shown in Fig. 10.6, the transmission line per-unit reactances are

$$X_{12} = 0.75 \qquad X_{13} = 0.25 \qquad X_{23} = 0.50$$

Bus 1 is selected as swing bus, and its voltage is used as reference and set at $V_1 = 1.00\underline{/0°}$ pu. Bus 2 is a generator bus and its voltage magnitude is set at $V_2 = 1.02$ pu and its real power generation specified as $P_{G2} = 0.6$ pu. Bus 3 is a load bus with real and reactive power demands of $P_{D3} = 0.8$ pu and $Q_{D3} = 0.6$ pu. Using the Gauss–Seidel iterative technique, obtain the power flow solution of this system within a tolerance of 0.01 pu.

10.5 Using the solution to Problem 10.4, calculate the power flow through line 1–3 at both ends.

10.6 Repeat Problem 10.5 for line 2–3.

10.7 Outline the solution of the Newton–Raphson power flow method as it applies to the power system of Problem 10.4.

10.8 For the four-bus power system shown in Fig. 10.8, choose bus 1 as swing bus.

a. For each bus k, specify the bus type and determine which of the variables V_k, δ_k, P_k, and Q_k are the known input data and which are the unknowns.

b. In terms of the partial derivatives, identify the elements of the Jacobian matrix for this particular power system.

c. Set up the linearized system of equations that are solved at each iteration of the Newton's method of power flow solution. *Do not solve* the system of equations.

10.9 For the three-bus power system shown in Fig. 10.6, generator G_1 is chosen as swing generator. The line reactances are given as

$$X_{12} = 0.20 \text{ pu} \qquad X_{13} = 0.25 \text{ pu} \qquad X_{23} = 0.40 \text{ pu}$$

  a. Find the bus admittance matrix $\mathbf{Y}_{bus}$.

  b. For each bus k, specify the bus type and determine which of the variables V_k, δ_k, P_k, and Q_k are input data and which are unknowns.

  c. Set up the linearized system of equations that are solved at each iteration of the Newton power flow solution method. *Do not solve* the system of equations.

10.10 For the three-bus power system shown in Fig. 10.9, form the $\mathbf{B}'$ and $\mathbf{B}''$ matrices used in the fast-decoupled power flow method.

10.11 The transmission line parameters of a four-bus power system are given as follows:

Line 1–2: $Z = (0.020 + j0.060)$ pu; $Y = j0.080$ pu
Line 1–3: $Z = (0.015 + j0.045)$ pu; $Y = j0.060$ pu
Line 1–4: $Z = (0.008 + j0.024)$ pu; $Y = j0.030$ pu
Line 2–3: $Z = (0.010 + j0.030)$ pu; $Y = j0.040$ pu
Line 3–4: $Z = j0.100$ pu

Bus 2 is a tie-line bus and is chosen as swing bus. Bus 1 is a generator bus, and its voltage magnitude is set to $V_1 = 1.0$ pu. Find the $\mathbf{B}'$ and $\mathbf{B}''$ matrices for use in the fast-decoupled power flow technique.

Eleven

Faults, Protection, and Stability

11.1   INTRODUCTION

The normal operation of the power system at steady state that was described in the previous chapter is affected, sometimes dramatically, by the occurrence of such disturbances as overloads and short circuits. Of primary concern are short circuits, not only because they can cause large damage to the affected system component but also because they may lead to instability of the whole power system. Thus, there is a need to design protection schemes to minimize the risks involved with the occurrences of disturbances. Fault analysis forms the basis for designing such protection systems. The proper coordination of the protective relays and the correct specification of circuit breaker ratings are based on the results of fault calculations. The performance of the power system is simulated in what is called transient stability analysis under a variety of disturbances, such as short circuits, sudden large load changes, and switching operations.

Fault calculations provide currents and voltages in a power system under faulted conditions. The modeling of various power system components, symmetrical components, the interconnections of the sequence networks, and symmetrical and unsymmetrical faults are presented in the second section of this chapter.

Power system protection is an important concern because short circuits present danger of damage to the equipment and loss of synchronism of the synchronous machines. The different types of protection systems and their applications for protecting the various power system components are discussed in the third section of this chapter.

Transient stability studies investigate the ability of the power system to remain in synchronism during major disturbances, such as equipment failure, major load changes, or momentary faults. Basic stability concepts are presented in the last section, along with a brief description of the models for generators, excitation systems, and governor-turbine systems.

11.2   FAULT ANALYSIS

The normal mode of operation of a power system is balanced three-phase AC. However, there are undesirable but unavoidable incidents that may temporarily disrupt normal conditions, as when the insulation of the system fails at any point or when a conducting material comes in contact with a bare conductor. Then we say a fault has occurred. A fault may be caused by lightning, trees falling on the electric wires, vehicular collision with the poles or towers, vandalism, and so forth.

Faults may be classified under four types. The different types of fault are listed here in the order of the frequency of their occurrence.

1. Single line-to-ground fault (SLG)
2. Line-to-line fault (L-L)
3. Double line-to-ground fault (2LG)
4. Balanced three-phase fault

Fault calculations provide information on currents and voltages in a power system during fault conditions. Short-circuit currents are computed for each relay and circuit breaker location and for various system contingency conditions, such as lines or generating units out of service, in order to determine minimum and maximum fault currents. This information is useful to the engineer in selecting circuit breakers for fault interruption, selecting relays for fault detection, and determining the relay settings, which is referred to as relay coordination. The proper selection and setting of protective devices ensure minimum disruption of electrical service and limit possible damage to the faulted equipment.

11.2.1   Three-Phase Fault Analysis

Sufficient accuracy in fault studies can be obtained with certain simplifications in the model of the power system. These assumptions include the following:

a. Shunt elements in the transformer model are neglected; that is, magnetizing currents and core losses are omitted.

b. Shunt capacitances in the transmission line model are neglected.

c. Transformers are set at nominal tap positions.

d. All internal voltage sources are set equal to $1.0\underline{/0°}$. This is equivalent to neglecting prefault load currents.

Three-phase fault calculations can be performed on a per-phase basis because the power system remains effectively balanced, or symmetrical, during a three-phase fault. Thus, the various power system components are represented by single-phase equivalent circuits wherein all three-phase connections are assumed to be converted to their equivalent wye connections. Calculations are performed using impedances per phase, phase currents, and line-to-neutral voltages.

Fault analysis, like any other power system calculation, is more conveniently performed using per-unit representation. The power system components are described in terms of their per-unit impedances or per-unit admittances. The power base is selected and is used for all parts of the power system. The voltage bases are different for various parts of the system, and they are selected so that the per-unit turns ratio of the transformers is equal to unity. The current base and the impedance base are computed using the specified power base and voltage bases.

EXAMPLE 11.1

A three-phase synchronous generator is rated 50 MVA, 13.2 kV, 0.8 power factor lagging and has a synchronous reactance of 20%. The generator is connected to a 60-MVA, 13.4/138-kV, three-phase transformer having a reactance of 15%. The generator is initially operating at no load and rated terminal voltage. A three-phase fault suddenly occurs on the transformer's high-voltage terminals. Determine the short-circuit current supplied by the generator and its terminal voltage.

Solution Choose the generator ratings as base quantities; thus, $S_{base} = 50$ MVA and $V_{base} = 13.2$ kV (line-to-line). The base current is calculated as follows.

$$I_{base} = 50{,}000/(13.2\sqrt{3}) = 2187 \text{ A}$$

The terminal voltage of the generator is taken as reference phasor; thus,

$$\mathbf{V}_t = 1.0\underline{/0°} \text{ pu}$$

Because the generator is initially operating at no load, the internal voltage is equal to the terminal voltage.

$$\mathbf{E}_a = \mathbf{V}_t = 1.0\underline{/0°} \text{ pu}$$

When a three-phase fault occurs at the high side of the transformer, the internal voltage of the generator is assumed to remain constant in determining the short-circuit current. The transformer reactance is converted to per unit referred to the chosen bases as follows.

$$X_t = (0.15)(50/60)(13.4/13.2)^2 = 0.129 \text{ pu}$$

Therefore, the total series reactance X_T is given by

$$X_T = X_s + X_t = 0.20 + 0.129 = 0.329 \text{ pu}$$

Hence, the short-circuit current is found by dividing the internal voltage by the total reactance.

$$\mathbf{I}_{sc} = \frac{\mathbf{E}_a}{jX_T} = \frac{1.0 \angle 0°}{j0.329} = 3.04 \angle -90° \text{ pu}$$
$$= (3.04)(2187) = 6648 \text{ A}$$

The terminal voltage of the generator is calculated as

$$\mathbf{V}_t = jX_t\mathbf{I}_{sc} = (j0.129)(3.04 \angle -90°) = 0.39 \angle 0° \text{ pu}$$
$$= (0.39)(13.2) = 5.15 \text{ kV} \quad \text{(line-to-line)}$$

DRILL PROBLEMS

D11.1 A step-up transformer is rated 1000 MVA, 26/345 kV and has a series impedance of $(0.004 + j0.085)$ pu based on its own ratings. It is connected to a 26-kV, 800-MVA generator that can be represented as a constant-voltage source in series with a reactance of $j1.65$ pu based on the generator ratings. The system is initially operating at no load when a three-phase fault occurs at the high-voltage terminals of the transformer. Choose a system power base of 1000 MVA and a voltage base in the generator circuit of 26 kV.

    a. Convert the per-unit generator reactance to the system base.

    b. Find the short-circuit current, in amperes, at the fault.

    c. Find the short-circuit current, in amperes, supplied by the generator.

D11.2 Two 50-MVA, 12-kV, 60-Hz generators G_1 and G_2 are connected to bus A. Each generator has a synchronous reactance of 0.15 pu. A three-phase, 100-MVA, 12/69-kV transformer of reactance 0.10 pu is connected to A. The transformer is operating at rated voltage and no load when a three-phase fault occurs on its high-voltage side. Calculate

a. The short-circuit current

b. The short-circuit current supplied by each generator

c. The voltage at bus A

D11.3 A 2000-kVA, 13.2-kV synchronous motor is connected through a distribution feeder to a 2000-kVA, 13.2-kV synchronous generator. The subtransient reactance of each machine is equal to 20% based on its own ratings, and the feeder has a reactance of 8.5 Ω. The motor is initially drawing 1500 kW at 13 kV and 0.8 PF leading when a three-phase fault occurs at the motor terminals. Determine

a. The subtransient short-circuit current

b. The short-circuit current supplied by the generator

c. The short-circuit current contribution of the motor

11.2.2   Symmetrical Components

The method of symmetrical components is used to solve power system problems involving unbalanced polyphase voltages and currents. It is analogous to the Fourier analysis of nonsinusoidal wave shapes wherein a nonsine wave is resolved into a number of sine waves of various frequencies. In symmetrical components, the unbalanced set of polyphase phasors is resolved into a number of balanced sets of phasors. After the unbalanced sets of voltage and current phasors are resolved into their symmetrical components, the power system may be solved using per-phase analysis.

A balanced, or symmetrical, set of phasors is defined as a set of phasors that have equal magnitudes and are separated by equal angles. Thus, the phasors in a three-phase balanced set are separated from each other by an angle of 120°.

In a three-phase system, an unbalanced set of phasors is resolved by using three sets of balanced phasors, namely positive sequence, negative sequence, and zero sequence. The positive (or abc) sequence consists of three phasors of equal magnitude and separated from each other by an angle of 120°. The negative (or cba) sequence also consists of three phasors of equal magnitude and separated from each other by an angle of 120°. The zero sequence also consists of three phasors of equal magnitude, but they are all in phase. These sequence components are illustrated in Fig. 11.1.

The a phase is considered as the principal phase, and its sequence components are used to represent the other phases. Thus, the sequence components of phases b and c are given by Eqs. 11.1, 11.2, and 11.3.

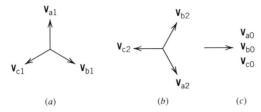

FIGURE 11.1   Sequence components of voltages.

$$\mathbf{V}_{b1} = \mathbf{V}_{a1}e^{j240°}$$
$$\mathbf{V}_{c1} = \mathbf{V}_{a1}e^{j120°} \qquad (11.1)$$
$$\mathbf{V}_{b2} = \mathbf{V}_{a2}e^{j120°}$$
$$\mathbf{V}_{c2} = \mathbf{V}_{a2}e^{j240°} \qquad (11.2)$$
$$\mathbf{V}_{b0} = \mathbf{V}_{a0}e^{j360°} = \mathbf{V}_{a0}$$
$$\mathbf{V}_{c0} = \mathbf{V}_{a0}e^{j360°} = \mathbf{V}_{a0} \qquad (11.3)$$

By introducing the operator $a = e^{j120°} = 1\underline{/120°}$, Eqs. 11.1 and 11.2 may be written as

$$\mathbf{V}_{b1} = a^2\mathbf{V}_{a1}$$
$$\mathbf{V}_{c1} = a\mathbf{V}_{a1} \qquad (11.4)$$
$$\mathbf{V}_{b2} = a\mathbf{V}_{a2}$$
$$\mathbf{V}_{c2} = a^2\mathbf{V}_{a2} \qquad (11.5)$$

The phase voltages are resolved into their sequence components and are expressed in terms of the sequence components of the principal phase a as follows:

$$\mathbf{V}_a = \mathbf{V}_{a0} + \mathbf{V}_{a1} + \mathbf{V}_{a2} \qquad (11.6)$$
$$\mathbf{V}_b = \mathbf{V}_{b0} + \mathbf{V}_{b1} + \mathbf{V}_{b2}$$
$$= \mathbf{V}_{a0} + a^2\mathbf{V}_{a1} + a\mathbf{V}_{a2} \qquad (11.7)$$
$$\mathbf{V}_c = \mathbf{V}_{c0} + \mathbf{V}_{c1} + \mathbf{V}_{c2}$$
$$= \mathbf{V}_{a0} + a\mathbf{V}_{a1} + a^2\mathbf{V}_{a2} \qquad (11.8)$$

In matrix form, Eqs. 11.6 to 11.8 may be expressed as

$$\begin{bmatrix} \mathbf{V}_a \\ \mathbf{V}_b \\ \mathbf{V}_c \end{bmatrix} = \begin{bmatrix} 1 & 1 & 1 \\ 1 & a^2 & a \\ 1 & a & a^2 \end{bmatrix} \begin{bmatrix} \mathbf{V}_{a0} \\ \mathbf{V}_{a1} \\ \mathbf{V}_{a2} \end{bmatrix} \qquad (11.9)$$

Adding Eqs. 11.6, 11.7, and 11.8 yields

$$\mathbf{V}_a + \mathbf{V}_b + \mathbf{V}_c = 3\mathbf{V}_{a0} + (1 + a^2 + a)\mathbf{V}_{a1} + (1 + a + a^2)\mathbf{V}_{a2} \qquad (11.10)$$

The quantity inside the parentheses on the right-hand side of Eq. 11.10 may be shown to reduce to zero; that is,

$$\begin{aligned} 1 + a + a^2 &= 1 + e^{j120°} + e^{j240°} \\ &= 1\underline{/0°} + 1\underline{/120°} + 1\underline{/240°} \\ &= 0.0 + j0.0 \end{aligned} \qquad (11.11)$$

Therefore, the zero-sequence component of the voltage of the principal phase a is found as

$$\mathbf{V}_{a0} = \tfrac{1}{3}(\mathbf{V}_a + \mathbf{V}_b + \mathbf{V}_c) \qquad (11.12)$$

Multiplying Eqs. 11.7 and 11.8 by the operators a and a^2, respectively, and adding their products to Eq. 11.6 yields

$$\mathbf{V}_a + a\mathbf{V}_b + a^2\mathbf{V}_c = (1 + a + a^2)\mathbf{V}_{a0} + 3\mathbf{V}_{a1} + (1 + a + a^2)\mathbf{V}_{a2} \qquad (11.13)$$

By using Eq. 11.11, the positive-sequence component of the voltage is determined.

$$\mathbf{V}_{a1} = \tfrac{1}{3}(\mathbf{V}_a + a\mathbf{V}_b + a^2\mathbf{V}_c) \qquad (11.14)$$

Similarly, Eqs. 11.7 and 11.8 may be multiplied by the operators a^2 and a, respectively, and added to Eq. 11.6. Upon simplifying, the expression for the negative-sequence component of the voltage is obtained.

$$\mathbf{V}_{a2} = \tfrac{1}{3}(\mathbf{V}_a + a^2\mathbf{V}_b + a\mathbf{V}_c) \qquad (11.15)$$

Equations 11.13, 11.14, and 11.15 may be grouped together and written in matrix form as follows:

$$\begin{bmatrix} \mathbf{V}_{a0} \\ \mathbf{V}_{a1} \\ \mathbf{V}_{a2} \end{bmatrix} = \frac{1}{3} \begin{bmatrix} 1 & 1 & 1 \\ 1 & a & a^2 \\ 1 & a^2 & a \end{bmatrix} \begin{bmatrix} \mathbf{V}_a \\ \mathbf{V}_b \\ \mathbf{V}_c \end{bmatrix} \qquad (11.16)$$

EXAMPLE 11.2

Find the sequence components of the unbalanced power system whose phase voltages are given by

$$\mathbf{V}_{an} = 100 \underline{/0^\circ} \text{ V}$$

$$\mathbf{V}_{bn} = 80 \underline{/-110^\circ} \text{ V}$$

$$\mathbf{V}_{cn} = 90 \underline{/130^\circ} \text{ V}$$

Solution   By using Eqs. 11.12, 11.14, and 11.15, the sequence components of the voltage are computed as follows:

$$\mathbf{V}_{a0} = \tfrac{1}{3}(\mathbf{V}_{an} + \mathbf{V}_{bn} + \mathbf{V}_{cn})$$
$$= \tfrac{1}{3}(100 \underline{/0^\circ} + 80 \underline{/-110^\circ} + 90 \underline{/130^\circ}) = 5.35 \underline{/-22.85^\circ} \text{ V}$$

$$\mathbf{V}_{a1} = \tfrac{1}{3}(\mathbf{V}_{an} + a\mathbf{V}_{bn} + a^2\mathbf{V}_{cn})$$
$$= \tfrac{1}{3}(100 \underline{/0^\circ} + 1 \underline{/120^\circ}\, 80 \underline{/-110^\circ} + 1 \underline{/240^\circ}\, 90 \underline{/130^\circ})$$
$$= 89.68 \underline{/6.3^\circ} \text{ V}$$

$$\mathbf{V}_{a2} = \tfrac{1}{3}(\mathbf{V}_{an} + a^2\mathbf{V}_{bn} + a\mathbf{V}_{cn})$$
$$= \tfrac{1}{3}(100 \underline{/0^\circ} + 1 \underline{/240^\circ}\, 80 \underline{/-110^\circ} + 1 \underline{/120^\circ}\, 90 \underline{/130^\circ})$$
$$= 9.77 \underline{/-52.6^\circ} \text{ V}$$

The transformation matrix in Eq. 11.9 relating the phase voltages to their sequence components and the matrix used in Eq. 11.13 to obtain the sequence components from the original unbalanced voltages equally apply to a system of unbalanced three-phase currents. Thus, the phase currents are expressed in terms of their sequence components as follows:

$$\begin{bmatrix} \mathbf{I}_a \\ \mathbf{I}_b \\ \mathbf{I}_c \end{bmatrix} = \begin{bmatrix} 1 & 1 & 1 \\ 1 & a^2 & a \\ 1 & a & a^2 \end{bmatrix} \begin{bmatrix} \mathbf{I}_{a0} \\ \mathbf{I}_{a1} \\ \mathbf{I}_{a2} \end{bmatrix} \qquad (11.17)$$

Similarly, the sequence components of the current are obtained from the original unbalanced phase currents by using the following equation:

$$\begin{bmatrix} \mathbf{I}_{a0} \\ \mathbf{I}_{a1} \\ \mathbf{I}_{a2} \end{bmatrix} = \frac{1}{3}\begin{bmatrix} 1 & 1 & 1 \\ 1 & a & a^2 \\ 1 & a^2 & a \end{bmatrix} \begin{bmatrix} \mathbf{I}_a \\ \mathbf{I}_b \\ \mathbf{I}_c \end{bmatrix} \qquad (11.18)$$

EXAMPLE 11.3

The sequence components of the current in a portion of a power system are given as

$$\mathbf{I}_{a1} = 2.5\,\underline{/-90^\circ}\text{ pu}$$
$$\mathbf{I}_{a2} = 1.65\,\underline{/90^\circ}\text{ pu}$$
$$\mathbf{I}_{a0} = 0.85\,\underline{/90^\circ}\text{ pu}$$

Obtain the three phase currents.

Solution By using Eq. 11.17, the phase currents are found from the sequence components as follows:

$$\mathbf{I}_a = \mathbf{I}_{ao} + \mathbf{I}_{a1} + \mathbf{I}_{a2} = 0.85\,\underline{/90^\circ} + 2.5\,\underline{/-90^\circ} + 1.65\,\underline{/90^\circ} = 0$$
$$\mathbf{I}_b = \mathbf{I}_{ao} + a^2\mathbf{I}_{a1} + a\mathbf{I}_{a2}$$
$$= 0.85\,\underline{/90^\circ} + 1\,\underline{/240^\circ}\ 2.5\,\underline{/-90^\circ} + 1\,\underline{/120^\circ}\ 1.65\,\underline{/90^\circ}$$
$$= 3.81\,\underline{/160.5^\circ}\text{ pu}$$
$$\mathbf{I}_c = \mathbf{I}_{ao} + a\mathbf{I}_{a1} + a^2\mathbf{I}_{a2}$$
$$= 0.85\,\underline{/90^\circ} + 1\,\underline{/120^\circ}\ 2.5\,\underline{/-90^\circ} + 1\,\underline{/240^\circ}\ 1.65\,\underline{/90^\circ}$$
$$= 3.81\,\underline{/19.5^\circ}\text{ pu}$$

DRILL PROBLEMS

D11.4 The phase current in a wye-connected, unbalanced load are

$$\mathbf{I}_a = 50 - j40\text{ A}$$
$$\mathbf{I}_b = -30 - j20\text{ A}$$
$$\mathbf{I}_c = -40 + j30\text{ A}$$

Determine the sequence components of the currents.

D11.5 The line-to-line voltages across a three-phase, wye-connected load consisting of $Z = 100\,\underline{/30^\circ}\ \Omega$ in each phase are as follows:

$$\mathbf{V}_{ab} = 205\,\underline{/0^\circ}\text{ V}$$
$$\mathbf{V}_{bc} = 250\,\underline{/-125^\circ}\text{ V}$$

$$\mathbf{V}_{ca} = 214\underline{/107°} \text{ V}$$

Determine the sequence components of the voltages.

D11.6 The sequence components of the current in phase a are

$$\mathbf{I}_{a1} = 6\underline{/-15°} \text{ pu}$$
$$\mathbf{I}_{a2} = 8\underline{/225°} \text{ pu}$$
$$\mathbf{I}_{a0} = 5\underline{/-165°} \text{ pu}$$

Determine the phase currents $\mathbf{I}_a$, $\mathbf{I}_b$, and $\mathbf{I}_c$.

11.2.3 Unsymmetrical Fault Analysis

Consider the portion of the power system shown in Fig. 11.2. The system is said to be symmetrical if it satisfies the following conditions:

1. Phase conductor impedances are equal, that is,

$$Z_{aa} = Z_{bb} = Z_{cc}$$

2. Mutual impedances between phase conductors are equal, that is,

$$Z_{ab} = Z_{bc} = Z_{ca}$$

3. Mutual impedances between phase conductors and the neutral wire are equal, that is,

$$Z_{an} = Z_{bn} = Z_{cn}$$

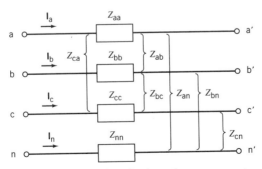

FIGURE 11.2 A portion of a three-phase power system.

The voltages on the right terminals (marked a', b', and c') of the portion of the system shown in Fig. 11.2 are identified as V_a, V_b, and V_c, and those on the left terminals (marked a, b, and c) are denoted as E_a, E_b, and E_c. The line currents may be expressed in terms of their sequence components as in Eq. 11.17. If the three currents are added, their sum is given by

$$I_a + I_b + I_c = 3I_{a0} = -I_n \tag{11.19}$$

A Kirchhoff's voltage equation may be written around each loop containing one phase with the neutral as return.

$$E_a = I_a Z_{aa} + I_b Z_{ab} + I_c Z_{ca} + I_n Z_{an} + V_a$$
$$- I_a Z_{an} - I_b Z_{bn} - I_c Z_{cn} - I_n Z_{nn} \tag{11.20}$$
$$E_b = I_b Z_{bb} + I_a Z_{ab} + I_c Z_{bc} + I_n Z_{bn} + V_b$$
$$- I_a Z_{an} - I_b Z_{bn} - I_c Z_{cn} - I_n Z_{nn} \tag{11.21}$$
$$E_c = I_c Z_{cc} + I_a Z_{ca} + I_b Z_{bc} + I_n Z_{cn} + V_c$$
$$- I_a Z_{an} - I_b Z_{bn} - I_c Z_{cn} - I_n Z_{nn} \tag{11.22}$$

Assuming that the power system is a symmetrical network and using the relation given in Eq. 11.19, Eqs. 11.20 to 11.22 may be expressed as

$$E_a = I_a Z_{aa} + (I_b + I_c)Z_{ab} + V_a + I_n(2Z_{an} - Z_{nn}) \tag{11.23}$$
$$E_b = I_b Z_{aa} + (I_a + I_c)Z_{ab} + V_b + I_n(2Z_{an} - Z_{nn}) \tag{11.24}$$
$$E_c = I_c Z_{aa} + (I_a + I_b)Z_{ab} + V_c + I_n(2Z_{an} - Z_{nn}) \tag{11.25}$$

The zero-sequence component of the voltage is found by adding Eqs. 11.23 to 11.25 and dividing the sum by 3. Thus,

$$E_{a0} = \tfrac{1}{3}(E_a + E_b + E_c)$$
$$= \tfrac{1}{3}[(I_a + I_b + I_c)(Z_{aa} + 2Z_{ab}) + I_n(6Z_{an} - 3Z_{nn}) + (V_a + V_b + V_c)]$$
$$= I_{a0}(Z_{aa} + 2Z_{ab} - 6Z_{an} + 3Z_{nn}) + V_{a0} \tag{11.26}$$

The positive-sequence component of the voltage is obtained by multiplying Eq. 11.24 by the operator a, multiplying Eq. 11.25 by the operator a^2, and adding the two products to Eq. 11.23. The sum is divided by 3; hence,

$$E_{a1} = \tfrac{1}{3}(E_a + aE_b + a^2E_c)$$
$$= \tfrac{1}{3}[(I_a + aI_b + a^2I_c)(Z_{aa} - Z_{ab}) + (V_a + aV_b + a^2V_c)]$$
$$= I_{a1}(Z_{aa} - Z_{ab}) + V_{a1} \tag{11.27}$$

Similarly, the negative-sequence component is obtained by multiplying Eq. 11.24 by a^2, multiplying Eq. 11.25 by a, adding the two products to

Eq. 11.23, and dividing the sum by 3; thus,

$$\mathbf{E}_{a2} = \tfrac{1}{3}(\mathbf{E}_a + a^2\mathbf{E}_b + a\mathbf{E}_c)$$
$$= \tfrac{1}{3}[(\mathbf{I}_a + a^2\mathbf{I}_b + a\mathbf{I}_c)(Z_{aa} - Z_{ab}) + (\mathbf{V}_a + a^2\mathbf{V}_b + a\mathbf{V}_c)]$$
$$= \mathbf{I}_{a2}(Z_{aa} - Z_{ab}) + \mathbf{V}_{a2} \qquad (11.28)$$

It is generally assumed that the voltages from the generators are of positive sequence only. Therefore, the zero-sequence voltage $\mathbf{E}_{a0}$ and the negative-sequence voltage $\mathbf{E}_{a2}$ on the source side are not present, and they can be set equal to zero in Eqs. 11.26 and 11.28, respectively. Thus, Eqs. 11.26 to 11.28 may be rewritten as

$$\mathbf{V}_{a0} = -\mathbf{I}_{a0}(Z_{aa} + 2Z_{ab} - 6Z_{an} + 3Z_{nn})$$
$$= -\mathbf{I}_{a0}Z_0 \qquad (11.29)$$
$$\mathbf{V}_{a1} = -\mathbf{I}_{a1}(Z_{aa} - Z_{ab}) + \mathbf{E}_{a1}$$
$$= -\mathbf{I}_{a1}Z_1 + \mathbf{E}_{a1} \qquad (11.30)$$
$$\mathbf{V}_{a2} = -\mathbf{I}_{a2}(Z_{aa} - Z_{ab})$$
$$= -\mathbf{I}_{a2}Z_2 \qquad (11.31)$$

The expression for the zero-sequence component of the voltage as given by Eq. 11.29 contains no positive- or negative-sequence terms. Also, the expression given by Eq. 11.30 contains only the positive-sequence components and no negative- or zero-sequence terms. Similarly, Eq. 11.31 contains only negative-sequence and no positive- or zero-sequence terms. In other words, Eqs. 11.29 to 11.31 are completely uncoupled.

It is also seen from Eq. 11.29 that the zero-sequence voltage $\mathbf{V}_{a0}$ depends only on the zero-sequence current $\mathbf{I}_{a0}$ and the impedance $(Z_{aa} + 2Z_{ab} - 6Z_{an} + 3Z_{nn})$, which is designated as Z_0 and is referred to as the *zero-sequence impedance*. Also, from Eq. 11.30, the positive-sequence voltage $\mathbf{V}_{a1}$ depends only on the positive-sequence current $\mathbf{I}_{a1}$ and the impedance $Z_1 = (Z_{aa} - Z_{ab})$, which is called the *positive-sequence impedance*. Furthermore, the negative-sequence voltage $\mathbf{V}_{a2}$ depends only on the negative-sequence current $\mathbf{I}_{a2}$ and the *negative-sequence impedance* $Z_2 = (Z_{aa} - Z_{ab})$.

In Eqs. 11.29 to 11.31, it is assumed that the voltage of the source $\mathbf{E}_{a1}$ and all the impedances are known; thus, there are six unknown variables to determine. These are the sequence components of the current and the sequence components of the voltage. There are exactly three equations (11.29–11.31) relating the current and voltage sequence components, and three more are needed for a unique solution. The additional three relationships are provided by the interconnection of the sequence networks for the various types of faults.

Sequence Networks and Sequence Impedances   The three equations 11.29 to 11.31 can be represented by three networks with no mutual coupling as shown in Fig. 11.3. These networks are called the *positive-sequence, negative-sequence,* and *zero-sequence networks.* They are constructed by using the positive-sequence, negative-sequence, and zero-sequence impedances of the various power system components.

The series impedance of a transmission line derived in Chapter 9 and also used in Section 10.1.3 represents the positive-sequence and negative-sequence impedances. The zero-sequence impedance is not the same as the positive- or negative-sequence impedance. The zero-sequence impedance is affected by several factors, including the characteristics of the earth as a return path and the type and number of ground wires used. A good discussion of the derivation of the zero-sequence impedance of a transmission line is given in Ref. 5, *Electrical Transmission and Distribution Reference Book.*

When the analysis involves calculation of the steady-state or sustained short-circuit current, the direct-axis synchronous reactance X_d of the synchronous machine is taken as the positive-sequence reactance. If subtransient or transient currents are required, X_d'' or X_d' is used instead, respectively.

The rotating magnetic field due to positive-sequence currents in a synchronous machine rotates in the same direction as the rotor. However, the negative-sequence currents produce a rotating magnetic field at synchronous speed but in the opposite direction; thus, this negative-sequence field passes over the poles, field winding, and damper windings at twice synchronous speed, resulting in higher opposition and effectively lower reactance.

The zero-sequence currents, although in phase, are physically displaced 120° from each other because of the distribution of the windings; thus, the sum of the magnetic fields is equal to zero except for the effects of unbalances. Therefore, the zero-sequence reactance due to the small flux produced by the zero-sequence currents and slot and end connection leakage is still smaller than the negative-sequence reactance. Typical values of reactances of synchronous machines are given in Table 10 of Appendix A.

The impedance of a transformer to both positive- and negative-sequence currents is the same, and the equivalent circuits are those given in Section 10.1.3. The phase shift, if there is any, has the same magnitude but opposite direction. Phase shifts will be neglected in the following discussions.

The zero-sequence equivalent circuit of a three-phase transformer bank depends on the type of connection and the method of grounding. In a delta-

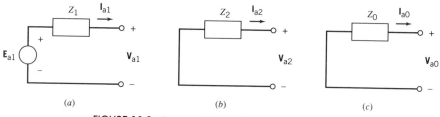

(a)   (b)   (c)

FIGURE 11.3   Representation of the sequence networks.

connected winding, zero-sequence currents can flow, but because they are equal in each phase of the delta, they do not leave the terminals of the transformer; thus, it appears as an open circuit as viewed from the external circuit. If the neutral of a wye-connected winding is not connected to ground, the three phase currents add to zero and there is no zero-sequence current (i.e., $I_{a0} = 0$); thus, it appears as an open circuit to the external circuit. When the neutral of a wye-connected winding is grounded, the sum of the three phase currents is equal to $3I_{a0}$ flowing to or from ground. Since the zero-sequence network represents only one phase with current I_{a0} flowing, any ground impedance Z_g is included as $3Z_g$ to provide the proper neutral voltage. Some of the typical transformer connections and their zero-sequence equivalent circuits are shown in Fig. 11.4.

Single Line-to-Ground Fault   A single line-to-ground (SLG) fault is the most commonly occurring unsymmetrical fault. It may be caused by a vehicular accident causing one of the phase conductors to fall and come in contact with the earth, or it may be caused by tree branches, or it could be caused by flashovers across dusty insulators during rainshowers. An SLG fault is illustrated in Fig. 11.5.

Assuming that prefault currents are negligible, the SLG fault is described by the following voltage and current relationships.

$$V_a = 0 \tag{11.32}$$

$$I_b = I_c = 0 \tag{11.33}$$

The sequence components of the short-circuit current are found by using Eq. (11.18). Thus,

$$\begin{bmatrix} I_{a0} \\ I_{a1} \\ I_{a2} \end{bmatrix} = \frac{1}{3} \begin{bmatrix} 1 & 1 & 1 \\ 1 & a & a^2 \\ 1 & a^2 & a \end{bmatrix} \begin{bmatrix} I_a \\ 0 \\ 0 \end{bmatrix} = \frac{1}{3} \begin{bmatrix} I_a \\ I_a \\ I_a \end{bmatrix} \tag{11.34}$$

Therefore, it is seen that the sequence components of the current are all equal to each other; that is,

$$I_{a0} = I_{a1} = I_{a2} = \tfrac{1}{3} I_a \tag{11.35}$$

The voltage of the principal phase a may be expressed in terms of its sequence components as

$$V_a = V_{a0} + V_{a1} + V_{a2} = 0 \tag{11.36}$$

The expressions for the sequence components of the voltage are given by Eqs. 11.29 to 11.31. Substituting these expressions in Eq. 11.36 and rearranging

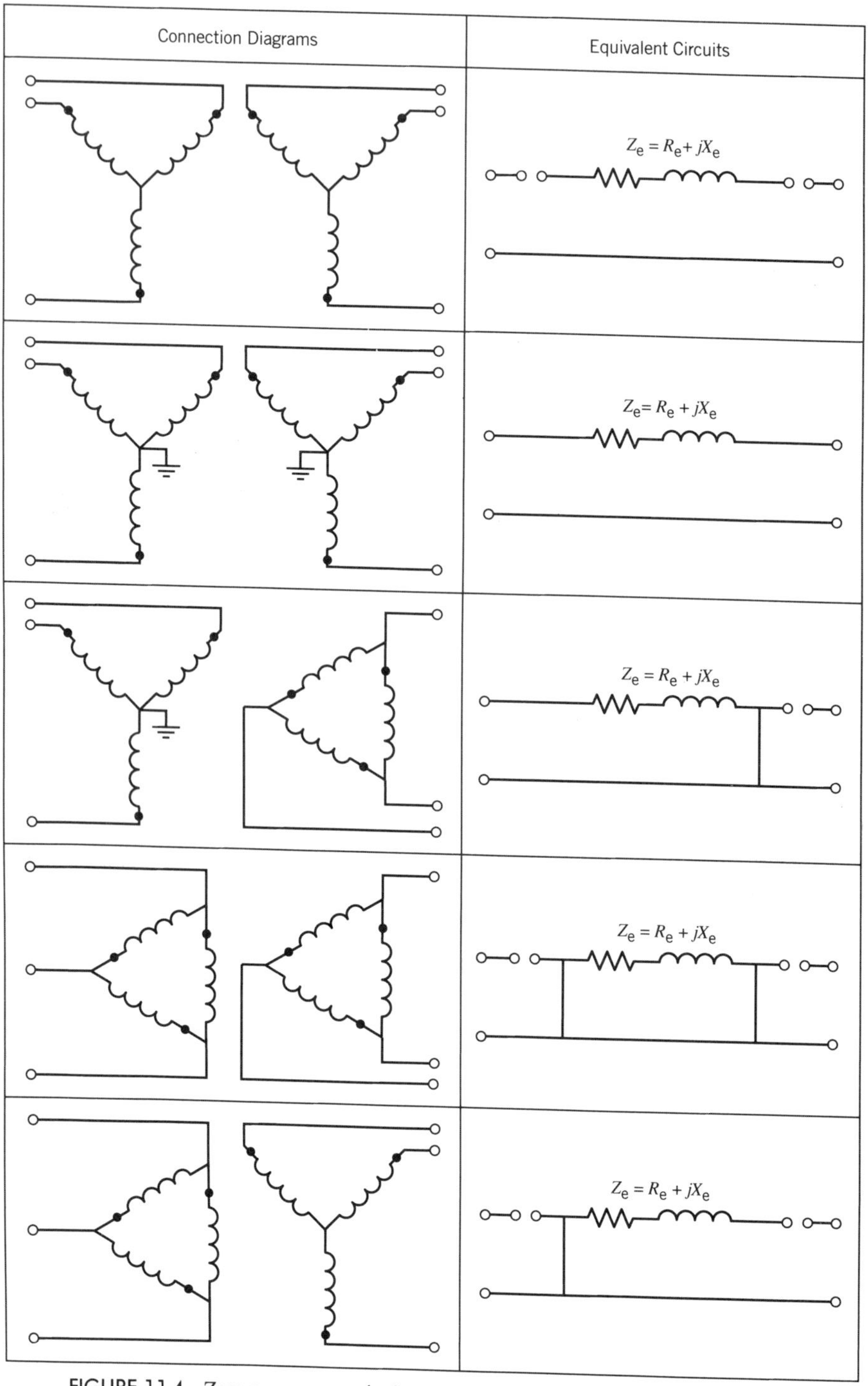

Connection Diagrams	Equivalent Circuits
	$Z_e = R_e + jX_e$
	$Z_e = R_e + jX_e$
	$Z_e = R_e + jX_e$
	$Z_e = R_e + jX_e$
	$Z_e = R_e + jX_e$

FIGURE 11.4   Zero-sequence equivalent circuits of three-phase transformer banks.

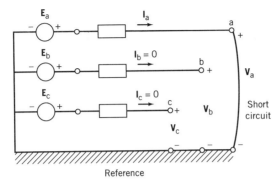

FIGURE 11.5 A single line-to-ground fault.

the terms yield

$$E_{a1} - I_{a1}(Z_1 + Z_2 + Z_0) = 0 \qquad (11.37)$$

The expression for I_{a1} is obtained from Eq. 11.37, and using Eq. 11.35 provides the expressions for I_{a2} and I_{a0} as follows:

$$\begin{aligned}
I_{a1} &= \frac{E_{a1}}{Z_1 + Z_2 + Z_0} \\
&= I_{a2} \\
&= I_{a0}
\end{aligned} \qquad (11.38)$$

The short-circuit current I_a is found as the sum of its sequence components; thus,

$$\begin{aligned}
I_a &= I_{a0} + I_{a1} + I_{a2} \\
&= 3\frac{E_{a1}}{Z_1 + Z_2 + Z_0}
\end{aligned} \qquad (11.39)$$

Equations 11.35 and 11.36 indicate that the sequence networks are to be interconnected in series for a single line-to-ground fault. This is illustrated in Fig. 11.6.

EXAMPLE 11.4

A three-phase generator is rated 150 MVA, 13.8 kV and has positive-, negative-, and zero-sequence reactances of 20%, 10%, and 5%, respectively. The generator is connected to a transmission line having positive- and negative-sequence reactances of 10% and a zero-sequence reactance of 30%. The

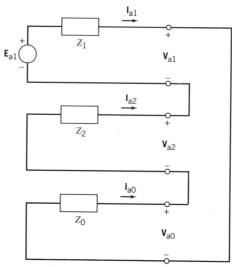

FIGURE 11.6 Interconnection of sequence networks for a single line-to-ground fault.

generator is operating at rated voltage and no-load conditions when a single line-to-ground short circuit occurs at the end of the line. Determine the phase currents at the fault and the phase voltages at the terminals of the generator. A one-line diagram of the power system is shown in Fig. 11.7.

Solution Assume that the fault is a phase-a-to-ground fault. The sequence networks are interconnected as shown in Fig. 11.8.

Since the generator is initially operating at no load and rated voltage, the generator voltage is taken as 1.0 per unit and used as reference. Thus,

$$\mathbf{E}_{a1} = 1.0 \underline{/0°} \text{ pu}$$

The sequence components of the current at the fault are computed as follows:

$$\mathbf{I}_{a1} = (1.0\underline{/0°})/(j0.3 + j0.2 + j0.35) = (1.0\underline{/0°})/(j0.85)$$

$$= 1.176\underline{/-90°} \text{ pu} = -j1.176 \text{ pu}$$

$$= \mathbf{I}_{a0} = \mathbf{I}_{a2}$$

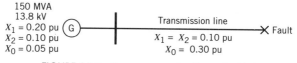

FIGURE 11.7 Power system for Example 10.8.

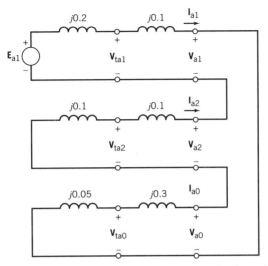

FIGURE 11.8   Interconnection of sequence networks for Example 11.4.

The phase currents are computed by using Eq. 11.17 as follows:

$$
\begin{bmatrix} \mathbf{I}_a \\ \mathbf{I}_b \\ \mathbf{I}_c \end{bmatrix} = \begin{bmatrix} 1 & 1 & 1 \\ 1 & a^2 & a \\ 1 & a & a^2 \end{bmatrix} \begin{bmatrix} 1.176\underline{/-90°} \\ 1.176\underline{/-90°} \\ 1.176\underline{/-90°} \end{bmatrix} = \begin{bmatrix} 3.528\underline{/-90°} \\ 0.0 \\ 0.0 \end{bmatrix} \text{pu}
$$

The base current is computed as follows:

$$
I_{base} = \frac{150,000}{13.8\sqrt{3}} = 6276 \text{ A}
$$

The per-unit short-circuit current is multiplied by the base current to give the actual current in amperes. Thus,

$$
\mathbf{I}_a = (3.528)(6276) = 22,140 \text{ A}
$$

The sequence components of the voltage at the terminals of the generator are computed as follows:

$\mathbf{V}_{ta0} = -j0.05\mathbf{I}_{a0} = -(j0.05)(-j1.176) = -0.0588 \text{ pu}$

$\mathbf{V}_{ta1} = \mathbf{E}_{a1} - j0.20\mathbf{I}_{a1} = 1.0\underline{/0°} - (j0.20)(-j1.176) = 0.7648 \text{ pu}$

$\mathbf{V}_{ta2} = -j0.10\mathbf{I}_{a2} = -(j0.10)(-j1.176) = -0.1176 \text{ pu}$

The phase voltages at the generator terminals are computed by using Eq. 11.9 as follows:

$$
\begin{bmatrix} \mathbf{V}_{ta} \\ \mathbf{V}_{tb} \\ \mathbf{V}_{tc} \end{bmatrix} = \begin{bmatrix} 1 & 1 & 1 \\ 1 & a^2 & a \\ 1 & a & a^2 \end{bmatrix} \begin{bmatrix} -0.0588 \\ 0.7648 \\ -0.1176 \end{bmatrix} = \begin{bmatrix} 0.588\,\underline{/0°} \\ 0.855\,\underline{/-117°} \\ 0.855\,\underline{/\;117°} \end{bmatrix} \text{pu}
$$

The line-to-line voltages at the generator terminals are found by multiplying the per-unit values by the base voltage.

$$
\begin{bmatrix} \mathbf{V}_{ta} \\ \mathbf{V}_{tb} \\ \mathbf{V}_{tc} \end{bmatrix} = \begin{bmatrix} 0.588\,\underline{/0°} \\ 0.855\,\underline{/-117°} \\ 0.855\,\underline{/\;117°} \end{bmatrix} (13.8 \text{ kV}) = \begin{bmatrix} 8.10\,\underline{/0°} \\ 11.80\,\underline{/-117°} \\ 11.80\,\underline{/\;117°} \end{bmatrix} \text{kV}
$$

Line-to-Line Fault   A line-to-line (L-L) fault involves a short circuit between two phase conductors that are assumed to be phases b and c. Therefore, there is symmetry with respect to the principal phase a. An L-L fault is illustrated in Fig. 11.9.

Assuming that prefault currents are negligible, the voltage and current relationships describing the line-to-line fault are given by

$$\mathbf{V}_b = \mathbf{V}_c \tag{11.40}$$

$$\mathbf{I}_b = -\mathbf{I}_c; \qquad \mathbf{I}_a = 0 \tag{11.41}$$

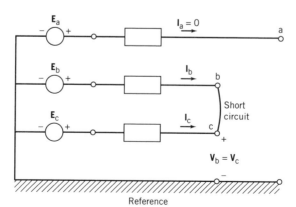

FIGURE 11.9   A line-to-line fault.

The sequence components of the short-circuit current are found by using Eq. 11.18. Thus,

$$\begin{bmatrix} \mathbf{I}_{a0} \\ \mathbf{I}_{a1} \\ \mathbf{I}_{a2} \end{bmatrix} = \frac{1}{3} \begin{bmatrix} 1 & 1 & 1 \\ 1 & a & a^2 \\ 1 & a^2 & a \end{bmatrix} \begin{bmatrix} \mathbf{I}_a \\ \mathbf{I}_b \\ \mathbf{I}_c \end{bmatrix} = \frac{1}{3} \begin{bmatrix} 0 \\ (a - a^2)\mathbf{I}_b \\ (a^2 - a)\mathbf{I}_b \end{bmatrix} = \begin{bmatrix} 0 \\ j(\mathbf{I}_b/\sqrt{3}) \\ -j(\mathbf{I}_b/\sqrt{3}) \end{bmatrix} \quad (11.42)$$

Therefore, it is seen that the negative-sequence component of the current is equal to the negative of its positive-sequence component; that is,

$$\mathbf{I}_{a2} = -\mathbf{I}_{a1} \quad (11.43)$$

The voltages of the shorted phases, b and c, may be expressed in terms of the sequence components as follows:

$$\mathbf{V}_b = \mathbf{V}_{a0} + a^2\mathbf{V}_{a1} + a\mathbf{V}_{a2} \quad (11.44)$$
$$\mathbf{V}_c = \mathbf{V}_{a0} + a\mathbf{V}_{a1} + a^2\mathbf{V}_{a2} \quad (11.45)$$

Subtracting Eq. 11.45 from 11.44 gives

$$\mathbf{V}_b - \mathbf{V}_c = (a^2 - a)\mathbf{V}_{a1} + (a - a^2)\mathbf{V}_{a2} = 0 \quad (11.46)$$

Because $\mathbf{V}_b$ and $\mathbf{V}_c$ are equal, Eq. 11.46 reduces to zero and yields

$$\mathbf{V}_{a1} = \mathbf{V}_{a2} \quad (11.47)$$

Equations 11.43 and 11.47 indicate that the positive- and negative-sequence networks are to be interconnected in parallel for a line-to-line fault. This is illustrated in Fig. 11.10.

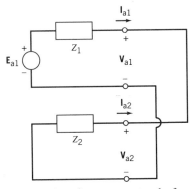

FIGURE 11.10 Interconnection of sequence networks for a line-to-line fault.

EXAMPLE 11.5

Repeat Example 11.4 assuming a phase-b-to-phase-c fault occurs at the end of the line in the given sample power system.

Solution   The sequence networks are interconnected as shown in Fig. 11.11. The generator voltage is taken as reference phasor. Thus,

$$\mathbf{E}_{a1} = 1.0\angle 0° \text{ pu}$$

The sequence components of the current at the fault are computed as follows:

$$\mathbf{I}_{a1} = (1.0\angle 0°)/(j0.3 + j0.2) = 2.0\angle -90° \text{ pu} = -j2.0 \text{ pu}$$

$$\mathbf{I}_{a2} = -\mathbf{I}_{a1} = 2.0\angle 90° \text{ pu} = j2.0 \text{ pu}$$

$$\mathbf{I}_{a0} = 0$$

The phase currents are computed by using Eq. 11.13 as follows:

$$\begin{bmatrix} \mathbf{I}_a \\ \mathbf{I}_b \\ \mathbf{I}_c \end{bmatrix} = \begin{bmatrix} 1 & 1 & 1 \\ 1 & a^2 & a \\ 1 & a & a^2 \end{bmatrix} \begin{bmatrix} 0 \\ -j2 \\ j2 \end{bmatrix} = \begin{bmatrix} 0 \\ -3.46 \\ 3.46 \end{bmatrix} \text{ pu} = \begin{bmatrix} 0 \\ -21.74 \\ 21.74 \end{bmatrix} \text{ kA}$$

The sequence components of the voltage at the terminals of the generator are computed as follows:

$$\mathbf{V}_{ta0} = 0$$

$$\mathbf{V}_{ta1} = \mathbf{E}_{a1} - j(0.20)\mathbf{I}_{a1} = 1.0\angle 0° - (j0.20)(-j2.0) = 0.6\angle 0° \text{ pu}$$

$$\mathbf{V}_{ta2} = -j0.10\mathbf{I}_{a2} = -(j0.10)(j2.0) = 0.2\angle 0° \text{ pu}$$

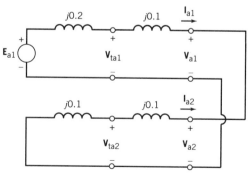

FIGURE 11.11   Interconnection of sequence networks for Example 11.5.

The phase voltages at the generator terminals are computed by using Eq. 11.9 as follows:

$$\begin{bmatrix} V_{ta} \\ V_{tb} \\ V_{tc} \end{bmatrix} = \begin{bmatrix} 1 & 1 & 1 \\ 1 & a^2 & a \\ 1 & a & a^2 \end{bmatrix} \begin{bmatrix} 0.0 \\ 0.6 \\ 0.2 \end{bmatrix} = \begin{bmatrix} 0.80 \angle 0° \\ 0.53 \angle -139° \\ 0.53 \angle 139° \end{bmatrix} \text{pu} = \begin{bmatrix} 11.0 \angle 0° \\ 7.3 \angle -139° \\ 7.3 \angle 139° \end{bmatrix} \text{kV}$$

Double Line-to-Ground Fault   A double line-to-ground (2LG) fault involves a short circuit between two phase conductors b and c and ground. As with the line-to-line fault, there is symmetry with respect to the principal phase a. A 2LG fault is illustrated in Fig. 11.12.

For a 2LG fault with prefault currents assumed negligible, the voltage and current relationships are given by the following:

$$V_b = V_c = 0 \tag{11.48}$$

$$I_a = 0 \tag{11.49}$$

The sequence components of the voltage at the fault location are found by using Eq. 11.16. Thus,

$$\begin{bmatrix} V_{a0} \\ V_{a1} \\ V_{a2} \end{bmatrix} = \frac{1}{3} \begin{bmatrix} 1 & 1 & 1 \\ 1 & a & a^2 \\ 1 & a^2 & a \end{bmatrix} \begin{bmatrix} V_a \\ 0 \\ 0 \end{bmatrix} = \frac{1}{3} \begin{bmatrix} V_a \\ V_a \\ V_a \end{bmatrix} \tag{11.50}$$

Therefore, it is seen that the sequence components of the voltage at the fault are all equal; that is,

$$V_{a1} = V_{a2} = V_{a0} \tag{11.51}$$

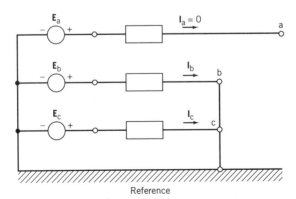

FIGURE 11.12   A double line-to-ground fault.

Neglecting prefault load currents, the current in phase a may be expressed as

$$\mathbf{I}_a = \mathbf{I}_{a0} + \mathbf{I}_{a1} + \mathbf{I}_{a2} = 0 \qquad (11.52)$$

Equations 11.51 and 11.52 indicate that the three sequence networks are to be interconnected in parallel for a double line-to-ground fault. This is illustrated in Fig. 11.13.

EXAMPLE 11.6

Repeat Example 11.4 assuming a double line-to-ground fault occurs at the end of the line in the given sample power system.

Solution   The sequence networks are interconnected in parallel as shown in Fig. 11.14.

The generator voltage is taken as reference phasor. Thus,

$$\mathbf{E}_{a1} = 1.0\underline{/0°} \text{ pu}$$

The sequence components of the current at the fault are computed as follows:

$$\mathbf{I}_{a1} = \frac{1.0\underline{/0°}}{j0.3 + (j0.2)(j0.35)/(j0.2 + j0.35)} = \frac{1}{j0.427}$$
$$= 2.34\underline{/-90°} = -j2.34 \text{ pu}$$

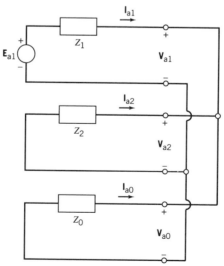

FIGURE 11.13   Interconnection of sequence networks for a double line-to-ground fault.

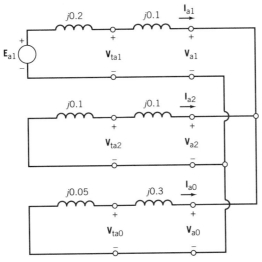

FIGURE 11.14 Interconnection of sequence networks for Example 11.6.

$$\mathbf{I}_{a2} = -\frac{j0.35}{j0.2 + j0.35}(2.34\underline{/-90°}) = 1.49\underline{/90°} = j1.49 \text{ pu}$$

$$\mathbf{I}_{a0} = -\frac{j0.20}{j0.2 + j0.35}(2.34\underline{/-90°}) = 0.85\underline{/90°} = j0.85 \text{ pu}$$

The phase currents are computed by using Eq. 11.17 as follows:

$$\begin{bmatrix} \mathbf{I}_a \\ \mathbf{I}_b \\ \mathbf{I}_c \end{bmatrix} = \begin{bmatrix} 1 & 1 & 1 \\ 1 & a^2 & a \\ 1 & a & a^2 \end{bmatrix} \begin{bmatrix} j0.85 \\ -j2.34 \\ j1.49 \end{bmatrix} = \begin{bmatrix} 0 \\ 3.55\underline{/159°} \\ 3.55\underline{/\,21°} \end{bmatrix} \text{ pu}$$

$$= \begin{bmatrix} 0 \\ 22.28\underline{/159°} \\ 22.28\underline{/\,21°} \end{bmatrix} \text{ kA}$$

The sequence components of the voltage at the terminals of the generator are computed as follows:

$$\mathbf{V}_{ta0} = -j0.05\mathbf{I}_{a0} = -(j0.05)(0.85\underline{/90°}) = 0.043 \text{ pu}$$

$$\mathbf{V}_{ta1} = \mathbf{E}_{a1} - j0.20\mathbf{I}_{a1} = 1.0\underline{/0°} - (j0.20)(2.34\underline{/-90°}) = 0.532 \text{ pu}$$

$$\mathbf{V}_{ta2} = -j0.10\mathbf{I}_{a2} = -(j0.10)(1.49\underline{/90°}) = 0.149 \text{ pu}$$

The phase voltages are computed by using Eq. 11.9; that is,

$$
\begin{bmatrix} \mathbf{V}_{ta} \\ \mathbf{V}_{tb} \\ \mathbf{V}_{tc} \end{bmatrix} = \begin{bmatrix} 1 & 1 & 1 \\ 1 & a^2 & a \\ 1 & a & a^2 \end{bmatrix} \begin{bmatrix} 0.043 \\ 0.532 \\ 0.149 \end{bmatrix} = \begin{bmatrix} 0.724 \underline{/0^\circ} \\ 0.514 \underline{/-133^\circ} \\ 0.514 \underline{/\ 133^\circ} \end{bmatrix} \text{pu}
$$

$$
= \begin{bmatrix} 10.0 \underline{/0^\circ} \\ 7.1 \underline{/-133^\circ} \\ 7.1 \underline{/\ 133^\circ} \end{bmatrix} \text{kV}
$$

Three-Phase Fault   A three-phase fault, although it is a symmetrical fault, may also be analyzed using the method of symmetrical components. Consider the three-phase fault illustrated in Fig. 11.15. Assuming the prefault currents are negligible, the voltage and current relationships describing this fault are

$$\mathbf{V}_a = \mathbf{V}_b = \mathbf{V}_c \qquad (11.53)$$

$$\mathbf{I}_a + \mathbf{I}_b + \mathbf{I}_c = 3\mathbf{I}_{a0} = 0 \qquad (11.54)$$

Equation 11.54 confirms that there is no zero-sequence current for a three-phase fault; that is, $\mathbf{I}_{a0}$ is identically equal to zero. The positive- and negative-sequence components of the voltage at the fault location are found as follows:

$$\mathbf{V}_{a1} = \tfrac{1}{3}(\mathbf{V}_a + a\mathbf{V}_b + a^2\mathbf{V}_c) = \tfrac{1}{3}(1 + a + a^2)\mathbf{V}_a = 0 \qquad (11.55)$$

$$\mathbf{V}_{a2} = \tfrac{1}{3}(\mathbf{V}_a + a^2\mathbf{V}_b + a\mathbf{V}_c) = \tfrac{1}{3}(1 + a^2 + a)\mathbf{V}_a = 0 \qquad (11.56)$$

Since the positive- and negative-sequence components of the voltage at the location of the fault are both equal to zero, the sequence components of the

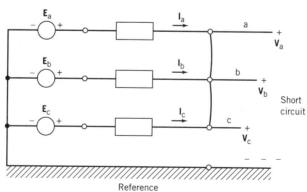

FIGURE 11.15   A three-phase fault.

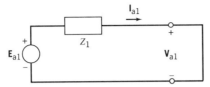

FIGURE 11.16   Sequence network for a three-phase fault.

current are obtained from Eqs. 11.30 and 11.31 as follows:

$$\mathbf{I}_{a1} = \frac{\mathbf{E}_{a1}}{Z_1}$$

$$\mathbf{I}_{a2} = 0 \qquad\qquad (11.57)$$

It has been found that no zero- and negative-sequence currents are flowing during a three-phase fault. Therefore, the analysis of a three-phase fault involves only the positive-sequence network, which is shown in Fig. 11.16.

EXAMPLE 11.7

Repeat Example 11.4 assuming a three-phase fault occurs at the end of the line in the given sample power system.

Solution   Only the positive-sequence network is used, and it is shown in Fig. 11.17.

At no load, the generator voltage is 1.0 per unit and is taken as reference. Thus,

$$\mathbf{E}_{a1} = 1.0\,\underline{/0°}\ \text{pu}$$

The sequence components of the current at the fault are computed as follows:

$$\mathbf{I}_{a1} = (1.0\,\underline{/0°})/(j0.3) = 3.33\,\underline{/-90°}\ \text{pu} = -j3.33\ \text{pu}$$

$$\mathbf{I}_{a2} = 0$$

$$\mathbf{I}_{a0} = 0$$

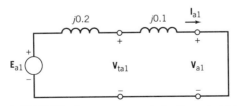

FIGURE 11.17   Sequence network for Example 11.7.

The phase currents are computed by using Eq. 11.17 as follows:

$$\begin{bmatrix} I_a \\ I_b \\ I_c \end{bmatrix} = \begin{bmatrix} 1 & 1 & 1 \\ 1 & a^2 & a \\ 1 & a & a^2 \end{bmatrix} \begin{bmatrix} 0.0 \\ -j3.33 \\ 0.0 \end{bmatrix} = \begin{bmatrix} 3.33\angle-90° \\ 3.33\angle150° \\ 3.33\angle 30° \end{bmatrix} pu = \begin{bmatrix} 20.9\angle-90° \\ 20.9\angle150° \\ 20.9\angle 30° \end{bmatrix} kA$$

The sequence components of the voltage at the terminals of the generator are computed as follows:

$$V_{ta0} = 0$$
$$V_{ta1} = E_{a1} - j0.20 I_{a1} = 1.0\angle0° - (j0.20)(-j3.33) = 0.333 \text{ pu}$$
$$V_{ta2} = 0$$

The phase voltages at the generator terminals are computed by using Eq. 11.9 as follows:

$$\begin{bmatrix} V_{ta} \\ V_{tb} \\ V_{tc} \end{bmatrix} = \begin{bmatrix} 1 & 1 & 1 \\ 1 & a^2 & a \\ 1 & a & a^2 \end{bmatrix} \begin{bmatrix} 0 \\ 0.333 \\ 0 \end{bmatrix} = \begin{bmatrix} 0.333\angle0° \\ 0.333\angle-120° \\ 0.333\angle120° \end{bmatrix} pu$$

$$= \begin{bmatrix} 4.6\angle0° \\ 4.6\angle-120° \\ 4.6\angle120° \end{bmatrix} kV$$

DRILL PROBLEMS

D11.7 A three-phase, 100-MVA, 12-kV, wye-connected synchronous generator has the following reactances: $X'' = X_2 = 20\%$ and $X_0 = 8\%$. The generator is connected to a 100-MVA, 12/115-kV, Δ/Y, three-phase transformer with a reactance of 10%. The neutrals of the generator and transformer windings are solidly grounded. The terminal voltage of the generator is 12 kV, and the transformer is at open circuit. A single line-to-ground fault occurs on the high-voltage side of the transformer.

a. Find the short-circuit current at the fault.

b. Find the short-circuit current supplied by the generator.

D11.8 Repeat Problem D11.7 if the fault is a line-to-line fault.

D11.9 Repeat Problem D11.7 if the fault is a double line-to-ground fault.

11.3   POWER SYSTEM PROTECTION

The operation of a power system is affected by disturbances that could be due to natural occurrences such as lightning, wind, trees, animals, and human errors or accidents. These disturbances could lead to abnormal system conditions such as short circuits, overloads, and open circuits.

Short circuits, which are also referred to as faults, are of the greatest concern because they could lead to damage to equipment or system elements and other operating problems including voltage drops, decrease in frequency, loss of synchronism, and complete system collapse. There is, therefore, a need for a device or a group of devices that is capable of recognizing a disturbance and acting automatically to alleviate any ill effects on the system element or on the operator. Such capability is provided by the protection system.

The protection system is designed to disconnect the faulted system element automatically when the short circuit currents are high enough to present a direct danger to the element or to the system as a whole. When the fault results in overloads or short-circuit currents that do not present any immediate danger, the protection system will initiate an alarm so that measures can be implemented to remedy the situation.

11.3.1   Components of Protection Systems

There are three principal components of a protection system:

1. Transducer
2. Protective relay
3. Circuit breaker

These components are described briefly in the following paragraphs.

Transducers   The transducer serves as a sensor to detect abnormal system conditions and to transform the high values of short-circuit current and voltage to lower levels. The main sensors used are the current transformer (CT) and the potential transformer (PT).

The current transformer is designed to provide a standard continuous secondary current of 5 A. Standard CT ratios available include 50/5, 100/5, 150/5, 200/5, 250/5, 300/5, 400/5, 500/5, 600/5, 800/5, 900/5, 1000/5, 1200/5, 1500/5, and 2000/5. During fault conditions, the short-circuit currents could reach over 10 times normal for short periods of time without damaging the CT windings.

The current transformer has a primary winding that usually consists of one turn and a secondary winding of several turns. It is, therefore, unsafe to open-circuit the secondary of a CT whose primary is energized.

The potential transformer is designed to operate at a constant standard secondary voltage of 120 V. For low-voltage applications, the PT is just like any other two-winding voltage transformer. For primary voltages in the HV and EHV levels, a capacitor voltage-divider circuit is used together with the PT. The primary voltage is impressed across the series-connected capacitors. The PT is used to measure the voltage of a few kilovolts across the capacitor of the smaller capacitance value.

Protective Relays   A protective relay is a device that processes the signals provided by the transducers, which may be in the form of a current, a voltage, or a combination of current and voltage. These signals arise as a result of a faulted condition such as a short circuit, defective equipment or lines, lightning strikes, or surges. The protective relay can initiate or permit the opening of various interrupting devices or sound an alarm. There are two main classifications of protective relays based on their construction: electromechanical and solid state.

The electromechanical relay develops an electromagnetic force or torque from the signal provided by the transducer; this force or torque is used to physically open, or close, a set of contacts to permit or initiate the tripping of circuit breakers or actuate an alarm.

The solid-state, or static, relay is energized by the same signals as in an electromechanical relay. However, there is no physical opening, or closing, of the relay contacts. Instead, the switching of the relay contacts is simulated by causing a solid-state device to change its status from conducting (closed position) to nonconducting (open position).

Electromechanical relays predate the solid-state relays. A majority of power system installations still use electromechanical relays. The improved reliability, versatility, and faster response (as low as one-quarter cycle) of solid-state relays have made them more attractive. Some electromechanical relays have been replaced by solid-state relays, and in newer installations a mixture of both types would usually be found.

Circuit Breakers   A circuit breaker is a mechanical device used to energize and interrupt an electric circuit. It should be able to open and close quickly, maybe in the order of a few milliseconds. It should be able to carry the rated current continuously at the nominal voltage, and it must be able to withstand the large short-circuit current (called its momentary rating) that flows during the first cycle after a fault occurs. The circuit breaker must be able to interrupt a large short-circuit current called its interrupting rating. The momentary rating is about 1.6 times the interrupting rating because the former includes the effect of the DC component of the transient short-circuit current. The actual value of current interrupted by the circuit breaker depends on its speed, which could be 1/2, 3, 5, or 8 cycles.

When the current-carrying contacts of the circuit breaker are opened, an electric field appears across the contacts that ionizes the medium between them, and an arc is established between the contacts. The circuit breaker must be able to extinguish, or interrupt, this arc as quickly as possible. The arc is made to take an elongated path, cooled, and finally extinguished when the AC current feeding the arc passes through its zero value. Sometimes, the arc is extinguished in air, oil, sulfur hexachloride (SF_6), or a vacuum.

11.3.2   Types of Protection Systems

The fundamental principle in designing protection systems is to divide the power system into zones that can be provided with the appropriate protection and can be disconnected or isolated as quickly as possible in order to minimize the effect on the rest of the power system. Each zone of protection is provided with two types of protection: primary and backup protection.

Primary or Main Protection   Primary protection is provided to ensure fast and reliable tripping of the circuit breakers to clear faults occurring within the boundary of its own zone of protection. In general, primary protection is provided for each transmission line segment, major piece of equipment, and switchgear.

A separate zone of protection is established around each system element as shown in Fig. 11.18. There is overlapping of adjacent zones of protection around the circuit breakers. Thus, if a fault occurs within the overlap region, more breakers are tripped than is actually necessary. On the other hand, if there is no overlap, a fault that occurs in the region between two zones of protection will not be in either zone, and no protective relay will operate and no circuit breaker will be tripped.

Although each major piece of equipment and transmission line segment is provided with primary protection and the primary zones of protection are made

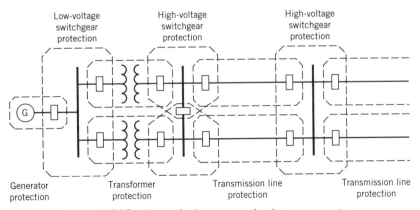

FIGURE 11.18   Zones of primary protection in a power system.

to overlap, it is still possible for the protection to fail. Hence, there is a need for backup protection.

Backup Protection Backup protection is provided in case the primary protection fails to operate or is under repair or maintenance. The protective relays used for backup protection have longer time delays; that is, they are slower acting than the relays used for primary protection to give the latter the opportunity to perform their function.

The backup protection should be designed in such a way that the cause of failure of the primary protection is not going to cause the same failure in the backup protection. Therefore, the backup protection should be located at a different station from the primary protection.

Consider a fault on line segment 3–4 of the power system shown in Fig. 11.19. If circuit breaker 3 fails to trip for some reason, the backup relay would trip circuit breaker 1; or, if the circuit breaker 4 fails to trip, the backup relays would trip circuit breakers 7 and 8. Because circuit breaker 1 is located differently from 3 and circuit breakers 7 and 8 are located differently from 4, the cause of failure of the primary protection would probably not cause failure of the backup protection, which might happen if backup protection were provided at 2 or 5 and 6 instead.

If a fault occurred at station C, primary protection would be provided by circuit breakers at 4, 5, and 6. Backup protection would be located at 3, 7, and 8.

11.3.3 Requirements of Protection Systems

In order to perform its functions properly, the protection system must have the following characteristics: speed, selectivity, reliability, and sensitivity.

Speed The speed of the protection system refers to the operating times of the protective relays. The potential damage to the faulted element depends on

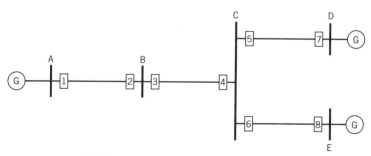

FIGURE 11.19 Location of backup protection.

the length of time the short-circuit currents are allowed to flow. The speed of clearing, or isolating, the faulted system component also affects the stability of the whole system.

Protective relays may be characterized as instantaneous with an operating time of about 0.10 s, or as high speed with an operating time of less than 0.05 s. Solid-state, or static, relays can have operating times as low as one quarter of a cycle.

Selectivity   Selectivity is the ability of the protection system to detect a fault, identify the point at which the fault occurred, and isolate the faulted circuit element by tripping the minimum number of circuit breakers. Selectivity of the protection system is obtained by proper coordination of the operating currents and time delays of the protective relays.

Reliability   The reliability of the protection system is its ability to operate upon the occurrence of any fault for which it was designed to protect. In other words, the protection system should operate when it is supposed to and not operate when it is not supposed to.

Sensitivity   Sensitivity refers to the characteristic of a protective relay that it operates reliably, when required, in response to a fault that produces the minimum short-circuit current flowing through the relay.

11.3.4   Types of Protective Relays

The protective relay is the device that responds to signals from the transducers by quickly initiating or allowing a control action to be implemented in order to prevent damage to the faulted equipment and to restore service as soon as possible. The operating characteristics of the more commonly used protective relays are described in this section.

A relay is said to *pick up* when it operates to open its normally closed (NC) contact or to close its normally open (NO) contact in response to a disturbance to produce a desired control action. The smallest value of the actuating quantity for the relay to operate is called its *pickup value*.

A relay is said to *reset* when it operates to close an open contact that is normally closed (NC) or to open a closed contact that is normally open (NO). The largest value of the actuating quantity for this to happen is called the *reset value*.

Overcurrent Relays   The actuating quantity of an overcurrent relay is a current. The relay is designed to operate when the actuating quantity equals, or exceeds, its pickup value. An overcurrent relay can be either of two types: instantaneous or time-delay type. Both relay types are frequently provided in one relay case and are actuated by the same current; however, their individual pickup values can be adjusted separately. The pickup values may be adjusted by varying the tap settings in the input.

The instantaneous relay element has no intentional time delay, and it operates quickly from 1/2 to 3 cycles depending on the value of the fault current. A typical operating characteristic of this relay type is shown in Fig. 11.20.

The time-delay relay element is characterized by having an operating time that varies inversely as the fault current flowing through the relay. A typical inverse time characteristic is shown in Fig. 11.20. The time-delay characteristic may be shifted up or down by adjusting the time-dial setting so that the relay operates with a different time delay for the same value of fault current.

The difficulty in using overcurrent relays is that they are inherently nonselective. They detect overcurrents (faults) not only in their own zones of protection but also in adjacent zones. The selectivity can be improved by proper coordination of the relay pickup values and time-delay settings. As the electric load grows and the power system configuration changes, operating conditions and magnitudes of short-circuit currents will vary. The pickup values of the overcurrent relays have to be readjusted continually in response to these changes.

Overcurrent relays are popular especially on low-voltage circuits because of their low cost. They are also used in specific applications on high-voltage systems.

Directional Relays   A directional relay is able to distinguish between current flowing in one direction and current flowing in the opposite direction. The relay responds to the phase angle difference between the actuating current and a reference current (or voltage) called the polarizing quantity.

Directional relaying is typically used in conjunction with some other relay, usually the overcurrent relay to improve its selectivity.

Differential Relays   The operation of a differential relay is based on the vector difference of two or more similar electrical actuating quantities. The most

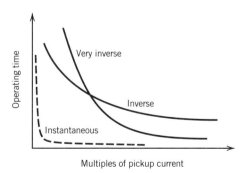

FIGURE 11.20   Time characteristics of overcurrent relays.

common application is current differential relaying, in which the current entering and the current leaving the protected element are compared. If the difference exceeds the pickup value of the relay, it operates to trip the breakers to isolate the element.

Typical differential relaying employing an overcurrent relay is shown in Fig. 11.21. Identical current transformers are placed at both ends of the protected element, and the CT secondaries are connected in parallel with an overcurrent relay. The directions of current flows shown in Fig. 11.21 are those corresponding to normal load conditions or to a fault external to the protected element. Thus, it is seen that the CT secondary currents merely circulate between the CTs, and no current flows in the overcurrent relay.

Suppose a fault occurs on the protected element as shown in Fig. 11.22. The short-circuit currents flow into the fault, and the CT secondary currents no longer circulate. The vectorial sum of the CT secondary currents flows through the overcurrent relay and causes the relay to disconnect the element from the system.

Even though the current transformers used for the differential relay are identical, the secondary currents may not be identical because of CT transformation inaccuracies. Thus, the secondary currents will no longer merely circulate for normal load conditions or for external faults. The differential current that will flow through the overcurrent relay may be sufficient for the relay to pick up and cause false tripping of the circuit breakers.

Percentage-Differential Relays   The difficulty encountered in differential relaying due to CT errors is eased by the use of a percentage-differential relay. This type of relay has an operating coil and two restraining coils. The operating current is proportional to $(I_A - I_B)$ and must exceed a certain percentage of the restraining current, which is proportional to $\frac{1}{2}(I_A + I_B)$, before the relay will operate.

Distance Relays   In a distance relay, a voltage and a current are balanced against each other and the relay responds to the ratio of the voltage to the current, which is the impedance of the transmission line from the relay location

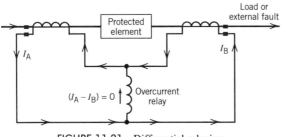

FIGURE 11.21   Differential relaying.

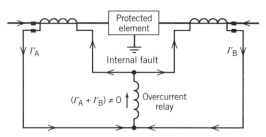

FIGURE 11.22 Fault currents in a differential relay.

to the point of interest. The impedance may be used to measure distance along a transmission line, hence the name distance relay.

This distance relay is useful because it is able to differentiate between a fault and normal operating conditions and to differentiate between faults in a specific area and a fault in a different part of the system. The operation of the distance relay is limited to a certain range of pickup values of impedance. The distance relay picks up whenever the measured impedance is less than or equal to the selected pickup value of impedance.

There are several types of distance relays, including impedance, reactance, and mho relays. The mho relay has an inherent directional characteristic; that is, it responds to or "sees" faults in only one direction. On the other hand, the impedance and the reactance relays see faults in all directions. Thus, a directional relay is commonly used together with the impedance relay, and a mho relay is used as a starting unit for the reactance relay.

Pilot Relays Pilot relaying is a means of communicating information from the end of a protected line to the protective relays at both line terminals. The relays determine whether a fault is internal or external to the protected line. The communication channel, or pilot, is used to transmit this information between line terminals. If the fault is internal to the protected line, all the circuit breakers at the terminals of the line are tripped in high speed. If the fault is external to the protected line, the tripping of the circuit breakers is prevented, or blocked.

Three types of pilots are commonly used for protective relaying: wire, power line carrier, and microwave pilot.

A wire pilot consists of a twisted pair of copper wires of the telephone line type. It may be leased from the telephone company, or it may be owned by the electric utility.

The power line carrier is the most commonly used pilot for protective relaying. In this type of pilot, a low-voltage, high-frequency current (30 to 300 kHz) is transmitted along one phase of the high-voltage power line to a receiver at the other end of the protected line. Line traps, located at both line terminals, serve to contain a carrier signal inside the zone of the protected line.

The microwave pilot is an ultra-high-frequency radio system operating above 900 MHz. In this pilot, transmitters and receivers operate the same way as in power line carrier pilot; however, line traps are replaced by a line-of-sight antenna.

11.3.5   Applications of Protection Systems

The applications of the different types of protection systems for the protection of various types of equipment and transmission lines are described in this section. These discussions are confined to protection for the high-voltage, bulk-power system components.

Generator Protection   The protection system provided to the synchronous generator must be able to detect any abnormal condition immediately and act quickly to prevent damage to the generator and minimize the effect on the power system. Synchronous generators are provided with protection against various disturbances, including short circuits in the stator windings, loss of field excitation, stator and rotor overheating, and overspeed. Short-circuit protection of the stator windings is of primary concern and is discussed in this section.

When a short circuit occurs between stator windings of a synchronous generator, or between a stator winding and ground, the protection system should quickly trip the main circuit breaker to disconnect the machine from the rest of the system and at the same time disconnect the field winding from the exciter.

The best stator winding short-circuit protection is provided by a percentage-differential relay, which is shown in Fig. 11.23. It may be noted that the

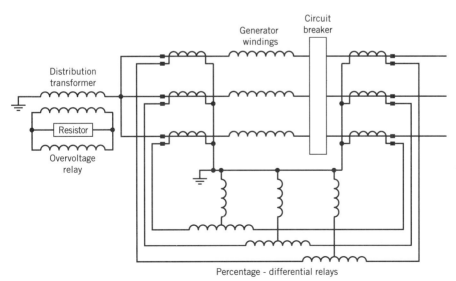

FIGURE 11.23   Generator protection.

relaying protects against a three-phase fault, or line-to-line fault, in the stator windings.

When the synchronous generator neutral is grounded through a high resistance, the short-circuit current for a single line-to-ground fault is much less than the short-circuit currents for faults involving the phase windings. Thus, the ground fault current may not be detected by the differential relaying protection. An overvoltage relay connected across the grounding resistor would be able to detect the increased voltage across the resistor in the presence of a ground fault, and the overvoltage relay will operate.

Transformer Protection   The protection system provided for a power transformer depends primarily on its capacity and its voltage rating. Thus, for small transformers with capacities up to about 2 MVA, power fuses are deemed to be adequate. For larger transformers, with capacities greater than 10 MVA, percentage-differential relays with harmonic restraint are recommended. The one-line diagram of the protection of a three-winding transformer using a three-winding percentage-differential relay is illustrated in Fig. 11.24.

The differential relaying protection must satisfy two basic requirements:

1. The protection must not operate for normal load conditions and faults external to the transformer.

2. The relays must operate to trip the circuit breakers for an internal fault that is severe enough to cause direct damage to the transformer.

Three-phase transformers with Y-Δ–connected windings require further consideration. The primary and secondary currents of such transformers normally

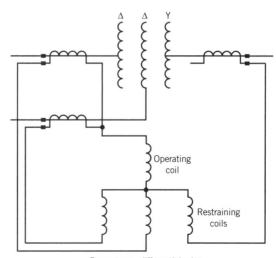

Percentage - differential relay

FIGURE 11.24   Three-winding transformer protection.

differ not only in magnitude but also in phase angles because of the inherent phase shifts in Y-Δ or Δ-Y connections. The current transformers must, therefore, be connected in such a manner that the CT secondary line currents as seen by the protective relays are equal under normal operating conditions or for external faults. The correct magnitude relationship is obtained by the proper choice of CT ratios and, if necessary, the use of an autotransformer in the CT secondary circuit. The correct phase-angle relationship is obtained by connecting the CTs on the wye-connected side of the transformer in delta and the CTs on the delta-connected side in wye. In this way, the CT connections are able to compensate for the phase shift introduced by the Y-Δ or Δ-Y connection. The design of the protection for a Δ-Y transformer using percentage-differential relays is illustrated in Fig. 11.25 and the following example.

EXAMPLE 11.8

Design the protection of a three-phase, 50-MVA, 230/34.5-kV power transformer using available standard CT ratios. The high-voltage side is Y connected

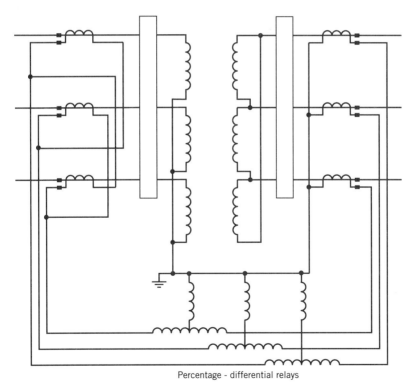

Percentage - differential relays

FIGURE 11.25 Y-Δ transformer protection.

and the low-voltage side is Δ connected. Specify the CT ratios, and show the three-phase wiring diagram indicating the CT polarities. Determine the currents in the transformer and the CTs. Specify the rating of an autotransformer, if one is needed.

Solution When the transformer is carrying rated load, the line currents on the high-voltage side and low-voltage side are

$$I_{HV} = \frac{50,000}{\sqrt{3}(230)} = 125.5 \text{ A}$$

$$I_{LV} = \frac{50,000}{\sqrt{3}(34.5)} = 836.7 \text{ A}$$

The CTs on the low-voltage side are connected in wye, and the CT ratio selected for this side is 900/5. The current in the leads flowing to the percentage-differential relay on this side is equal to the CT secondary current and is given by

$$I_{LV \text{ lead}} = 836.7\left(\frac{5}{900}\right) = 4.65 \text{ A}$$

The current in the leads to the relay from the low-voltage side must be balanced by an equal current in the leads connected to the Δ-connected CTs on the high-voltage side. This requires a CT secondary current equal to

$$I_{CT \text{ sec}} = \frac{4.65}{\sqrt{3}} = 2.68 \text{ A}$$

To obtain a CT secondary current of 2.68 A, the CT ratio of the high-voltage CTs is chosen as

$$\text{CT ratio} = \frac{125.5}{2.68} = 46.8$$

The nearest available standard CT ratio is 250/5. If this CT ratio is selected, the CT secondary currents will actually be

$$I_{CT \text{ sec}} = 125.5\left(\frac{5}{250}\right) = 2.51 \text{ A}$$

Therefore, the currents in the leads to the Δ-connected CTs from the percentage-differential relays will be

$$I_{HV \text{ lead}} = \sqrt{3}(2.51) = 4.35 \text{ A}$$

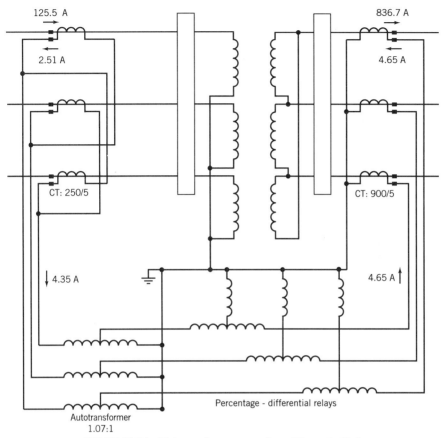

FIGURE 11.26 Y-Δ transformer protection of Example 11.8.

It is seen that the currents in the leads on both sides of the percentage-differential relay are not balanced. This condition cannot just be ignored because it could lead to improper tripping of the circuit breaker for an external fault. This problem can be solved by using an autotransformer as shown in Fig. 11.26. The autotransformer should have a turns ratio of

$$N_{\text{autotransformer}} = \frac{4.65}{4.35} = 1.07$$

In the design of the transformer protection of Example 11.8, the magnetizing current of the transformer has been assumed to be negligible. This is a reasonable assumption during normal operating conditions because the magnetizing current is a small percentage of the rated load current. However, when a transformer is being energized, it may draw a large magnetizing inrush current that soon decays with time to its normal value. The inrush current flows only in the primary, causing an unbalance in current, and the differential relay will interpret this an internal fault and will pick up to trip the circuit breakers.

To prevent the protection system from operating and tripping the transformer during its energization, percentage-differential relaying with harmonic restraint is recommended. This is based on the fact that the magnetizing inrush current has high harmonic content, whereas the fault current consists mainly of fundamental frequency sinusoid. Thus, the current supplied to the restraining coil consists of the fundamental and harmonic components of the normal restraining current of $(I_A + I_B)/2$, plus another signal proportional to the harmonic content of the differential current $(I_A - I_B)/2$. Only the fundamental frequency of the differential current is supplied to the operating coil of the relay.

Bus Protection   The sum of the currents flowing through the lines connected to the same bus is equal to zero for normal operating conditions or for a fault external to the bus. Hence, differential relaying using overcurrent relays can be used to provide protection against bus faults. This is illustrated in the one-line diagram shown in Fig. 11.27.

All of the current transformers are connected in parallel and an overcurrent relay is connected across their output. When there is no fault on the bus, the line currents are indicated by the solid arrows, and the current through the overcurrent relay is given by

$$I_R = \frac{1}{\text{CT ratio}}(I_1 + I_2 - I_3 - I_4) = 0$$

Therefore, no current flows through the relay. On the other hand, when there is a fault at the bus, currents I_3 and I_4 reverse directions as shown by the broken

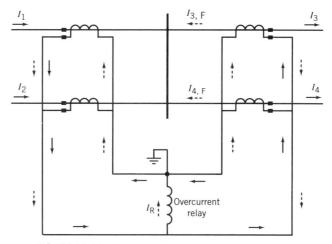

FIGURE 11.27   Bus protection using an overcurrent relay.

arrows; thus, the current through the relay is now given by

$$I_R = \frac{1}{CT \text{ ratio}}(I_1 + I_2 + I_3 + I_4) > 0$$

Hence, the overcurrent relay picks up to trip all circuit breakers.

Transmission Line Protection   Transmission lines provide the means for bringing the electric energy from the generating plants to the major substations, from these substations to the load center substations, and ultimately to the individual consumers. Transmission lines travel over wide-open spaces, as well as thickly populated metropolitan areas. Because of these long distances and extensive exposure to nature and human accidents, most of the short circuits that occur in power systems are on overhead transmission and distribution lines.

For the low-voltage distribution circuits, the protection system usually provided consists of circuit reclosers and power fuses that act as relays and circuit breakers combined. For the protection of medium-voltage and high-voltage transmission lines, separate relays and circuit breakers are employed. Since HV and EHV transmission lines are the major means of bulk-power transmission, the protection of these transmission lines is designed to be more reliable and more selective, and it is also more expensive.

Transmission Line Protection by Overcurrent Relays   Consider that the portion of the power system to be protected is radial as shown in Fig. 11.28. The generator connected to bus 1 represents the rest of the power system, and it supplies loads at buses 1, 2, 3, 4, and 5 through four transmission lines. Since the short-circuit current comes from the left side of each line, it is sufficient to provide only one circuit breaker at the sending end for each line. So that service disruption is minimized, for a fault on line 4–5, only circuit breaker CB_4 should be tripped; for a fault on the line 3–4, only breaker CB_3 should be opened; for a fault on the line 2–3, only CB_2 should be tripped; and so on.

The short-circuit current due to a fault on any of the lines depends on the fault location and the type of fault. Since the total impedance increases with the distance from the generator to the fault, the short-circuit current is inversely proportional to this distance.

Overcurrent relays are used mainly to provide protection for subtransmission and distribution lines. Two forms of overcurrent protection are

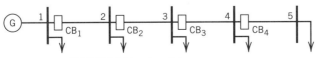

FIGURE 11.28 A radial power system.

provided: primary protection for the line itself and backup protection for an adjacent line. Two types of overcurrent relay units are used: time overcurrent relay and instantaneous relay.

The time overcurrent relays at each of the four buses 1, 2, 3, and 4 provide primary protection for their own line segment and provide remote backup protection to adjacent line downstream from the relay location. Thus, the relay at bus 1 provides primary protection for line 1–2 and also provides backup protection for line 2–3; the relay at bus 2 provides primary protection for line 2–3 and backup protection for line 3–4; and so forth.

When the relay at bus 1 provides backup protection for line 2–3, it must be adjusted to be selective with the primary relaying at bus 2. The relay at bus 2 is expected to operate first for a fault on line 2–3 before the relay at bus 1 operates. The operating times of the relays at the different buses are shown in Fig. 11.29. Thus, for the fault shown, the relay at bus 4 picks up to trip circuit breaker CB_4. If CB_4 fails to open for any reason, the fault current persists, and after a certain time delay the relay at bus 3 picks up to trip CB_3. The relay at bus 2 is selective with the relay at bus 3; thus, if both CB_3 and CB_4 failed to open, the relay at bus 2 would pick up and trip CB_2 after a longer time delay.

For setting the pickup values and the selectivity clearances between the time overcurrent relays for backup protection, there are four criteria to consider:

1. The relay must be able to pick up for the minimum short-circuit currents for which the relay is designed to protect.

2. The relay should not pick up for normal load conditions. A common practice is to set the minimum pickup value equal to twice the peak load current in order to carry emergency peak loads and pick up cold load.

3. The relay nearer the source providing backup protection for the downstream relay must pick up for one third of the minimum current seen by the latter.

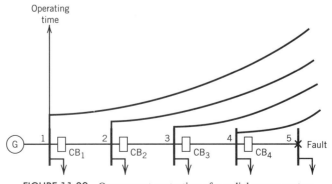

FIGURE 11.29 Overcurrent protection of a radial power system.

4. The relay nearer the source providing backup protection for the downstream relay should pick up for the maximum current seen by the latter but no sooner than (a selectivity clearance of) at least 0.3 s.

The instantaneous overcurrent relay unit provides primary protection for its own protected line segment to supplement the time-delay overcurrent relay unit. The instantaneous unit operates quickly with no intentional time delay, so it should be adjusted such that it does not operate for a fault on neighboring lines. The pickup value is usually adjusted so that the instantaneous relay is able to detect a fault occurring at a distance of up to 80% of the line segment. The operating times of the instantaneous relay and the time-delay units are shown in Fig. 11.30.

The complete protection system for a line consists of three overcurrent relays for phase fault protection and one overcurrent relay for ground fault protection. This protection system is shown in Fig. 11.31.

Transmission Line Protection by Distance Relays   To improve reliability, the generating plants, transmission lines, and substations of a high-voltage power system are interconnected to form a network. Overcurrent relay protection can no longer be used because coordinating the relays becomes a difficult, if not impossible, task.

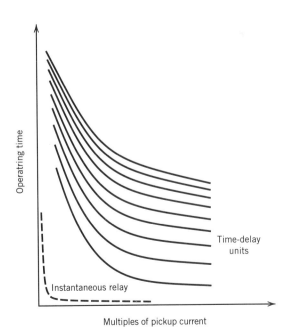

FIGURE 11.30   Typical time characteristics of an overcurrent relay.

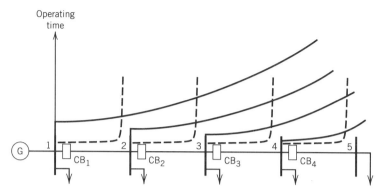

FIGURE 11.31 Complete overcurrent protection of a radial system.

Protection of transmission lines connected as a network can be provided by distance relays. These distance relays provide phase fault protection for the line, while an overcurrent relay provides ground fault protection.

Distance relays provide primary protection for a line section and backup protection for an adjacent line. The value of impedance at the farthest fault location for which a distance relay picks up is called its "reach." One distance relay contains three distance relay units responding to three independently adjustable pickup impedance values (or zones of protection) with three independently adjustable time delays.

The primary impedance corresponding to a particular fault location, or relay unit reach, is converted to a secondary value that is used to adjust the phase or ground distance relay. This secondary impedance value is given by

$$Z_{\text{sec}} = Z_{\text{pri}} \left(\frac{\text{CT ratio}}{\text{PT ratio}} \right)$$

The first-zone, or high-speed zone, unit of the distance relay has a reach of up to 80% to 90% of the length of the transmission line. This relay unit operates with no time delay.

The second-zone unit of the distance relay provides protection for the rest of the line beyond the reach of the first-zone unit. This second-zone unit has a reach of at least 20% of the adjoining line section. Because this unit "sees" faults in the adjoining line, it must be selective with the first-zone unit of the adjacent line. Therefore, this second-zone unit is provided with a time delay of about 0.2 s to 0.5 s.

The third-zone unit of the distance relay provides backup protection for the rest of the adjoining line section. This unit is adjusted to reach beyond the end of the adjoining line to ensure that backup protection is provided for the full line section. The time delay provided for this unit is usually about 0.4 s to 1.0 s. The protection of a portion of a power system by using distance relays is illustrated in Fig. 11.32.

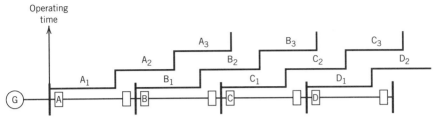

FIGURE 11.32 Line protection by distance relays.

*11.4 TRANSIENT STABILITY

Power system stability refers to the ability of the various synchronous machines in the system to remain in synchronism, or stay in step, with each other following a disturbance. Stability may be classified as steady-state, dynamic, or transient stability.

Steady-state stability refers to the ability of the various machines to regain and maintain synchronism after a small and slow disturbance, such as a gradual change in load. Transient stability is stability after a sudden large disturbance such as a fault, loss of a generator, a switching operation, and a sudden load change. Dynamic stability is the case between steady-state and transient stability, and the period of study is much longer so that the effects of regulators and governors may be included.

Consider the two-machine power system shown in Fig. 11.33. Each machine may be represented as a constant emf in series with a synchronous reactance and negligible resistance. Assume that the machine on the left acts as a generator and the other machine acts as a motor. The emfs may be expressed in polar form as follows:

$$\mathbf{E_g} = E_g \underline{/\delta}$$
$$\mathbf{E_m} = E_m \underline{/0^\circ} \tag{11.58}$$

The current flowing in the circuit is given by

$$\mathbf{I} = \frac{\mathbf{E_g} - \mathbf{E_m}}{jX_T} = \frac{E_g \underline{/\delta} - E_m \underline{/0^\circ}}{jX_T} \tag{11.59}$$

where $X_T = X_g + X_e + X_m$.

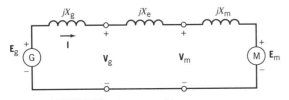

FIGURE 11.33 A two-machine power system.

The real power delivered by the generator to the motor is found as follows:

$$P = \text{Re}[\mathbf{E_g I}^*]$$

$$= \text{Re}\left[(E_g \, \underline{/\delta})\left(\frac{E_g \, \underline{/\delta} - E_m \, \underline{/0°}}{jX_T}\right)^*\right]$$

$$= \text{Re}\left[\frac{E_g^2}{X_T} \, \underline{/90°} - \frac{E_g E_m}{X_T} \, \underline{/\delta + 90°}\right]$$

$$= \frac{E_g E_m}{X_T} \sin \delta \qquad\qquad (11.60)$$

Thus, it is seen that the power supplied by the generator to the motor varies with the sine of the phase angle difference of the two emfs, which is also the displacement angle between the two rotors.

The maximum power P_{max} that can be transmitted at steady state from the generator to the motor with the total reactance X_T is found from Eq. 11.60 for the case $\delta = 90°$. That is,

$$P_{max} = \frac{E_g E_m}{X_T} \qquad\qquad (11.61)$$

P_{max} is called the steady-state stability limit. Its value can be raised only by increasing either of the two internal voltages by adjusting their respective excitations or by decreasing the series reactance of the transmission line connecting the two machines.

The plot of the transmitted power versus the displacement angle is called the power-angle curve, and it is illustrated in Fig. 11.34. The graph indicates that the system is stable over the operating region $-90° < \delta < 90°$, where

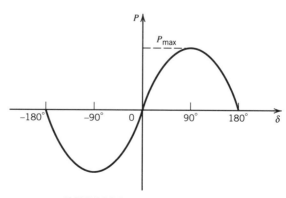

FIGURE 11.34   Power-angle curve.

the derivative of P with respect to δ is positive, that is, $dP/d\delta > 0$; this condition implies that an increase in displacement angle results in an increase in transmitted power.

Consider the two-machine system operating at steady state at the point A on the power-angle curve of Fig. 11.35. Assume that the generator is delivering electrical power P_0 at an angle δ_0 to the motor, which is driving a mechanical load connected to its shaft.

Case 1. Suppose that a small increment of shaft load is added to the motor. This net torque tends to retard the motor and its speed decreases, causing an increase in δ. The input power increases correspondingly until equilibrium is obtained at a new operating point B, higher than A.

Case 2. Suppose that motor load is increased gradually until point C of maximum power is reached. If additional load is applied to the motor, δ increases as before and goes beyond 90°. This time, however, instead of an increase in the input power, the input power decreases, which further increases the net retarding torque. This torque retards the motor even faster until it pulls out of step.

Case 3. Suppose a large increment of load is suddenly applied to the motor. The deficiency in input will be supplied temporarily by the decrease in kinetic energy. The motor will slow down, increasing δ and the power input. Assuming that the new load is less than P_{max}, δ will increase to a new value such that the motor input equals the motor load. When this is reached, the motor is still running slow, thus increasing δ beyond its proper value and producing an accelerating torque that increases the motor speed. However, when the motor regains normal speed, δ may have gone beyond point D so that the motor input is less than the load. This leads to motor pull-out.

Case 4. Assume for this case that the sudden incremental load is not too great. The motor will regain its normal speed before δ becomes too large. The motor speed will continue to increase because of the net accelerating torque

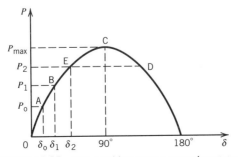

FIGURE 11.35 Two machine system operating states.

and will become greater than normal. When this happens, δ will decrease and again approach its proper value. These oscillations will later die out because of damping torques, and the motor will come to a stable operating condition at point E.

The upper limit to the sudden increment in load that the rotor can carry without pulling out of step is called the transient stability limit. This is always lower than the steady-state limit and can have different values depending on the nature and magnitude of the disturbance.

*11.4.1   Swing Equation

The motion of a synchronous machine is governed by Newton's law of rotation, which states that the product of the moment of inertia times the angular acceleration is equal to the net accelerating torque. Mathematically, this may be expressed as follows:

$$J\alpha = T_a = T_m - T_e \qquad (11.62)$$

Equation 11.62 may also be written in terms of the angular position as follows:

$$J\frac{d^2\theta_m}{dt^2} = T_a = T_m - T_e \qquad (11.63)$$

where

J = moment of inertia of the rotor

T_a = net accelerating torque or algebraic sum of all torques acting on the machine

T_m = shaft torque corrected for the rotational losses including friction and windage and core losses

T_e = electromagnetic torque

By convention, the values of T_m and T_e are taken as positive for generator action and negative for motor action.

For stability studies, it is necessary to find an expression for the angular position of the machine rotor as a function of time t. However, because the displacement angle and relative speed are of greater interest, it is more convenient to measure angular position and angular velocity with respect to a synchronously rotating reference frame with a synchronous speed of ω_{sm}. Thus, the rotor position may be described by the following:

$$\theta_m = \omega_{sm}t + \delta_m \qquad (11.64)$$

The derivatives of θ_m may be expressed as

$$\frac{d\theta_m}{dt} = \omega_{sm} + \frac{d\delta_m}{dt} \tag{11.65}$$

$$\frac{d^2\theta_m}{dt^2} = \frac{d^2\delta_m}{dt^2} \tag{11.66}$$

Substituting Eq. 11.66 into Eq. 11.63 yields

$$J\frac{d^2\delta_m}{dt^2} = T_a = T_m - T_e \tag{11.67}$$

Multiplying Eq. 11.67 by the angular velocity of the rotor transforms the torque equation into a power equation. Thus,

$$J\omega_m \frac{d^2\delta}{dt^2} = \omega_m T_a = \omega_m T_m - \omega_m T_e \tag{11.68}$$

Replacing $\omega_m T$ by P and $J\omega_m$ by M, the so-called swing equation is obtained. The swing equation describes how the machine rotor moves, or swings, with respect to the synchronously rotating reference frame in the presence of a disturbance, that is, when the net accelerating power is not zero.

$$M\frac{d^2\delta_m}{dt^2} = P_a = P_m - P_e \tag{11.69}$$

where

$M = J\omega$ = inertia constant

$P_a = P_m - P_e$ = net accelerating power

$P_m = \omega T_m$ = shaft power input corrected for the rotational losses

$P_e = \omega T_e$ = electrical power output corrected for the electrical losses

It may be noted that the inertia constant was taken equal to the product of the moment of inertia J and the angular velocity ω_m, which actually varies during a disturbance. Provided the machine does not lose synchronism, however, the variation in ω_m is quite small. Thus, M is usually treated as a constant.

Another constant, which is often used because its range of values for particular types of rotating machines is quite narrow, is the so-called normalized inertia constant H. It is related to M as follows:

$$H = \frac{1}{2}\frac{M\omega_{sm}}{S_{rated}} \text{ MJ/MVA} \tag{11.70}$$

Solving for M from Eq. 11.70 and substituting into 11.69 yields the swing equation expressed in per unit. Thus,

$$\frac{2H}{\omega_{sm}} \frac{d^2\delta_m}{dt^2} = \frac{P_a}{S_{rated}} = \frac{P_m - P_e}{S_{rated}} \tag{11.71}$$

It may be noted that the angle δ_m and angular velocity ω_m in Eq. 11.71 are expressed in mechanical radians and mechanical radians per second, respectively. For a synchronous generator with p poles, the electrical power angle and radian frequency are related to the corresponding mechanical variables as follows:

$$\delta(t) = \frac{p}{2}\delta_m(t)$$

$$\omega(t) = \frac{p}{2}\omega_m(t) \tag{11.72}$$

Similarly, the synchronous electrical radian frequency is related to synchronous angular velocity as follows:

$$\omega_s = \frac{p}{2}\omega_{sm} \tag{11.73}$$

Therefore, the per-unit swing equation of Eq. 11.71 may be expressed in electrical units and takes the form of Eq. 11.74.

$$\frac{2H}{\omega_s} \frac{d^2\delta}{dt^2} = P_a = P_m - P_e \tag{11.74}$$

Depending on the unit of the angle δ, Eq. 11.74 takes the form of either Eq. 11.75 or Eq. 11.76. Thus, the per-unit swing equation takes the form

$$\frac{H}{\pi f} \frac{d^2\delta}{dt^2} = P_a = P_m - P_e \tag{11.75}$$

when δ is in electrical radians, or

$$\frac{H}{180f} \frac{d^2\delta}{dt^2} = P_a = P_m - P_e \tag{11.76}$$

when δ is in electrical degrees.

When a disturbance occurs, an unbalance in the power input and power output ensues, producing a net accelerating torque. The solution of the swing equation in the form of the differential equation of (11.75) or (11.76) is appropriately called the swing curve $\delta(t)$.

*11.4.2 Single Machine-to-Infinite Bus

Consider the system shown in Fig. 11.36, which consists of a single machine connected to an infinite bus through an external reactance. The synchronous machine is driven by a constant-speed prime mover.

The electrical power output of the synchronous generator may be found in the same manner as in Eq. 11.60 and may be expressed as follows:

$$P_e = P_{\max} \sin \delta \qquad (11.77)$$

Substituting the expression for the electrical power output of the generator and rearranging the terms in the swing equation yield the following:

$$\frac{2H}{\omega_s} \frac{d^2\delta}{dt^2} + P_{\max} \sin \delta = P_m \qquad (11.78)$$

It is seen that the resulting differential equation is nonlinear. Except for a few cases, it may not be possible to solve the equation analytically to obtain a closed form of the solution $\delta(t)$. It is, therefore, usually necessary to use a numerical technique to solve the differential equation (11.75) or (11.76).

EXAMPLE 11.9

A 500-MVA, 20-kV, 60-Hz, four-pole synchronous generator is connected to an infinite bus through a purely reactive network. The generator has an inertia constant $H = 6.0$ MJ/MVA and is delivering power of 1.0 per unit to the infinite bus at steady state. The maximum power that can be delivered is 2.5 per unit. A fault occurs that reduces the generator output power to zero.

a. Find the angular acceleration.
b. Find the speed in rev/min at the end of 15 cycles.
c. Find the change in the angle δ at the end of 15 cycles.

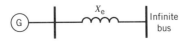

FIGURE 11.36 Single machine-to-infinite bus system.

Solution

a. Because the generator is initially operating at steady state, the mechanical input power is equal to the electrical power output prior to the fault. That is,

$$P_m = P_e = 1.0 \text{ pu}$$

Therefore, the accelerating torque is found by using Eq. 11.75.

$$\alpha = \frac{d^2\delta}{dt^2} = \frac{180f}{H}(P_m - P_e)$$

$$= \frac{(180)(60)}{6.0}(1.0 - 0.0) = 1800 \text{ elec. degrees/s}^2$$

For a four-pole machine,

$$\alpha = \frac{d^2\delta}{dt^2} = \frac{180f}{H}(P_m - P_e) = \frac{(180)(60)}{6.0}(1.0 - 0.0)$$

$$= 1800 \text{ elec. degrees/s}^2$$

$$= \left(\frac{2}{p}\right)1800 = 900 \text{ mech. degrees/s}^2$$

$$= 900\left(\frac{60 \text{ s/min}}{360°/\text{rev}}\right) = 150 \text{ rpm/s}$$

b. A 15-cycle interval is equivalent to a time interval of

$$t = (15)(1/60) = 0.25 \text{ s}$$

The synchronous speed of the machine is found as follows:

$$\omega_{sm} = 120f/p = (120)(60)/4 = 1800 \text{ rpm}$$

Integrating Eq. 11.76 starting from $t = 0$ up to a final time $t = 0.25$ s yields

$$\omega_m = \omega_{sm} + \alpha t = 1800 + (150)(0.25) = 1837.5 \text{ rpm}$$

c. The machine is initially operating at the angle δ_0, which is found from the following expression:

$$P_0 = P_{max} \sin \delta_0$$

Therefore, the initial angle is

$$\delta_0 = \sin^{-1}(P_0/P_{max}) = \sin^{-1}(1.0/2.5) = 23.58°$$

Integrating the angular acceleration expressed in electrical degrees/s² twice from $t = 0$ up to the final time of $t = 0.25$ s gives the value of the angle. That is,

$$\delta = \delta_0 + \tfrac{1}{2}\alpha t^2 = 23.58° + \tfrac{1}{2}(1800)(0.25)^2 = 79.83°$$

DRILL PROBLEMS

D11.10 A three-phase, 350-MVA, 13.8-kV, 60-Hz, four-pole, synchronous generator has an inertia constant of $H = 6$ MJ/MVA. The machine is connected to an infinite bus and is operating at steady state. Determine the kinetic energy stored in the rotor at synchronous speed.

D11.11 The prime mover of the generator of Problem D11.10 is supplying an input power (net of the rotational losses) of 500,000 hp. The electrical power developed suddenly decreases to 280 MW.

  a. Determine the angular acceleration.

  b. If the acceleration computed in part (a) remains constant for a period of 20 cycles, find the change in the rotor angle δ in electrical degrees and the speed in rpm at the end of 20 cycles.

D11.12 The generator of Problem D11.10 is initially operating at steady state and is delivering rated MVA at 0.8 power factor lagging. An external fault suddenly occurs that reduces the generator power output by 60%. Neglect the machine losses, and assume that the power input to the shaft remains constant. Find the accelerating torque in newton-meters when the fault occurs.

D11.13 For the generator of Problem D11.10, the mechanical power input P_m is initially 0.8 pu and is assumed to remain constant. The maximum power P_{max} that the generator can develop is 2.25 pu. An external fault suddenly occurs that reduces the generator output power to zero. Find the change in angle δ at the end of 30 cycles.

*11.4.3   Multimachine Systems

Consider two finite synchronous machines swinging with respect to each other. The swing equations are given by the following:

$$M_1 \frac{d^2\delta_1}{dt^2} = P_{a1} = P_{m1} - P_{e1} \tag{11.79}$$

$$M_2 \frac{d^2\delta_2}{dt^2} = P_{a2} = P_{m2} - P_{e2} \tag{11.80}$$

Subscripts 1 and 2 pertain to machines 1 and 2, respectively. P_{m1} and P_{m2} are the mechanical power inputs from the respective prime movers of the two machines. P_{e1} and P_{e2} are the electrical power outputs of the two machines, which may be expressed in terms of the power flow equations of Section 10.2 as follows:

$$P_{e1} = V_1 V_2 Y_{12} \cos(\delta_1 - \delta_2 - \theta_{12}) + V_1^2 Y_{11} \cos \theta_{11} \tag{11.81}$$

$$P_{e2} = V_2 V_1 Y_{21} \cos(\delta_2 - \delta_1 - \theta_{21}) + V_2^2 Y_{22} \cos \theta_{22} \tag{11.82}$$

It is seen, therefore, that the swing equations are nonlinear differential equations that are coupled through the electrical power that each synchronous machine delivers. The nonlinearities involve trigonometric functions, and the coupling is dependent on the electrical network connecting the two machines.

In a multimachine system, the electrical outputs of the synchronous machines are functions of the angular positions and, possibly, the angular velocities of all of the synchronous machines in the system. For example, if there were four machines in the system, the swing equations are given by the following:

$$M_1 \frac{d^2\delta_1}{dt^2} = P_{m1} - P_{e1}(\delta_1, \delta_2, \delta_3, \delta_4, \omega_1, \omega_2, \omega_3, \omega_4) \tag{11.83}$$

$$M_2 \frac{d^2\delta_2}{dt^2} = P_{m2} - P_{e2}(\delta_1, \delta_2, \delta_3, \delta_4, \omega_1, \omega_2, \omega_3, \omega_4) \tag{11.84}$$

$$M_3 \frac{d^2\delta_3}{dt^2} = P_{m3} - P_{e3}(\delta_1, \delta_2, \delta_3, \delta_4, \omega_1, \omega_2, \omega_3, \omega_4) \tag{11.85}$$

$$M_4 \frac{d^2\delta_4}{dt^2} = P_{m4} - P_{e4}(\delta_1, \delta_2, \delta_3, \delta_4, \omega_1, \omega_2, \omega_3, \omega_4) \tag{11.86}$$

Equations 11.83 to 11.86 constitute a system of nonlinear differential equations whose solution requires any of the known numerical integration techniques such as the Runge–Kutta method and various predictor-corrector methods.

* 11.4.4   Computer Solutions

Earlier stability studies analyzed the behavior of a single machine–infinite bus system that involved the simulation of a few ordinary differential equations.

Subsequent studies included a simple representation of the prime mover in which a constant torque is assumed to be applied to the generator shaft. The exciter, the speed governor, and saturation effects were not modeled.

Present transient stability programs allow representation of some of the synchronous machines in greater detail. In the vicinity of the fault, it is a common practice to use a two-axis generator model and to include the excitation and speed-governing systems. These models provide an accurate description of the postfault conditions for a longer time span. Where the simple models may be adequate for "first swing" calculations of about 1.0 s, the more detailed models may be valid for as long as 5.0 s.

Model Formulation The interactions among the various models that are usually included in a modern transient stability program are depicted in Fig. 11.37.

The electrical network assigns the exciter a voltage magnitude that it tries to maintain. The exciter supplies the voltage V_f to the generator field winding. A supplementary signal V_s may be included as part of the input to the exciter to provide positive damping. This input signal may be derived from the angular frequency deviation ω and other auxiliary signals.

The generator is coupled to the electrical network through its bus phasor voltage $\mathbf{V}$ and stator phasor current $\mathbf{I}$. The dynamic motion of the machine rotor is given in terms of the swing angle δ, the angular frequency deviation ω, the mechanical input power P_m, and the electrical output power P_e. Mathematically, this is expressed as follows:

$$\frac{d\delta}{dt} = \omega \tag{11.87}$$

$$\frac{d\omega}{dt} = \frac{1}{M}(P_m - P_e - D\omega) \tag{11.88}$$

where

$M = J\omega =$ inertia constant

$J =$ generator inertia

$D =$ damping coefficient

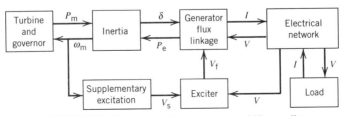

FIGURE 11.37 System models in modern stability studies.

The electrical power delivered by the generator to the external network is given by

$$P_{elec} = \text{Re}[\mathbf{VI}^*] \qquad (11.89)$$

where $\mathbf{V}$ and $\mathbf{I}$ are the generator terminal voltage and current phasors, respectively.

The stator current is decomposed into components along the direct and quadrature axes, $\mathbf{I}_d$ and $\mathbf{I}_q$, respectively. The direct axis is taken along the centerline of the machine pole and the quadrature axis is perpendicular to the direct axis. Thus, the stator current may be expressed as

$$\mathbf{I} = \mathbf{I}_d + \mathbf{I}_q \qquad (11.90)$$

The excitation voltage of the synchronous generator at steady state is found as follows:

$$\begin{aligned} \mathbf{E}_a &= \mathbf{V} + jX_q\mathbf{I}_q + jX_d\mathbf{I}_d \qquad (11.91) \\ &= \mathbf{E}_q + j(X_d - X_q)\mathbf{I}_d \end{aligned}$$

where

$$\mathbf{E}_q = \mathbf{V} + jX_q\mathbf{I} \qquad (11.92)$$

During transient conditions, the d-axis transient reactance X_d' is used instead of the synchronous reactance X_d. The q-axis reactance X_q' is the same as X_q because the field is on the direct axis only. Although X_d is normally greater than X_q, X_q is also normally larger than X_d'. The phasor diagram during transient conditions is shown in Fig. 11.38 with $\mathbf{E}_q$ as reference.

From the phasor diagram, the voltage $\mathbf{E}_q'$ behind transient reactance is given by

$$\mathbf{E}_q' = \mathbf{E}_q + j(X_d' - X_q)\mathbf{I}_d \qquad (11.93)$$

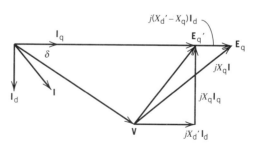

FIGURE 11.38 Synchronous generator transient phasor diagram.

E'_q is the voltage proportional to the resultant field flux linkages due to the interaction between the field and armature magnetic fields. Since the field flux linkages cannot change instantaneously after a disturbance, E'_q also does not change instantaneously. If X'_d is assumed to be equal to X_q, then E'_q reduces to E_q and can be calculated as follows:

$$E'_q = V + jX'_d I \qquad (11.94)$$

The synchronous generator is thus modeled as a constant-voltage source in series with its transient reactance.

In a typical transient stability program, the exciters of the synchronous machines being studied may be represented in varying complexities. The block diagram of what has come to be known as the IEEE type 1 excitation control system is shown in Fig. 11.39.

The input to the exciter control system is the generator terminal voltage V_t. The first transfer function represents the regulator input filtering; since its time constant T_R is very small, it may be neglected. At the first summing point, the terminal voltage is compared with the reference voltage V_{ref}, and to this error are added the supplementary signals V_s. The error voltage serves as the input to the voltage regulator, which is represented by a transfer function with a gain K_A and a time constant T_A. Maximum and minimum limits are imposed on the regulator output.

The exciter is modeled as a transfer function $1/(K_E + sT_E)$. The saturation function $S(E_f)$, representing the increase in field excitation requirements, may be modeled if it is desired to include the effects of saturation. The transfer function $K_F s/(1 + sT_F)$ in the feedback circuit from the exciter output to the second summing point provides the major loop damping.

The network equations of power flow are used to describe the performance of the electrical network for transient stability studies. The expression for the phasor voltage at bus k was derived in Section 10.2 and is repeated here as Eq. 11.95.

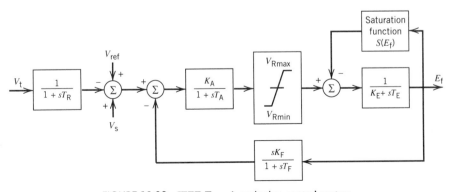

FIGURE 11.39   IEEE Type 1 excitation control system.

$$\mathbf{V}_k = \frac{1}{Y_{kk}}\left[\frac{P_k - jQ_k}{\mathbf{V}_k^*} - \Sigma_{m\neq k}\, Y_{km}\mathbf{V}_m\right] \qquad (11.95)$$

Power system loads are represented as either static impedance (static admittance) to ground, or constant power (constant current) at fixed power factor, or a combination of these types. Values of static-impedance and constant-current models are obtained from the scheduled bus loads and the bus voltages found in a base case power flow solution.

Summarizing, the mathematical expressions describing the power system may be divided into two categories:

1. Differential equations involving the vector of state variables $\mathbf{X}$ representing machine rotor angles and angular velocities, vector of bus voltages $\mathbf{V}$, and vector of bus currents $\mathbf{I}$.

$$\mathbf{F}(\dot{\mathbf{X}}, \mathbf{X}, \mathbf{V}, \mathbf{I}) = 0 \qquad (11.96)$$

2. Algebraic power flow equations.

$$\mathbf{G}(\mathbf{X}, \mathbf{V}, \mathbf{I}) = 0 \qquad (11.97)$$

Runge–Kutta Integration Method   The usual procedure for solving transient stability problems is to integrate the differential equations (11.96) for time step n using the voltage and current values from time step $(n-1)$. The Runge–Kutta method is commonly used for the numerical integration of nonlinear differential equations. Once the integration step is done, the new values of the voltages are used in the solution of the algebraic equations (11.97) to compute the values of the bus voltages and currents for time step n. The process of alternating integration and network solution is repeated as time progresses to step $(n+1)$. A fixed time increment of 0.01 s is frequently used in the integration.

In this Runge–Kutta computational method, the generators are effectively treated independently of each other during the integration step and are merely coupled to each other during the solution of the algebraic network equations. The solution of Eq. 11.96 involves the solution of a set of at least two and up to several differential equations for each generator. These are typically solved as a block; that is, integration of the machine equations proceeds for one machine at a time. It may be noted that there is a time skew between the integration step and the network solution step; that is, integration is performed using values of the other parameters one time step behind. For small integration steps, this can still give acceptable results.

Predictor-Corrector Integration Method   The time skew problem that is characteristic of the Runge–Kutta integration method may be minimized by solving the differential and algebraic equations simultaneously. Such a simultaneous solution is performed using predictor-corrector methods. Larger integration steps are allowed, and integration step sizes are easily varied. When the power system is quiescent, long step sizes are used; and when it is rapidly changing, much shorter step sizes are automatically chosen. This results in a minimum number of time steps.

REFERENCES

1. Anderson, Paul M. *Analysis of Faulted Power Systems*. Iowa State Press, Ames, 1973.

2. Arrillaga, J., C. P. Arnold, and B. J. Harker. *Computer Modelling of Electrical Power Systems*. Wiley, New York, 1986.

3. Blackburn, J. Lewis, ed. *Applied Protective Relaying*. Westinghouse Electric Corporation, Newark, N.J., 1976.

4. Del Toro, Vincent. *Electric Power Systems*. Prentice Hall, Englewood Cliffs, N.J., 1992.

5. *Electrical Transmission and Distribution Reference Book*. 4th ed. ABB Power T&D Co., Raleigh, NC, 1964.

6. El-Hawary, Mohamed E. *Electric Power Systems: Design and Analysis*. Reston Publishing, Reston, Va., 1983.

7. Glover, J. Duncan, and Mulukutla Sarma. *Power System Analysis and Design*. PWS Publishers, Boston, 1987.

8. Gönen, Turan. *Electric Power Transmission System Engineering Analysis and Design*. Wiley, New York, 1988.

9. ———. *Modern Power System Analysis*. Wiley, New York, 1988.

10. Gross, Charles A. *Power System Analysis*. 2nd ed. Wiley, New York, 1986.

11. Gungor, Behic R. *Power Systems*. Harcourt Brace Jovanovich, New York, 1988.

12. Heydt, G. T. *Computer Analysis Methods for Power Systems*. Macmillan, New York, 1986.

13. Mason, C. Russell. *The Art and Science of Protective Relaying*. Wiley, New York, 1956.

14. Neuenswander, John R. *Modern Power Systems*. International Textbook, Scranton, Pa., 1971.

15. Rustebakke, Homer M., ed. *Electric Utility Systems and Practices*. 4th ed. Wiley, New York, 1983.

16. Shipley, Randall B. *Introduction to Matrices and Power Systems*. Wiley, New York, 1976.

17. Stagg, Glenn A., and A. H. El-Abiad. *Computer Methods in Power System Analysis*. McGraw-Hill, New York, 1968.

18. Stevenson, William D., Jr. *Elements of Power System Analysis*. 4th ed. McGraw-Hill, New York, 1982.

19. Wallach, Y. *Calculations and Programs for Power System Networks*. Prentice Hall, Englewood Cliffs, N.J., 1986.

PROBLEMS

11.1 A three-phase synchronous generator is connected to a step-up three-phase transformer T_1, which is connected to a 60-km-long transmission line. At the far end of the line, a step-down transformer bank T_2 is connected. The secondary of T_2 supplies two motor loads M_1 and M_2. The ratings of the various types of equipment are

Generator: 10 MVA; 12 kV; $X = 20\%$; wye
T_1: 5 MVA; 12/69 kV; $X = 10\%$; delta-wye
T_2: 5 MVA; 69/4.16 kV; $X = 10\%$; wye-delta
M_1: 2000 kVA; 4.16 kV; $X = 20\%$; wye
M_2: 1000 kVA; 4.16 kV; $X = 20\%$; wye
Transmission line: $X = 0.5 \ \Omega/\text{km}$

a. Draw the per-phase equivalent circuit of the system showing all reactances in per unit and the base voltages used at various parts of the network. Choose the generator ratings as bases in the generator circuit.

b. For a three-phase fault on the low-voltage terminals of transformer T_2, calculate the short-circuit current in amperes supplied by the generator assuming that all internal voltages are $1.0 \underline{/0°}$ pu.

11.2 A three-phase, wye-connected synchronous generator is rated 350 MVA, 13.2 kV, with reactances of $X_d'' = 14\%$, $X_d' = 24\%$, and $X_d = 120\%$. The generator is connected to a step-up transformer rated 400 MVA, 13.2/115 kV, $X = 8\%$. The generator is operating at no load and rated voltage when a three-phase short circuit occurs on the high-voltage terminals of the transformer. Find the following currents expressed in per unit and in amperes.

a. The subtransient short-circuit current at the fault

b. The transient short-circuit current supplied by the generator

c. The steady-state short-circuit current at the fault

11.3 A synchronous generator is rated 300 MVA, 12 kV, 60 Hz. It is wye connected, and its neutral is solidly grounded. The machine reactances are $X_d'' = X_2 = 0.15$ pu and $X_0 = 0.05$ pu. The generator is operating at rated voltage at no load when a fault occurs at the generator terminals.

a. Find the ratio of the short-circuit current for a single line-to-ground fault to the short-circuit current for a three-phase fault.

b. Find the ratio of the short-circuit current for a line-to-line fault to the short-circuit current for a three-phase fault.

c. Find the ratio of the short-circuit current for a double line-to-ground fault to the short-circuit current for a three-phase fault.

11.4 An inductive reactance is to be inserted between the neutral of the generator of Problem 11.3 and ground in order to limit the short-circuit current for a single line-to-ground fault to that for a three-phase fault.

a. Determine the required value of reactance in ohms.

b. With the inductive reactance inserted between the neutral of the generator and ground, find the ratio of the short-circuit current for a line-to-line fault to the short-circuit current for a three-phase fault on the terminals of the generator.

c. With the inductive reactance computed in part (a) inserted between the neutral of the generator and ground, find the ratio of the short-circuit current for a double line-to-ground fault to the short-circuit current for a three-phase fault on the terminals of the generator.

11.5 A resistance is to be inserted between the neutral of the generator of Problem 11.3 and ground in order to limit the short-circuit current for a single line-to-ground fault to that for a three-phase fault.

a. Determine the required value of resistance in ohms.

b. With the resistance inserted between the neutral of the generator and ground, find the ratio of the short-circuit current for a line-to-line fault to the short-circuit current for a three-phase fault on the terminals of the generator.

c. With the resistance computed in part (a) inserted between the neutral of the generator and ground, find the ratio of the short-circuit current for a double line-to-ground fault to the short-circuit current for a three-phase fault on the terminals of the generator.

11.6 The power system shown in the one-line diagram of Fig. 11.40 has two synchronous generators each rated 100 MVA, 13.2 kV. The generator reactances are $X'' = X_2 = 20\%$ and $X_0 = 4\%$. The two transformers are each rated 100 MVA, 13.2/115 kV, with a reactance of 10%. The reactances of the transmission line are $X_1 = X_2 = 15\%$ and $X_0 = 50\%$. The system is operating at no load and rated voltage.

a. Find the short-circuit current for a double line-to-ground fault at bus 3.

b. Repeat part (a) for a single line-to-ground fault.

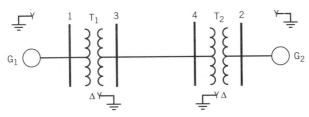

FIGURE 11.40 Power system of Problem 11.6.

11.7 Two generators G_1 and G_2 are connected through transformers T_1 and T_2 to a high-voltage bus to which a high-voltage transmission line is also connected. The line is open at the far end, wherein a three-phase fault occurs. The prefault voltage at the end of the transmission line is 350 kV. The equipment ratings and the per-unit reactances based on their respective ratings are as follows:

G_1: wye, 600 MVA, 12 kV, $X_d'' = X_2 = 0.10$, $X_0 = 0.05$
G_2: wye, 400 MVA, 12 kV, $X_d'' = X_2 = 0.13$, $X_0 = 0.07$

T_1: 600 MVA, 12/345 kV, Δ/Y, $X = 0.16$
T_2: 400 MVA, 12/345 kV, Y/Y, $X = 0.14$
Line: $X_1 = X_2 = 0.20$, $X_0 = 0.50$ on a base of 1000 MVA, 345 kV

a. Assume all wye-winding neutrals are solidly grounded. Determine the fault current and the contributions of each generator to the fault current.

b. Repeat part (a) for a single line-to-ground fault.

c. Repeat part (a) for a line-to-line fault.

d. Repeat part (a) for a double line-to-ground fault.

11.8 A 69-kV transmission line is supplied through a transformer by a 13.2-kV generator. The high-voltage side of the transformer is wye connected with its neutral solidly grounded, and the generator side is connected in delta. The positive-sequence reactances of the generator, transformer, and transmission line are 20 Ω, 10 Ω, and 50 Ω, respectively, and the negative-sequence reactance of the generator is 15 Ω. The zero-sequence reactances of the generator, and transmission line are 10 Ω and 150 Ω, respectively.

a. Calculate the short-circuit current for a single line-to-ground fault at the receiving end of the line.

b. Repeat part (a) for a double line-to-ground fault.

11.9 A single line-to-ground fault occurs at the far end of a radial transmission line with the following sequence impedances:

$$Z_1 = (0.5 + j1.50) \text{ pu}$$
$$Z_2 = (0.5 + j1.50) \text{ pu}$$
$$Z_0 = (1.0 + j3.00) \text{ pu}$$

The source is assumed to be an infinite bus of voltage 1.0 pu. Determine the fault current.

11.10 A four-bus power system is shown in Fig. 11.41. The ratings of the various types of equipment are given as follows:

G_1: 300 MVA, 12 kV, $X'' = 0.20$ per unit
G_2: 500 MVA, 20 kV, $X'' = 0.17$ per unit
G_3: 750 MVA, 20 kV, $X'' = 0.15$ per unit

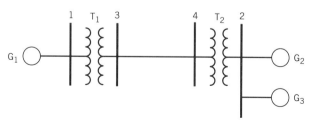

FIGURE 11.41 Power system of Problem 11.10.

T_1: 300 MVA, 12/230 kV, Δ-Y, $X = 0.10$ per unit
T_2: 500 MVA, 230/20 kV, Y-Δ, $X = 0.08$ per unit
Transmission line: 230-kV, $X = 75\ \Omega$

A three-phase short circuit occurs at bus 3, where the prefault voltage is 230 kV. Neglect prefault load currents.

a. Draw the impedance diagram in per unit using a power base of 1000 MVA and a voltage base of 20 kV for generator G_3.

b. Determine the fault current in amperes.

c. Calculate the contributions to the fault current from generator G_1 and from the transmission line.

d. Compute the fault current supplied by generators G_2 and G_3.

e. Determine the voltage at bus 2.

11.11 For the power system of Fig. 11.19, specify the location of the backup protection so that the cause of failure in the primary protection of the following system components will not cause the same failure in the backup.

a. Line 5–7

b. Line 6–8

c. Station B.

11.12 Consider the power system of Fig. 11.19. In response to a fault, the protection system tripped breakers 3, 7, and 8. Determine all possible locations of the fault. Determine what single failure had occurred, if any.

11.13 Repeat Problem 11.12 if the breakers tripped are 4, 6, and 7.

11.14 Repeat Problem 11.12 if the breakers tripped are 4, 6, 7, and 8.

11.15 For the three-phase high-voltage bus shown in Fig. 11.42, sketch the developed three-phase wiring diagram for the bus differential protection using overcurrent relays.

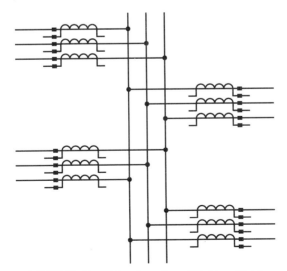

FIGURE 11.42   High-voltage bus of Problem 11.15.

11.16 A three-phase step-down transformer bank is rated 10 MVA and 69/13.8 kV. The high-voltage side is wye connected, and the low-voltage side is delta connected. Sketch the developed three-phase wiring diagram for the protection of the transformer bank using percentage-differential relays. Show all CT ratings, connections, and polarities. Also show the values of the currents in the lines, leads, relay windings, and transformer windings. Indicate the connections and ratings of any autotransformer that may be needed.

11.17 Repeat Problem 11.16 for the protection of a three-phase power transformer rated 100 MVA and 230/69 kV. Assume that the transformer windings are wye connected in both the primary and secondary sides.

11.18 Repeat Problem 11.17 when the transformer windings are connected in wye at the primary and in delta at the secondary.

11.19 A portion of a power system is shown in Fig. 11.43. Stations A and B have distance relays that are each adjusted for a first-zone reach of 100 ohms and a third-zone reach of 125 ohms. A fault occurs at F.

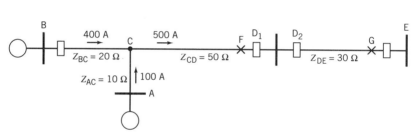

FIGURE 11.43 Portion of power system of Problem 11.19.

 a. What is the impedance seen by the relay at A and by the relay at B?

 b. Can the relay at A see the fault before the breaker at B has tripped?

 c. Can the relay at A see the fault after the breaker at B has tripped?

11.20 Repeat Problem 11.19 if the fault occurs at G and the breaker D2 fails to open. Assume that the fault current magnitudes are the same as those shown in Fig. 11.43.

11.21 A three-phase, 75-MVA, 13.2-kV, 60-Hz, eight-pole, synchronous generator has an inertia constant of $H = 4$ MJ/MVA. The machine is connected to an infinite bus, and it is operating at steady state. The prime mover is initially supplying an input power of 80,000 hp. The electric power output suddenly decreases to 40 MW.

 a. Calculate the angular acceleration.

 b. If the acceleration computed in part (a) remains constant for a period of 20 cycles, find the change in the rotor angle δ in electrical degrees and the speed in rpm at the end of 20 cycles.

11.22 A three-phase, 60-Hz synchronous generator has an inertia constant $H = 6$ MJ/MVA. It is connected to an infinite bus through a purely reactive network, and it delivers a real power of 1.0 pu. The maximum power that the generator could deliver is 2.5 per unit. A fault occurs at the terminals of the generator, and the output power is reduced to zero. The fault is cleared 0.30 s after it occurs. Determine the angle δ after the fault is cleared.

11.23 A three-phase synchronous generator has an inertia constant $H = 5$ MJ/MVA. It is connected to an infinite bus through a purely reactive network, and it delivers a real power of 0.8 pu. A fault occurs at the terminals of the generator, and the output power is reduced to zero. The maximum power that the generator could deliver is 2.0 pu. The fault is cleared 0.5 s after it occurs. Determine the angle δ after the fault is cleared.

11.24 The generator shown in Fig. 11.44 is delivering 1.0 pu current at 0.8 PF lagging to the infinite bus. The infinite-bus voltage is 1.0 pu. A fault occurs at point F. The fault is cleared by opening the circuit breakers at the ends of the faulted line section. Determine the prefault, on-fault, and postfault power-angle equations.

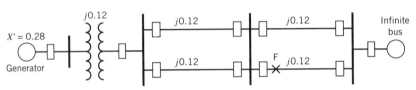

FIGURE 11.44   Power system of Problem 11.24.

11.25 A synchronous generator is connected to an infinite bus through two parallel transmission lines, each with a reactance of 0.5 pu. The excitation voltage of the generator is 1.2 pu, and its transient reactance is 0.20 pu. The generator is delivering a real power of 1.10 pu when a three-phase fault suddenly occurs near the end of one of the transmission lines where it connects to the bus. Determine the angle δ at the end of 10 cycles. Assume $H = 4$ pu-s.

11.26 A three-phase, 300-MVA, 13.8-kV, 60-Hz, 16-pole synchronous generator has an inertia constant $H = 2.5$ pu-s. The generator is initially operating at steady state with $P_m = P_e = 1.0$ pu and $\delta = 20°$. A three-phase fault occurs at the generator terminals. Determine the power angle eight cycles after the short occurs. Assume that P_m remains constant at 1.0 pu.

11.27 A hydroelectric power plant has two three-phase synchronous generators with the following ratings:

Unit 1: 400 MVA, 13.8 kV, 0.80 PF, 32 poles, $H_1 = 2.0$ pu-s
Unit 2: 200 MVA, 13.8 kV, 0.80 PF, 16 poles, $H_2 = 2.5$ pu-s

Choose a power base of 100 MVA.

a. Write the swing equation for each generator.

b. Derive the single equivalent machine swing equation.

11.28 A three-phase synchronous generator is connected to an infinite bus through a transformer and parallel transmission lines. The transformer reactance is 0.10 pu, and each transmission line has a reactance of 0.20 pu. The generator has an inertia constant $H = 3.0$ pu-s, and its transient reactance is 0.25. Assume that the mechanical power input P_m remains constant. The generator delivers a real power of 1.0 pu at 0.95 PF lagging to the infinite bus.

   a. Determine the excitation voltage of the generator.

   b. Write the expression for the electrical power delivered by the generator as a function of its power angle δ.

11.29 The generator of Problem 11.28 is initially operating at the given steady-state condition. A three-phase short circuit occurs on the bus interconnecting the transformer and the transmission lines. The fault is cleared after five cycles. Assume that none of the circuit breakers opened during the fault. Determine the angle δ after the fault is cleared.

11.30 The block diagram of an IEEE type 2 Excitation System is shown in Fig. 11.45. Derive the state equations in matrix form describing the dynamic performance of this excitation system. Assume that there is no exciter saturation.

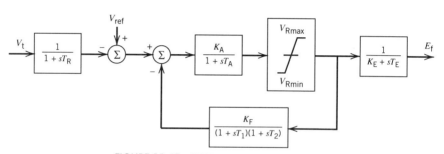

FIGURE 11.45   IEEE type 2 excitation system.

Appendices

Appendix A

Table of Conductor Characteristics

The following tables contain data on the electrical characteristics of various types of conductors, including copper, ACSR, hollow copper, Copperweld-copper, and Copperweld conductors. Other tables include skin effects and spacing factors for inductive reactances and capacitive reactances. Table 10 gives typical values of the different reactances of synchronous machines.

These tables are taken from *Electrical Transmission and Distribution Reference Book,* 4th ed., copyright © 1964, by ABB Power T&D Company Inc., Raleigh, NC, pp. 50–55. Reprinted by permission.

Table 1 Characteristics of Copper Conductors, Hard Drawn, 97.3 Percent Conductivity

Size of Conductor — Circular Mils	A.W.G. or B. & S.	No. of Strands	Diam. of Indiv. Strands (in)	Outside Diam. (in)	Breaking Strength (lb)	Weight (lb/mi)	Approx. Current Carrying Capacity Amps*	Geometric Mean Radius at 60 Cycles (ft)	r_a d-c 25°C	r_a 25 cyc 25°C	r_a 50 cyc 25°C	r_a 60 cyc 25°C	r_a d-c 50°C	r_a 25 cyc 50°C	r_a 50 cyc 50°C	r_a 60 cyc 50°C	x_a 25	x_a 50	x_a 60	x_a' 25	x_a' 50	x_a' 60
1 000 000	...	37	0.1644	1.151	43 830	16 300	1 300	0.0368	0.0585	0.0594	0.0620	0.0634	0.0640	0.0648	0.0672	0.0685	0.1666	0.333	0.400	0.216	0.1081	0.090
900 000	...	37	0.1560	1.092	39 510	14 670	1 220	0.0349	0.0650	0.0658	0.0682	0.0695	0.0711	0.0718	0.0740	0.0752	0.1693	0.339	0.406	0.220	0.1100	0.091
800 000	...	37	0.1470	1.029	35 120	13 040	1 130	0.0329	0.0731	0.0739	0.0760	0.0772	0.0800	0.0806	0.0826	0.0837	0.1722	0.344	0.413	0.224	0.1121	0.093
750 000	...	37	0.1424	0.997	33 400	12 230	1 090	0.0319	0.0780	0.0787	0.0807	0.0818	0.0853	0.0859	0.0878	0.0888	0.1739	0.348	0.417	0.226	0.1132	0.094
700 000	...	37	0.1375	0.963	31 170	11 410	1 040	0.0308	0.0836	0.0842	0.0861	0.0871	0.0914	0.0920	0.0937	0.0947	0.1759	0.352	0.422	0.229	0.1145	0.095
600 000	...	37	0.1273	0.891	27 020	9 781	940	0.0285	0.0975	0.0981	0.0997	0.1006	0.1066	0.1071	0.1086	0.1095	0.1799	0.360	0.432	0.235	0.1173	0.097
500 000	...	37	0.1162	0.814	22 510	8 151	840	0.0260	0.1170	0.1175	0.1188	0.1196	0.1280	0.1283	0.1296	0.1303	0.1845	0.369	0.443	0.241	0.1205	0.100
500 000	...	19	0.1622	0.811	21 590	8 151	840	0.0256	0.1170	0.1175	0.1188	0.1196	0.1280	0.1283	0.1296	0.1303	0.1853	0.371	0.445	0.241	0.1206	0.100
450 000	...	19	0.1539	0.770	19 750	7 336	780	0.0243	0.1300	0.1304	0.1316	0.1323	0.1422	0.1426	0.1437	0.1443	0.1879	0.376	0.451	0.245	0.1224	0.102
400 000	...	19	0.1451	0.726	17 560	6 521	730	0.0229	0.1462	0.1466	0.1477	0.1484	0.1600	0.1603	0.1613	0.1619	0.1909	0.382	0.458	0.249	0.1245	0.103
350 000	...	19	0.1357	0.679	15 590	5 706	670	0.0214	0.1671	0.1675	0.1684	0.1690	0.1828	0.1831	0.1840	0.1845	0.1943	0.389	0.466	0.254	0.1269	0.105
350 000	...	12	0.1708	0.710	15 140	5 706	670	0.0225	0.1671	0.1675	0.1684	0.1690	0.1828	0.1831	0.1840	0.1845	0.1918	0.384	0.460	0.251	0.1253	0.104
300 000	...	19	0.1257	0.629	13 510	4 891	610	0.01987	0.1950	0.1953	0.1961	0.1966	0.213	0.214	0.214	0.215	0.1982	0.396	0.476	0.259	0.1296	0.108
300 000	...	12	0.1581	0.657	13 170	4 891	610	0.0208	0.1950	0.1953	0.1961	0.1966	0.213	0.214	0.214	0.215	0.1957	0.392	0.470	0.256	0.1281	0.106
250 000	...	19	0.1147	0.574	11 360	4 076	540	0.01813	0.234	0.234	0.235	0.235	0.256	0.256	0.257	0.257	0.203	0.406	0.487	0.266	0.1329	0.111
250 000	...	12	0.1443	0.600	11 130	4 076	540	0.01902	0.234	0.234	0.235	0.235	0.256	0.256	0.257	0.257	0.200	0.401	0.481	0.263	0.1313	0.109
211 600	4/0	19	0.1055	0.528	9 617	3 450	480	0.01668	0.276	0.277	0.277	0.277	0.302	0.302	0.303	0.303	0.207	0.414	0.497	0.272	0.1359	0.113
211 600	4/0	12	0.1328	0.552	9 483	3 450	490	0.01750	0.276	0.277	0.277	0.278	0.302	0.303	0.303	0.303	0.205	0.409	0.491	0.269	0.1343	0.111
211 600	4/0	7	0.1739	0.522	9 154	3 450	420	0.01559	0.276	0.277	0.277	0.278	0.302	0.303	0.303	0.303	0.210	0.420	0.503	0.273	0.1363	0.113
167 800	3/0	12	0.1183	0.492	7 556	2 736	420	0.01559	0.349	0.349	0.349	0.350	0.381	0.381	0.382	0.382	0.210	0.421	0.505	0.277	0.1384	0.115
167 800	3/0	7	0.1548	0.464	7 366	2 736	360	0.01404	0.349	0.349	0.349	0.350	0.381	0.381	0.382	0.382	0.216	0.432	0.518	0.281	0.1405	0.117
133 100	2/0	7	0.1379	0.414	5 926	2 170	310	0.01252	0.440	0.440	0.440	0.440	0.481	0.481	0.481	0.481	0.222	0.443	0.532	0.289	0.1445	0.120
105 500	1/0	7	0.1228	0.368	4 752	1 720	270	0.01113	0.555	0.555	0.555	0.555	0.606	0.607	0.607	0.607	0.227	0.455	0.546	0.298	0.1488	0.124
83 690	1	7	0.1093	0.328	3 804	1 364	270	0.00992	0.699	0.699	0.699	0.699	0.765				0.233	0.467	0.560	0.306	0.1528	0.127
83 690	1	3	0.1670	0.360	3 620	1 351	270	0.01016	0.692	0.692	0.692	0.692	0.757				0.232	0.464	0.557	0.299	0.1495	0.124
66 370	2	7	0.0974	0.292	3 045	1 082	230	0.00883	0.881	0.882	0.882	0.882	0.964				0.239	0.478	0.574	0.314	0.1570	0.130
66 370	2	3	0.1487	0.320	2 913	1 071	240	0.00903	0.873				0.955				0.238	0.476	0.571	0.307	0.1537	0.128
66 370	2	1	......	0.258	3 003	1 061	220	0.00836	0.864				0.945				0.242	0.484	0.581	0.323	0.1614	0.134
52 630	3	7	0.0867	0.260	2 433	858	200	0.00787	1.112		Same as d-c		1.216		Same as d-c		0.245	0.490	0.588	0.316	0.1611	0.134
52 630	3	3	0.1325	0.285	2 359	850	200	0.00805	1.101		Same as d-c		1.204		Same as d-c		0.244	0.488	0.585	0.316	0.1578	0.131
52 630	3	1	......	0.229	2 439	841	190	0.00745	1.090		Same as d-c		1.192		Same as d-c		0.248	0.496	0.595	0.331	0.1656	0.138
41 740	4	3	0.1180	0.254	1 879	674	180	0.00717	1.388		Same as d-c		1.518		Same as d-c		0.250	0.499	0.599	0.324	0.1619	0.134
41 740	4	1	......	0.204	1 970	667	170	0.00663	1.374		Same as d-c		1.503		Same as d-c		0.254	0.507	0.609	0.339	0.1697	0.141
33 100	5	3	0.1050	0.226	1 505	534	150	0.00638	1.750		Same as d-c		1.914		Same as d-c		0.256	0.511	0.613	0.332	0.1661	0.138
33 100	5	1	......	0.1819	1 591	529	140	0.00591	1.733		Same as d-c		1.895		Same as d-c		0.260	0.519	0.623	0.348	0.1738	0.144
26 250	6	3	0.0935	0.201	1 205	424	130	0.00568	2.21		Same as d-c		2.41		Same as d-c		0.262	0.523	0.628	0.341	0.1703	0.141
26 250	6	1	......	0.1620	1 280	420	120	0.00526	2.18		Same as d-c		2.39		Same as d-c		0.265	0.531	0.637	0.356	0.1779	0.148
20 820	7	1	......	0.1443	1 030	333	110	0.00468	2.75		Same as d-c		3.01		Same as d-c		0.271	0.542	0.651	0.364	0.1821	0.151
16 510	8	1	......	0.1285	826	264	90	0.00417	3.47		Same as d-c		3.80		Same as d-c		0.277	0.554	0.665	0.372	0.1862	0.155

* For conductor at 75°C., air at 25°C., wind 1.4 miles per hour (2 ft/sec), frequency=60 cycles.

ble 2.A Characteristics of Aluminum Cable Steel Reinforced

Circular Mils or A.W.G. Aluminum	Aluminum Strands	Aluminum Layers	Aluminum Strand Dia. Inches	Steel Strands	Steel Strand Dia. Inches	Outside Diameter Inches	Copper Equivalent* Circular Mils or A.W.G.	Ultimate Strength Pounds	Weight Pounds per Mile	Geometric Mean Radius at 60 Cycles Feet	Approx. Current Carrying Capacity† Amps	r_a Resistance Ohms per Conductor per Mile, 25°C. (77°F.) Small Currents: d-c	25 cycles	50 cycles	60 cycles	50°C. (122°F.) Current Approx. 75% Capacity‡: d-c	25 cycles	50 cycles	60 cycles	x_a Inductive Reactance Ohms per Conductor per Mile at 1 Ft. Spacing All Currents: 25 cycles	50 cycles	60 cycles	x_a' Shunt Capacitive Reactance Megohms per Conductor per Mile at 1 Ft. Spacing: 25 cycles	50 cycles	60 cycles
00 000	54	3	0.1716	19	0.1030	1.545	1 000 000	56 000	10 777	0.0520	1 380	0.0587	0.0588	0.0590	0.0591	0.0646	0.0656	0.0675	0.0684	0.1495	0.299	0.359	0.1953	0.0977	0.0814
0 500	54	3	0.1673	19	0.1004	1.506	950 000	53 200	10 237	0.0507	1 340	0.0618	0.0619	0.0621	0.0622	0.0680	0.0690	0.0710	0.0720	0.1508	0.302	0.362	0.1971	0.0986	0.0821
31 000	54	3	0.1628	19	0.0977	1.465	900 000	50 400	9 699	0.0493	1 300	0.0652	0.0653	0.0655	0.0656	0.0718	0.0729	0.0749	0.0760	0.1522	0.304	0.365	0.1991	0.0996	0.0830
51 000	54	3	0.1582	19	0.0949	1.424	850 000	47 600	9 160	0.0479	1 250	0.0691	0.0692	0.0694	0.0695	0.0761	0.0771	0.0792	0.0803	0.1536	0.307	0.369	0.201	0.1006	0.0838
72 000	54	3	0.1535	19	0.0921	1.382	800 000	44 800	8 621	0.0465	1 200	0.0734	0.0735	0.0737	0.0738	0.0808	0.0819	0.0840	0.0851	0.1551	0.310	0.372	0.203	0.1016	0.0847
92 500	54	3	0.1486	19	0.0892	1.338	750 000	43 100	8 082	0.0450	1 160	0.0783	0.0784	0.0786	0.0788	0.0862	0.0872	0.0894	0.0906	0.1568	0.314	0.376	0.206	0.1028	0.0857
13 000	54	3	0.1436	19	0.0862	1.293	700 000	40 200	7 544	0.0435	1 110	0.0839	0.0840	0.0842	0.0844	0.0924	0.0935	0.0957	0.0969	0.1585	0.317	0.380	0.208	0.1040	0.0867
33 500	54	3	0.1384	7	0.1384	1.246	650 000	37 100	7 019	0.0420	1 060	0.0903	0.0905	0.0907	0.0909	0.0994	0.1005	0.1025	0.1035	0.1603	0.321	0.385	0.211	0.1053	0.0878
54 000	54	3	0.1329	7	0.1329	1.196	600 000	34 200	6 479	0.0403	1 010	0.0979	0.0980	0.0981	0.0982	0.1078	0.1088	0.1118	0.1128	0.1624	0.325	0.390	0.214	0.1068	0.0890
00 000	54	3	0.1291	7	0.1291	1.162	566 000	32 300	6 112	0.0391	970	0.104	0.104	0.104	0.104	0.1145	0.1155	0.1175	0.1185	0.1639	0.328	0.393	0.216	0.1078	0.0898
74 500	54	3	0.1273	7	0.1273	1.146	550 000	31 400	5 940	0.0386	950	0.107	0.107	0.107	0.108	0.1178	0.1188	0.1218	0.1228	0.1646	0.329	0.395	0.217	0.1083	0.0903
95 000	54	3	0.1214	7	0.1214	1.093	500 000	28 500	5 399	0.0368	900	0.117	0.118	0.118	0.119	0.1288	0.1308	0.1358	0.1378	0.1670	0.334	0.401	0.220	0.1100	0.0917
95 000	26	2	0.1749	7	0.1360	1.108	500 000	31 200	5 770	0.0375	900	0.117	0.117	0.117	0.117	0.1288	0.1288	0.1288	0.1288	0.1660	0.332	0.399	0.219	0.1095	0.0912
95 000	30	2	0.1628	19	0.0977	1.140	500 000	38 400	6 517	0.0393	910	0.117	0.117	0.117	0.117	0.1288	0.1288	0.1288	0.1288	0.1637	0.327	0.393	0.217	0.1085	0.0904
15 500	54	3	0.1151	7	0.1151	1.036	450 000	26 300	4 859	0.0349	830	0.131	0.131	0.131	0.132	0.1442	0.1452	0.1472	0.1482	0.1697	0.339	0.407	0.224	0.1119	0.0932
15 500	26	2	0.1659	7	0.1290	1.051	450 000	28 100	5 193	0.0355	840	0.131	0.131	0.131	0.131	0.1442	0.1442	0.1442	0.1442	0.1687	0.337	0.405	0.223	0.1114	0.0928
15 500	30	2	0.1544	19	0.0926	1.081	450 000	34 600	5 865	0.0372	840	0.131	0.131	0.131	0.131	0.1442	0.1442	0.1442	0.1442	0.1664	0.333	0.399	0.221	0.1104	0.0920
66 600	54	3	0.1111	7	0.1111	1.000	419 000	24 500	4 527	0.0337	800	0.140	0.140	0.141	0.141	0.1541	0.1571	0.1591	0.1601	0.1715	0.343	0.412	0.226	0.1132	0.0943
36 000	54	3	0.1085	7	0.1085	0.977	400 000	23 600	4 319	0.0329	770	0.147	0.147	0.148	0.148	0.1618	0.1638	0.1678	0.1688	0.1726	0.345	0.414	0.228	0.1140	0.0950
36 000	26	2	0.1564	7	0.1216	0.990	400 000	25 000	4 616	0.0335	780	0.147	0.147	0.147	0.147	0.1618	0.1618	0.1618	0.1618	0.1718	0.344	0.412	0.227	0.1135	0.0946
36 000	30	2	0.1456	19	0.0874	1.019	400 000	31 500	5 213	0.0351	780	0.147	0.147	0.147	0.147	0.1618	0.1618	0.1618	0.1618	0.1693	0.339	0.406	0.225	0.1125	0.0937
05 000	54	3	0.1059	7	0.1059	0.953	380 500	22 500	4 109	0.0321	750	0.154	0.155	0.155	0.155	0.1695	0.1715	0.1755	0.1775	0.1739	0.348	0.417	0.230	0.1149	0.0957
05 000	26	2	0.1525	7	0.1186	0.966	380 500	24 100	4 391	0.0327	760	0.154	0.154	0.154	0.154	0.1700	0.1720	0.1720	0.1720	0.1730	0.346	0.415	0.229	0.1144	0.0953
56 500	26	2	0.1463	7	0.1138	0.927	350 000	22 400	4 039	0.0313	730	0.168	0.168	0.168	0.168	0.1849	0.1859	0.1859	0.1859	0.1751	0.350	0.420	0.232	0.1159	0.0965
56 500	30	2	0.1362	7	0.1362	0.953	350 000	27 200	4 588	0.0328	730	0.168	0.168	0.168	0.168	0.1849	0.1859	0.1859	0.1859	0.1728	0.346	0.415	0.230	0.1149	0.0957
00 000	30	2	0.1291	7	0.1291	0.904	314 500	24 400	4 122	0.0311	690	0.187	0.187	0.187	0.187	0.206				0.1754	0.351	0.421	0.234	0.1167	0.0973
77 000	26	2	0.1355	7	0.1054	0.858	300 000	19 430	3 462	0.0290	670	0.196	0.196	0.196	0.196	0.216				0.1790	0.358	0.430	0.237	0.1186	0.0988
77 000	30	2	0.1261	7	0.1261	0.883	300 000	23 300	3 933	0.0304	670	0.196	0.196	0.196	0.196	0.216				0.1766	0.353	0.424	0.235	0.1176	0.0980
97 500	26	2	0.1236	7	0.0961	0.783	250 000	16 190	2 885	0.0265	590	0.235				0.259				0.1836	0.367	0.441	0.244	0.1219	0.1015
97 500	30	2	0.1151	7	0.1151	0.806	250 000	19 980	3 277	0.0278	600	0.235		Same as d-c		0.259		Same as d-c		0.1812	0.362	0.435	0.242	0.1208	0.1006
36 400	26	2	0.1138	7	0.0885	0.721	4/0	14 050	2 442	0.0244	530	0.278				0.306				0.1872	0.376	0.451	0.250	0.1248	0.1039
36 400	30	2	0.1059	7	0.1059	0.741	4/0	17 040	2 774	0.0255	530	0.278				0.306				0.1855	0.371	0.445	0.248	0.1238	0.1032
00 000	26	2	0.1074	7	0.0835	0.680	188 700	12 650	2 178	0.0230	490	0.311				0.342				0.1908	0.382	0.458	0.254	0.1269	0.1057
00 000	30	2	0.1000	7	0.1000	0.700	188 700	15 430	2 473	0.0241	500	0.311				0.342				0.1883	0.377	0.452	0.252	0.1258	0.1049
266 800	26	2	0.1013	7	0.0788	0.642	3/0	11 250	1 936	0.0217	460	0.350				0.385				0.1936	0.387	0.465	0.258	0.1289	0.1074

Circular Mils or A.W.G. Aluminum	Aluminum Strands	Aluminum Layers	Aluminum Strand Dia. Inches	Steel Strands	Steel Strand Dia. Inches	Outside Diameter Inches	Copper Equivalent*	Ultimate Strength Pounds	Weight Pounds per Mile	GMR at 60 Cycles Feet	For Current Approx. 75% Capacity ‡	25°C d-c	25 cycles	50 cycles	60 cycles	50°C d-c	25 cycles	50 cycles	60 cycles	x_a Single Layer Conductors, Small Currents: 25 cycles	50 cycles	60 cycles	Current Approx. 75% Capacity‡: 25 cycles	50 cycles	60 cycles	x_a' 25 cycles	50 cycles	60 cycles
266 800	6	1	0.2109	7	0.0703	0.633	3/0	9 645	1 802	0.00684	460	0.351	0.351	0.351	0.352	0.386	0.430	0.510	0.552	0.194	0.388	0.466	0.252	0.504	0.605	0.259	0.1294	0.1079
4/0	6	1	0.1878	1	0.1878	0.563	2/0	8 420	1 542	0.00814	340	0.441	0.442	0.444	0.445	0.485	0.514	0.567	0.592	0.218	0.437	0.524	0.242	0.484	0.581	0.267	0.1336	0.1113
3/0	6	1	0.1672	1	0.1672	0.502	1/0	6 675	1 223	0.00600	300	0.556	0.557	0.559	0.560	0.612	0.642	0.697	0.723	0.225	0.450	0.540	0.259	0.517	0.621	0.275	0.1377	0.1147
2/0	6	1	0.1490	1	0.1490	0.447	1	5 345	970	0.00510	270	0.702	0.702	0.704	0.706	0.773	0.806	0.866	0.895	0.231	0.462	0.554	0.267	0.534	0.641	0.284	0.1418	0.1182
1/0	6	1	0.1327	1	0.1327	0.398	2	4 280	769	0.00446	230	0.885	0.885	0.887	0.888	0.974	1.01	1.08	1.12	0.237	0.473	0.568	0.273	0.547	0.656	0.292	0.1460	0.1216
1	6	1	0.1182	1	0.1182	0.355	3	3 480	610	0.00418	200	1.12	1.12	1.12	1.12	1.23	1.27	1.34	1.38	0.242	0.483	0.580	0.277	0.554	0.665	0.300	0.1500	0.1250
2	6	1	0.1052	1	0.1052	0.316	4	2 790	484	0.00418	180	1.41	1.41	1.41	1.41	1.55	1.59	1.66	1.69	0.247	0.493	0.592	0.277	0.554	0.665	0.308	0.1542	0.1285
2	7	1	0.0974	1	0.1299	0.325	4	3 525	566	0.00504	180	1.41	1.41	1.41	1.41	1.55	1.59	1.62	1.65	0.247	0.493	0.592	0.267	0.535	0.642	0.306	0.1532	0.1276
3	6	1	0.0937	1	0.0937	0.281	5	2 250	384	0.00430	160	1.78	1.78	1.78	1.78	1.95	1.98	2.04	2.07	0.252	0.503	0.604	0.275	0.551	0.661	0.317	0.1583	0.1320
4	6	1	0.0834	1	0.0834	0.250	6	1 830	304	0.00437	140	2.24	2.24	2.24	2.24	2.47	2.50	2.54	2.57	0.257	0.514	0.611	0.274	0.549	0.659	0.325	0.1627	0.1355
4	7	1	0.0772	1	0.1029	0.257	6	2 288	356	0.00452	140	2.24	2.24	2.24	2.24	2.47	2.50	2.53	2.55	0.257	0.515	0.618	0.273	0.545	0.655	0.323	0.1615	0.1346
5	6	1	0.0743	1	0.0743	0.223	7	1 460	241	0.00416	120	2.82	2.82	2.82	2.82	3.10	3.12	3.16	3.18	0.262	0.525	0.630	0.279	0.557	0.665	0.333	0.1666	0.1388
6	6	1	0.0661	1	0.0661	0.198	8	1 170	191	0.00394	100	3.56	3.56	3.56	3.56	3.92	3.94	3.97	3.98	0.268	0.536	0.643	0.281	0.561	0.673	0.342	0.1708	0.1423

* Based on copper 97 percent, aluminum 61 percent conductivity.
† For conductor at 75°C., air at 25°C., wind 1.4 miles per hour (2 ft/sec), frequency = 60 cycles.
‡ "Current Approx. 75% Capacity" is 75% of the "Approx. Current Carrying Capacity in Amps." and is approximately the current which will produce 50°C. conductor temp. (25°C. rise) with 25°C. air temp., wind 1.4 miles per hour.

able 2.B Characteristics of "Expanded" Aluminum Cable Steel Reinforced

Circular Mils or A.W.G. Aluminum	Aluminum Strands	Aluminum Layers	Aluminum Strand Dia. Inches	Steel Strands	Steel Strand Dia. Inches	Filler Section Aluminum Strands	Filler Aluminum Strand Dia. Inches	Filler Paper Strands	Filler Paper Layers	Outside Diameter Inches	Copper Equivalent Circular Mils or A.W.G.	Ultimate Strength Pounds	Weight Pounds per Mile	Geometric Mean Radius at 60 Cycles Feet	Approx. Current Carrying Capacity Amps	r_a Resistance Ohms per Conductor per Mile, 25°C. (77°F.) Small Currents: d-c, 25, 50, 60 cycles	50°C. (122°F.) Current Approx. 75% Capacity: d-c, 25, 50, 60 cycles	x_a Inductive Reactance Ohms per Conductor per Mile at 1 Ft. Spacing All Currents: 25 cycles	50 cycles	60 cycles	x_a' Shunt Capacitive Reactance Megohms per Conductor per Mile at 1 Ft. Spacing: 25 cycles	50 cycles	60 cycles
850 000	54	2	0.1255	19	0.0834	4	0.1182	23	2	1.38	534 000	35 371	7 200	(1)	(1)	(1)	(1)	(1)	(1)	(1)	(1)	(1)	(1)
150 000	54	2	0.1409	19	0.0921	4	0.1353	24	2	1.55	724 000	41 900	9 070										
338 000	66	2	0.1350	19	0.100	4	0.184	18	2	1.75	840 000	49 278	11 340										

(1) Electrical Characteristics not available until laboratory measurements are completed.

Table 3.A Characteristics of Anaconda Hollow Copper Conductors

Design Number	Size of Conductor Circular Mils or A.W.G.	Wires: Number	Wires: Diameter Inches	Outside Diameter Inches	Breaking Strain Pounds	Weight Pounds per Mile	Geometric Mean Radius at 60 Cycles Feet	Approx. Current Carrying Capacity Amps‡	r_a Resistance Ohms per Conductor per Mile: 25°C. (77°F.) d-c 25 cycles	25°C. (77°F.) 50 cycles 60 cycles	50°C. (122°F.) d-c 25 cycles	50°C. (122°F.) 50 cycles 60 cycles	x_a Inductive Reactance Ohms per Conductor per Mile at 1 Ft. Spacing: 25 cycles	50 cycles	60 cycles	x_a' Shunt Capacitive Reactance Megohms per Conductor per Mile at 1 Ft. Spacing: 25 cycles	50 cycles	60 cycles
966	890 500	28	0.1610	1.650	36 000	15 085	0.0612	1395	0.0671	0.0676	0.0734	0.0739	0.1412	0.282	0.339	0.1907	0.0953	0.079
96R1	750 000	42	0.1296	1.155	34 200	12 345	0.0408	1160	0.0786	0.0791	0.0860	0.0865	0.1617	0.323	0.388	0.216	0.1080	0.090
939	650 000	50	0.1097	1.126	29 500	10 761	0.0406	1060	0.0909	0.0915	0.0994	0.1001	0.1621	0.324	0.389	0.218	0.1089	0.090
360R1	600 000	50	0.1053	1.007	27 500	9 905	0.0387	1020	0.0984	0.0991	0.1077	0.1084	0.1644	0.329	0.395	0.221	0.1105	0.092
938	550 000	50	0.1009	1.036	25 200	9 103	0.0373	960	0.1076	0.1081	0.1177	0.1183	0.1663	0.333	0.399	0.224	0.1119	0.093
4R5	510 000	50	0.0970	1.000	22 700	8 485	0.0360	910	0.1173	0.1178	0.1283	0.1289	0.1681	0.336	0.404	0.226	0.1131	0.094
892R3	500 000	18	0.1558	1.080	21 400	8 263	0.0394	900	0.1178	0.1184	0.1289	0.1296	0.1630	0.326	0.391	0.221	0.1104	0.092
933	450 000	21	0.1353	1.074	19 300	7 476	0.0398	850	0.1319	0.1324	0.1443	0.1448	0.1630	0.326	0.391	0.221	0.1106	0.092
924	400 000	21	0.1227	1.014	17 200	6 642	0.0376	810	0.1485	0.1491	0.1624	0.1631	0.1658	0.332	0.398	0.225	0.1126	0.093
925R1	380 500	22	0.1211	1.003	16 300	6 331	0.0373	780	0.1565	0.1572	0.1712	0.1719	0.1663	0.333	0.399	0.226	0.1130	0.094
565R1	350 000	21	0.1196	0.950	15 100	5 813	0.0353	750	0.1695	0.1700	0.1854	0.1860	0.1691	0.338	0.406	0.230	0.1150	0.095
936	350 000	15	0.1444	0.860	15 400	5 776	0.0311	740	0.1690	0.1695	0.1849	0.1854	0.1754	0.351	0.421	0.237	0.1185	0.098
378R1	350 000	30	0.1059	0.736	16 100	5 739	0.0253	700	0.1685	0.1690	0.1843	0.1849	0.1860	0.372	0.446	0.248	0.1241	0.103
954	321 000	22	0.1113	0.920	13 850	5 343	0.0340	700	0.1851	0.1856	0.202	0.203	0.1710	0.342	0.410	0.232	0.1161	0.096
935	300 000	18	0.1205	0.839	13 100	4 984	0.0307	670	0.1980	0.1985	0.216	0.217	0.1761	0.352	0.423	0.239	0.1194	0.099
903R1	300 000	15	0.1338	0.797	13 200	4 953	0.0289	660	0.1969	0.1975	0.215	0.216	0.1793	0.359	0.430	0.242	0.1212	0.101
178R2	300 000	12	0.1507	0.750	13 050	4 937	0.0266	650	0.1964	0.1969	0.215	0.216	0.1833	0.367	0.440	0.247	0.1234	0.102
926	250 000	18	0.1100	0.766	10 950	4 155	0.0279	600	0.238	0.239	0.260	0.261	0.1810	0.362	0.434	0.245	0.1226	0.102
915R1	250 000	15	0.1214	0.725	11 000	4 148	0.0266	590	0.237	0.238	0.259	0.260	0.1834	0.367	0.440	0.249	0.1246	0.103
24R1	250 000	12	0.1368	0.683	11 000	4 133	0.0245	580	0.237	0.238	0.259	0.260	0.1876	0.375	0.450	0.253	0.1267	0.106
923	4/0	18	0.1005	0.700	9 300	3 521	0.0255	530	0.281	0.282	0.307	0.308	0.1855	0.371	0.445	0.252	0.1258	0.104
922	4/0	15	0.1109	0.663	9 300	3 510	0.0238	520	0.281	0.282	0.307	0.308	0.1889	0.378	0.453	0.256	0.1278	0.106
50R2	4/0	14	0.1152	0.650	9 300	3 510	0.0234	520	0.280	0.281	0.306	0.307	0.1898	0.380	0.455	0.257	0.1285	0.107
158R1	3/0	16	0.0961	0.606	7 500	2 785	0.0221	460	0.354	0.355	0.387	0.388	0.1928	0.386	0.463	0.262	0.1310	0.109
495R1	3/0	15	0.0996	0.595	7 600	2 785	0.0214	460	0.353	0.354	0.386	0.387	0.1943	0.389	0.466	0.263	0.1316	0.109
570R2	3/0	12	0.1123	0.560	7 600	2 772	0.0201	450	0.352	0.353	0.385	0.386	0.1976	0.395	0.474	0.268	0.1338	0.111
909R2	2/0	15	0.0880	0.530	5 950	2 213	0.0191	370	0.446	0.446	0.487	0.487	0.200	0.400	0.481	0.271	0.1357	0.113
412R2	2/0	14	0.0913	0.515	6 000	2 207	0.0184	370	0.446	0.446	0.487	0.487	0.202	0.404	0.485	0.274	0.1368	0.114
937	2/0	13	0.0950	0.505	6 000	2 203	0.0181	370	0.446	0.446	0.487	0.487	0.203	0.406	0.487	0.275	0.1375	0.114
930	125 600	14	0.0885	0.500	5 650	2 083	0.0180	360	0.473	0.473	0.517	0.517	0.203	0.406	0.487	0.276	0.1378	0.114
934	121 300	15	0.0836	0.500	5 400	2 015	0.0179	350	0.491	0.491	0.537	0.537	0.203	0.407	0.488	0.276	0.1378	0.114
901	119 400	12	0.0936	0.470	5 300	1 979	0.0165	340	0.507	0.507	0.555	0.555	0.207	0.415	0.498	0.280	0.1400	0.116

‡For conductor at 75°C., air at 25°C., wind 1.4 miles per hour (2 ft/sec), frequency=60 cycles, average tarnished surface.

Table 3.B Characteristics of General Cable Type HH Hollow Copper Conductors

Conductor Size Circular Mils or A.W.G.	Outside[1] Diameter Inches	Wall Thickness Inches	Weight Pounds per Mile	Breaking Strength Pounds	Geometric Mean Radius Feet	Approx. Current Carrying Capacity[2] Amps	r_a Resistance Ohms per Conductor per Mile: 25°C. (77°F.) d-c	25°C. 25 cycles	25°C. 50 cycles	25°C. 60 cycles	50°C. (122°F.) d-c	50°C. 25 cycles	50°C. 50 cycles	50°C. 60 cycles	x_a Inductive Reactance Ohms per Conductor per Mile at 1 Foot Spacing: 25 cycles	50 cycles	60 cycles	x_a' Shunt Capacitive Reactance Megohms per Conductor per Mile at 1 Foot Spacing: 25 cycles	50 cycles	60 cycles
1 000 000	2.103	0.150*	16 160	43 190	0.0833	1620	0.0576	0.0576	0.0577	0.0577	0.0630	0.0630	0.0631	0.0631	0.1257	0.251	0.302	0.1734	0.0867	0.0722
950 000	2.035	0.147*	15 350	41 030	0.0805	1565	0.0606	0.0606	0.0607	0.0607	0.0663	0.0664	0.0664	0.0664	0.1274	0.255	0.306	0.1757	0.0879	0.0732
900 000	1.966	0.144*	14 540	38 870	0.0778	1505	0.0640	0.0640	0.0641	0.0641	0.0700	0.0701	0.0701	0.0701	0.1291	0.258	0.310	0.1782	0.0891	0.0742
850 000	1.901	0.140*	13 730	36 710	0.0751	1450	0.0677	0.0678	0.0678	0.0678	0.0741	0.0742	0.0742	0.0742	0.1309	0.262	0.314	0.1805	0.0903	0.0752
800 000	1.820	0.137*	12 920	34 550	0.0722	1390	0.0720	0.0720	0.0720	0.0721	0.0788	0.0788	0.0788	0.0788	0.1329	0.266	0.319	0.1833	0.0917	0.0764
790 000	1.650	0.131†	12 760	34 120	0.0646	1335	0.0729	0.0729	0.0730	0.0730	0.0797	0.0798	0.0799	0.0799	0.1385	0.277	0.332	0.1906	0.0953	0.0794
750 000	1.750	0.133*	12 120	32 390	0.0691	1325	0.0768	0.0768	0.0768	0.0769	0.0840	0.0840	0.0841	0.0841	0.1351	0.270	0.324	0.1864	0.0932	0.0777
700 000	1.686	0.130*	11 310	30 230	0.0665	1265	0.0822	0.0823	0.0823	0.0823	0.0900	0.0900	0.0901	0.0901	0.1370	0.274	0.329	0.1891	0.0945	0.0788
650 000	1.610	0.126*	10 500	28 070	0.0635	1200	0.0886	0.0886	0.0886	0.0887	0.0969	0.0970	0.0970	0.0970	0.1394	0.279	0.335	0.1924	0.0962	0.0802
600 000	1.558	0.123*	9 692	25 910	0.0615	1140	0.0959	0.0960	0.0960	0.0960	0.1050	0.1051	0.1051	0.1051	0.1410	0.282	0.338	0.1947	0.0974	0.0811
550 000	1.478	0.119*	8 884	23 750	0.0583	1075	0.1047	0.1048	0.1048	0.1048	0.1146	0.1146	0.1147	0.1147	0.1437	0.287	0.345	0.1985	0.0992	0.0827
512 000	1.400	0.115*	8 270	22 110	0.0551	1020	0.1124	0.1125	0.1125	0.1125	0.1230	0.1230	0.1231	0.1231	0.1466	0.293	0.352	0.202	0.1012	0.0843
500 000	1.390	0.115*	8 076	21 590	0.0547	1005	0.1151	0.1151	0.1152	0.1152	0.1259	0.1260	0.1260	0.1260	0.1469	0.294	0.353	0.203	0.1014	0.0845
500 000	1.268	0.109†	8 074	21 590	0.0494	978	0.1151	0.1152	0.1152	0.1152	0.1259	0.1260	0.1260	0.1261	0.1521	0.304	0.365	0.209	0.1047	0.0872
500 000	1.100	0.130†	8 068	21 590	0.0420	937	0.1150	0.1151	0.1152	0.1153	0.1258	0.1259	0.1260	0.1260	0.1603	0.321	0.385	0.219	0.1098	0.0915
500 000	1.020	0.144†	8 063	21 590	0.0384	915	0.1150	0.1150	0.1152	0.1152	0.1258	0.1259	0.1260	0.1261	0.1648	0.330	0.396	0.225	0.1124	0.0937
450 000	1.317	0.111*	7 268	19 430	0.0518	939	0.1279	0.1280	0.1280	0.1280	0.1400	0.1401	0.1401	0.1401	0.1496	0.299	0.359	0.207	0.1033	0.0861
450 000	1.188	0.105†	7 266	19 430	0.0462	910	0.1278	0.1279	0.1279	0.1280	0.1399	0.1400	0.1400	0.1401	0.1554	0.311	0.373	0.214	0.1070	0.0892
400 000	1.218	0.106*	6 460	17 270	0.0478	864	0.1439	0.1440	0.1440	0.1440	0.1575	0.1576	0.1576	0.1576	0.1537	0.307	0.369	0.212	0.1061	0.0884
400 000	1.103	0.100†	6 458	17 270	0.0428	838	0.1438	0.1439	0.1439	0.1440	0.1574	0.1575	0.1575	0.1576	0.1593	0.319	0.382	0.219	0.1097	0.0914
350 000	1.128	0.102*	5 653	15 110	0.0443	790	0.1644	0.1645	0.1645	0.1645	0.1799	0.1800	0.1800	0.1800	0.1576	0.315	0.378	0.218	0.1089	0.0907
350 000	1.014	0.096†	5 650	15 110	0.0393	764	0.1644	0.1645	0.1645	0.1646	0.1799	0.1800	0.1800	0.1801	0.1637	0.328	0.393	0.225	0.1127	0.0939
300 000	1.020	0.096*	4 845	12 950	0.0399	709	0.1918	0.1919	0.1919	0.1919	0.210	0.210	0.210	0.210	0.1628	0.326	0.391	0.225	0.1124	0.0937
300 000	0.919	0.091†	4 843	12 950	0.0355	687	0.1917	0.1918	0.1918	0.1919	0.210	0.210	0.210	0.210	0.1688	0.338	0.405	0.232	0.1162	0.0968
250 000	0.914	0.091*	4 037	10 790	0.0357	626	0.230	0.230	0.230	0.230	0.252	0.252	0.252	0.252	0.1685	0.337	0.404	0.233	0.1163	0.0970
250 000	0.818	0.086†	4 036	10 790	0.0315	606	0.230	0.230	0.230	0.230	0.252	0.252	0.252	0.252	0.1748	0.350	0.420	0.241	0.1203	0.1002
250 000	0.766	0.094†	4 034	10 790	0.0292	594	0.230	0.230	0.230	0.230	0.252	0.252	0.252	0.252	0.1787	0.357	0.429	0.245	0.1226	0.1022
214 500	0.650	0.098†	3 459	9 265	0.0243	524	0.268	0.268	0.268	0.268	0.293	0.293	0.293	0.294	0.1879	0.376	0.451	0.257	0.1285	0.1071
4/0	0.733	0.082†	3 415	9 140	0.0281	539	0.272	0.272	0.272	0.272	0.297	0.297	0.298	0.298	0.1806	0.361	0.433	0.248	0.1242	0.1035
3/0	0.608	0.080†	2 707	7 240	0.0230	454	0.343	0.343	0.343	0.343	0.375	0.375	0.375	0.375	0.1907	0.381	0.458	0.262	0.1309	0.1091
2/0	0.500	0.080†	2 146	5 750	0.0186	382	0.432	0.432	0.432	0.432	0.472	0.473	0.473	0.473	0.201	0.403	0.483	0.276	0.1378	0.1149

Notes: *Thickness at edges of interlocked segments. †Thickness uniform throughout.
(1) Conductors of smaller diameter for given cross-sectional area also available; in the naught sizes, some additional diameter expansion is possible.
(2) For conductor at 75°C., air at 25°C., wind 1.4 miles per hour (2 ft/sec), frequency=60 cycles.

Table 4.A Characteristics of Copperweld-Copper Conductors

Nominal Designation	Size of Conductor: Number and Diameter of Wires		Size of Conductor: Outside Diameter Inches	Copper Equivalent Circular Mils or A.W.G.	Rated Breaking Load Lbs.	Weight Lbs. per Mile	Geometric Mean Radius at 60 Cycles Feet	Approx. Current Carrying Capacity at 60 Cycles Amps*	r_a Resistance Ohms per Conductor per Mile at 25°C.(77°F.) Small Currents				r_a Resistance Ohms per Conductor per Mile at 50°C. (122°F.) Current Approx. 75% of Capacity**				x_a Inductive Reactance Ohms per Conductor per Mile One ft. Spacing Average Currents			x_a' Capacitive Reactance Megohms per Conductor per Mile One ft. Spacing		
	Copperweld	Copper							d-c	25 cycles	50 cycles	60 cycles	d-c	25 cycles	50 cycles	60 cycles	25 cycles	50 cycles	60 cycles	25 cycles	50 cycles	60 cycles
0 E	7x.1576"	12x.1576"	0.788	350 000	32 420	7 409	0.0220	660	0.1658	0.1728	0.1789	0.1812	0.1812	0.1915	0.201	0.204	0.1929	0.386	0.463	0.243	0.1216	0.1014
0 EK	4x.1470"	15x.1470"	0.735	350 000	23 850	6 536	0.0245	680	0.1658	0.1682	0.1700	0.1705	0.1812	0.1845	0.1873	0.1882	0.1875	0.375	0.450	0.248	0.1241	0.1034
0 V	3x.1751"	9x.1893"	0.754	350 000	23 480	6 578	0.0226	650	0.1655	0.1725	0.1800	0.1828	0.1809	0.1910	0.202	0.206	0.1915	0.383	0.460	0.246	0.1232	0.1027
0 E	7x.1459"	12x.1459"	0.729	300 000	27 770	6 351	0.0204	600	0.1934	0.200	0.207	0.209	0.211	0.222	0.232	0.235	0.1969	0.394	0.473	0.249	0.1244	0.1037
0 EK	4x.1361"	15x.1361"	0.680	300 000	20 960	5 602	0.0227	610	0.1934	0.1958	0.1976	0.1981	0.211	0.215	0.218	0.219	0.1914	0.383	0.460	0.254	0.1269	0.1057
0 V	3x.1621"	9x.1752"	0.698	300 000	20 730	5 639	0.0209	590	0.1930	0.200	0.208	0.210	0.211	0.222	0.233	0.237	0.1954	0.391	0.469	0.252	0.1259	0.1050
0 E	7x.1332"	12x.1332"	0.666	250 000	23 920	5 292	0.01859	540	0.232	0.239	0.245	0.248	0.254	0.265	0.275	0.279	0.202	0.403	0.484	0.255	0.1276	0.1064
0 EK	4x.1242"	15x.1242"	0.621	250 000	17 840	4 669	0.0207	540	0.232	0.235	0.236	0.237	0.254	0.258	0.261	0.261	0.1960	0.392	0.471	0.260	0.1301	0.1084
0 V	3x.1480"	9x.1600"	0.637	250 000	17 420	4 699	0.01911	530	0.232	0.239	0.246	0.249	0.253	0.264	0.276	0.281	0.200	0.400	0.480	0.258	0.1292	0.1077
0 E	7x.1225"	12x.1225"	0.613	4/0	20 730	4 479	0.01711	480	0.274	0.281	0.287	0.290	0.300	0.312	0.323	0.326	0.206	0.411	0.493	0.261	0.1306	0.1088
0 G	2x.1944"	5x.1944"	0.583	4/0	15 640	4 168	0.01409	460	0.273	0.284	0.294	0.298	0.299	0.318	0.336	0.342	0.215	0.431	0.517	0.265	0.1324	0.1103
0 EK	4x.1143"	15x.1143"	0.571	4/0	15 370	3 951	0.01903	490	0.274	0.277	0.278	0.279	0.300	0.304	0.307	0.308	0.200	0.401	0.481	0.266	0.1331	0.1109
0 V	3x.1361"	9x.1472"	0.586	4/0	15 000	3 977	0.01758	470	0.274	0.281	0.288	0.291	0.299	0.311	0.323	0.328	0.204	0.409	0.490	0.264	0.1322	0.1101
0 F	1x.1833"	6x.1833"	0.550	4/0	12 290	3 750	0.01558	470	0.273	0.280	0.285	0.287	0.299	0.309	0.318	0.322	0.210	0.421	0.505	0.269	0.1344	0.1120
0 E	7x.1091"	12x.1091"	0.545	3/0	16 800	3 552	0.01521	420	0.346	0.353	0.359	0.361	0.378	0.391	0.402	0.407	0.212	0.423	0.508	0.270	0.1348	0.1123
0 J	3x.1851"	4x.1851"	0.555	3/0	16 170	3 732	0.01156	410	0.344	0.356	0.367	0.372	0.377	0.398	0.419	0.428	0.225	0.451	0.541	0.268	0.1341	0.1118
0 G	2x.1731"	2x.1731"	0.519	3/0	12 860	3 305	0.01254	400	0.344	0.355	0.365	0.369	0.377	0.397	0.416	0.423	0.221	0.443	0.531	0.273	0.1365	0.1137
0 EK	4x.1018"	4x.1018"	0.509	3/0	12 370	3 134	0.01697	420	0.346	0.348	0.350	0.351	0.378	0.382	0.386	0.386	0.206	0.412	0.495	0.274	0.1372	0.1143
0 V	3x.1311"	9x.1311"	0.522	3/0	12 220	3 154	0.01566	410	0.345	0.352	0.360	0.362	0.377	0.390	0.403	0.408	0.210	0.420	0.504	0.273	0.1363	0.1136
0 F	1x.1632"	6x.1632"	0.490	3/0	9 980	2 974	0.01388	410	0.344	0.351	0.356	0.358	0.377	0.388	0.397	0.401	0.216	0.432	0.519	0.277	0.1385	0.1155
0 K	4x.1780"	3x.1780"	0.534	2/0	17 600	3 411	0.00912	360	0.434	0.447	0.459	0.466	0.475	0.499	0.524	0.535	0.237	0.475	0.570	0.271	0.1355	0.1129
0 J	3x.1648"	4x.1648"	0.494	2/0	13 430	2 960	0.01029	350	0.434	0.446	0.457	0.462	0.475	0.498	0.520	0.530	0.231	0.463	0.555	0.277	0.1383	0.1152
0 G	2x.1542"	5x.1542"	0.463	2/0	10 510	2 622	0.01119	350	0.434	0.445	0.456	0.459	0.475	0.497	0.518	0.525	0.227	0.454	0.545	0.281	0.1406	0.1171
0 V	3x.1080"	9x.1167"	0.465	2/0	9 846	2 502	0.01395	360	0.435	0.442	0.450	0.452	0.476	0.489	0.504	0.509	0.216	0.432	0.518	0.281	0.1404	0.1170
0 F	1x.1454"	6x.1454"	0.436	2/0	8 094	2 359	0.01235	350	0.434	0.441	0.446	0.448	0.475	0.487	0.497	0.501	0.222	0.444	0.533	0.285	0.1427	0.1189
0 K	4x.1585"	3x.1585"	0.475	1/0	14 490	2 703	0.00812	310	0.548	0.560	0.573	0.579	0.599	0.625	0.652	0.664	0.243	0.487	0.584	0.279	0.1397	0.1164
0 J	3x.1467"	4x.1467"	0.440	1/0	10 970	2 346	0.00917	310	0.548	0.559	0.570	0.576	0.599	0.624	0.648	0.659	0.237	0.474	0.569	0.285	0.1423	0.1186
0 G	2x.1373"	5x.1373"	0.412	1/0	8 563	2 078	0.00996	310	0.548	0.559	0.568	0.573	0.599	0.623	0.645	0.654	0.233	0.466	0.559	0.289	0.1447	0.1206
0 F	1x.1294"	6x.1294"	0.388	1/0	6 536	1 870	0.01099	310	0.548	0.554	0.559	0.562	0.599	0.612	0.622	0.627	0.228	0.456	0.547	0.294	0.1469	0.1224
N	5x.1546"	2x.1546"	0.464	1	15 410	2 541	0.00638	280	0.691	0.705	0.719	0.726	0.755	0.787	0.818	0.832	0.256	0.512	0.614	0.281	0.1405	0.1171
K	4x.1412"	3x.1412"	0.423	1	11 900	2 144	0.00723	270	0.691	0.704	0.716	0.722	0.755	0.784	0.813	0.825	0.249	0.498	0.598	0.288	0.1438	0.1198
J	3x.1307"	4x.1307"	0.392	1	9 000	1 861	0.00817	270	0.691	0.703	0.714	0.719	0.755	0.783	0.808	0.820	0.243	0.486	0.583	0.293	0.1465	0.1221
G	2x.1222"	5x.1222"	0.367	1	6 956	1 649	0.00887	260	0.691	0.702	0.712	0.716	0.755	0.781	0.805	0.815	0.239	0.478	0.573	0.298	0.1488	0.1240
F	1x.1153"	6x.1153"	0.346	1	5 266	1 483	0.00980	270	0.691	0.698	0.704	0.705	0.755	0.769	0.781	0.786	0.234	0.468	0.561	0.302	0.1509	0.1258
P	6x.1540"	1x.1540"	0.462	2	16 870	2 487	0.00501	250	0.871	0.886	0.901	0.909	0.952	0.988	1.024	1.040	0.268	0.536	0.643	0.281	0.1406	0.1172
N	5x.1377"	2x.1377"	0.413	2	12 680	2 015	0.00568	240	0.871	0.885	0.899	0.906	0.952	0.986	1.020	1.035	0.261	0.523	0.627	0.289	0.1446	0.1205
K	4x.1257"	3x.1257"	0.377	2	9 730	1 701	0.00644	240	0.871	0.884	0.896	0.902	0.952	0.983	1.014	1.028	0.255	0.510	0.612	0.296	0.1479	0.1232
J	3x.1164"	4x.1164"	0.349	2	7 322	1 476	0.00727	230	0.871	0.883	0.894	0.899	0.952	0.982	1.010	1.022	0.249	0.498	0.598	0.301	0.1506	0.1255
A	1x.1699"	2x.1699"	0.366	2	5 876	1 356	0.00763	240	0.869	0.875	0.880	0.882	0.950	0.962	0.973	0.979	0.247	0.493	0.592	0.298	0.1489	0.1241
G	2x.1089"	5x.1089"	0.327	2	5 626	1 307	0.00790	230	0.871	0.882	0.892	0.896	0.952	0.980	1.006	1.016	0.245	0.489	0.587	0.306	0.1529	0.1275
F	1x.1026"	6x.1026"	0.308	2	4 233	1 176	0.00873	230	0.871	0.878	0.884	0.885	0.952	0.967	0.979	0.985	0.230	0.479	0.575	0.310	0.1551	0.1292
P	6x.1371"	1x.1371"	0.411	3	13 910	1 973	0.00445	220	1.098	1.113	1.127	1.136	1.200	1.239	1.273	1.296	0.274	0.547	0.657	0.290	0.1448	0.1207
N	5x.1226"	2x.1226"	0.368	3	10 390	1 598	0.00506	210	1.098	1.112	1.126	1.133	1.200	1.237	1.273	1.289	0.267	0.534	0.641	0.298	0.1487	1.1239
K	4x.1120"	3x.1120"	0.336	3	7 910	1 349	0.00574	210	1.098	1.111	1.123	1.129	1.200	1.233	1.267	1.281	0.261	0.522	0.626	0.304	0.1520	0.1266
J	3x.1036"	4x.1036"	0.311	3	5 955	1 171	0.00648	200	1.098	1.110	1.121	1.126	1.200	1.232	1.262	1.275	0.255	0.509	0.611	0.309	0.1547	0.1289
A	1x.1513"	2x.1513"	0.326	3	4 810	1 075	0.00679	210	1.096	1.102	1.107	1.109	1.198	1.211	1.225	1.229	0.252	0.505	0.606	0.306	0.1531	0.1275
P	6x.1221"	1x.1221"	0.366	4	11 420	1 564	0.00397	190	1.385	1.400	1.414	1.423	1.514	1.555	1.598	1.616	0.280	0.559	0.671	0.298	0.1489	0.1241
N	5x.1092"	2x.1092"	0.328	4	8 460	1 267	0.00451	180	1.385	1.399	1.413	1.420	1.514	1.554	1.593	1.610	0.273	0.546	0.655	0.306	0.1528	0.1274
D	2x.1615"	1x.1615"	0.348	4	7 340	1 191	0.00566	190	1.382	1.389	1.396	1.399	1.511	1.529	1.544	1.542	0.262	0.523	0.628	0.301	0.1507	0.1256
A	1x.1347"	2x.1347"	0.290	4	3 938	853	0.00604	180	1.382	1.388	1.393	1.395	1.511	1.525	1.540	1.545	0.258	0.517	0.620	0.314	0.1572	0.1310
P	6x.1087"	1x.1087"	0.326	5	9 311	1 240	0.00353	160	1.747	1.762	1.776	1.785	1.909	1.954	2.00	2.02	0.285	0.571	0.685	0.306	0.1531	0.1275
D	2x.1438"	1x.1438"	0.310	5	6 035	944	0.00504	160	1.742	1.749	1.756	1.759	1.905	1.924	1.941	1.939	0.268	0.535	0.642	0.310	0.1548	0.1290
A	1x.1200"	2x.1200"	0.258	5	3 193	676	0.00538	160	1.742	1.748	1.753	1.755	1.905	1.920	1.936	1.941	0.264	0.528	0.634	0.323	0.1614	0.1345
D	2x.1281"	1x.1281"	0.276	6	4 942	749	0.00449	140	2.20	2.21	2.21	2.22	2.40	2.42	2.44	2.44	0.273	0.547	0.656	0.318	0.1590	0.1325
A	1x.1068"	2x.1068"	0.230	6	2 585	536	0.00479	140	2.20	2.20	2.21	2.21	2.40	2.42	2.44	2.44	0.270	0.540	0.648	0.331	0.1655	0.1379
C	1x.1046"	2x.1046"	0.225	6	2 143	514	0.00469	130	2.20	2.20	2.21	2.21	2.40	2.42	2.44	2.44	0.271	0.542	0.651	0.333	0.1663	0.1386
D	2x.1141"	1x.1141"	0.246	7	4 022	594	0.00400	120	2.77	2.78	2.79	2.79	3.03	3.05	3.07	3.07	0.279	0.558	0.670	0.326	0.1631	0.1359
A	1x.1266"	2x.0895"	0.223	7	2 754	495	0.00441	120	2.77	2.78	2.78	2.78	3.03	3.05	3.07	3.07	0.274	0.548	0.658	0.333	0.1666	0.1388
D	2x.1016"	1x.1016"	0.219	8	3 256	471	0.00356	110	3.49	3.50	3.51	3.51	3.82	3.84	3.86	3.86	0.285	0.570	0.684	0.334	0.1672	0.1393
A	1x.1127"	2x.0797"	0.199	8	2 233	392	0.00394	100	3.49	3.50	3.51	3.51	3.82	3.84	3.86	3.87	0.280	0.560	0.672	0.341	0.1706	0.1422
C	1x.0808"	2x.0834"	0.179	8	1 362	320	0.00373	100	3.49	3.50	3.51	3.51	3.82	3.84	3.86	3.86	0.283	0.565	0.679	0.349	0.1744	0.1453
½ D	2x.0808"	1x.0808"	0.174	9½	1 743	298	0.00283	85	4.91	4.92	4.92	4.93	5.37	5.39	5.42	5.42	0.297	0.593	0.712	0.351	0.1754	0.1462

*Based on a conductor temperature of 75°C. and an ambient of 25°C., wind 1.4 miles per hour (2 ft/sec.), frequency=60 cycles, average tarnished surface.

**Resistances at 50°C. total temperature, based on an ambient of 25°C. plus 25°C. rise due to heating effect of current. The approximate magnitude of current necessary to produce the 25° C. rise is 75% of the "Approximate Current Carrying Capacity at 60 cycles."

Table 4.B   Characteristics of Copperweld Conductors

Nominal Conductor Size	Number and Size of Wires	Outside Diameter Inches	Area of Conductor Circular Mils	Rated Breaking Load Pounds — High	Rated Breaking Load Pounds — Extra High	Weight Pounds per Mile	Geometric Mean Radius at 60 cycles Average Currents Feet	Approx. Current Carrying Capacity* Amps at 60 Cycles	r_a Resistance Ohms per Conductor per Mile at 25°C. (77°F.) Small Currents d-c	25 cycles	50 cycles	60 cycles	r_a Resistance Ohms per Conductor per Mile at 75°C. (167°F.) Current Approx. 75% of Capacity** d-c	25 cycles	50 cycles	60 cycles	x_a Inductive Reactance Ohms per Conductor per Mile One Ft. Spacing Average Currents 25 cycles	50 cycles	60 cycles	x_a' Capacitive Reactance Megohms per Conductor per Mile One Ft. Spacing 25 cycles	50 cycles	60 cycles

30% Conductivity

7/8″	19 No. 5	0.910	628 900	55 570	66 910	9 344	0.00758	620	0.306	0.316	0.326	0.331	0.363	0.419	0.476	0.499	0.261	0.493	0.592	0.233	0.1165	0.0971
13/16″	19 No. 6	0.810	498 800	45 830	55 530	7 410	0.00675	540	0.386	0.396	0.406	0.411	0.458	0.518	0.580	0.605	0.267	0.505	0.606	0.241	0.1206	0.1005
23/32″	19 No. 7	0.721	395 500	37 740	45 850	5 877	0.00601	470	0.486	0.496	0.506	0.511	0.577	0.643	0.710	0.737	0.273	0.517	0.621	0.250	0.1248	0.1040
21/32″	19 No. 8	0.642	313 700	31 040	37 690	4 660	0.00535	410	0.613	0.623	0.633	0.638	0.728	0.799	0.872	0.902	0.279	0.529	0.635	0.258	0.1289	0.1074
9/16″	19 No. 9	0.572	248 800	25 500	30 610	3 696	0.00477	360	0.773	0.783	0.793	0.798	0.917	0.995	1.075	1.106	0.285	0.541	0.649	0.266	0.1330	0.1108
5/8″	7 No. 4	0.613	292 200	24 780	29 430	4 324	0.00511	410	0.656	0.664	0.672	0.676	0.778	0.824	0.870	0.887	0.281	0.533	0.640	0.261	0.1306	0.1088
9/16″	7 No. 5	0.546	231 700	20 470	24 650	3 429	0.00455	360	0.827	0.835	0.843	0.847	0.981	1.030	1.080	1.099	0.287	0.545	0.654	0.269	0.1347	0.1122
1/2″	7 No. 6	0.486	183 800	16 890	20 460	2 719	0.00405	310	1.042	1.050	1.058	1.062	1.237	1.290	1.343	1.364	0.293	0.557	0.668	0.278	0.1388	0.1157
7/16″	7 No. 7	0.433	145 700	13 910	16 890	2 157	0.00361	270	1.315	1.323	1.331	1.335	1.560	1.617	1.675	1.697	0.299	0.569	0.683	0.286	0.1429	0.1191
3/8″	7 No. 8	0.385	115 600	11 440	13 890	1 710	0.00321	230	1.658	1.666	1.674	1.678	1.967	2.03	2.09	2.12	0.305	0.581	0.697	0.294	0.1471	0.1226
11/32″	7 No. 9	0.343	91 650	9 393	11 280	1 356	0.00286	200	2.09	2.10	2.11	2.11	2.48	2.55	2.61	2.64	0.311	0.592	0.711	0.303	0.1512	0.1260
5/16″	7 No. 10	0.306	72 680	7 758	9 196	1 076	0.00255	170	2.64	2.64	2.65	2.66	3.13	3.20	3.27	3.30	0.316	0.604	0.725	0.311	0.1553	0.1294
3 No. 5	3 No. 5	0.392	99 310	9 262	11 860	1 467	0.00457	220	1.926	1.931	1.936	1.938	2.29	2.31	2.34	2.35	0.289	0.545	0.654	0.293	0.1465	0.122
3 No. 6	3 No. 6	0.349	78 750	7 639	9 754	1 163	0.00407	190	2.43	2.43	2.44	2.44	2.88	2.91	2.94	2.95	0.295	0.556	0.668	0.301	0.1506	0.1258
3 No. 7	3 No. 7	0.311	62 450	6 291	7 922	922.4	0.00363	160	3.06	3.07	3.07	3.07	3.63	3.66	3.70	3.71	0.301	0.568	0.682	0.310	0.1547	0.1289
3 No. 8	3 No. 8	0.277	49 530	5 174	6 282	731.5	0.00323	140	3.86	3.87	3.87	3.87	4.58	4.61	4.65	4.66	0.307	0.580	0.696	0.318	0.1589	0.132
3 No. 9	3 No. 9	0.247	39 280	4 250	5 129	580.1	0.00288	120	4.87	4.87	4.88	4.88	5.78	5.81	5.85	5.86	0.313	0.591	0.710	0.326	0.1629	0.1358
3 No. 10	3 No. 10	0.220	31 150	3 509	4 160	460.0	0.00257	110	6.14	6.14	6.15	6.15	7.28	7.32	7.36	7.38	0.319	0.603	0.724	0.334	0.1671	0.1392

40% Conductivity

7/8″	19 No. 5	0.910	628 900	50 240	.....	9 344	0.01175	690	0.229	0.239	0.249	0.254	0.272	0.321	0.371	0.391	0.236	0.449	0.539	0.233	0.1165	0.0971
13/16″	19 No. 6	0.810	498 800	41 600	.....	7 410	0.01046	610	0.289	0.299	0.309	0.314	0.343	0.396	0.450	0.472	0.241	0.461	0.553	0.241	0.1206	0.1005
23/32″	19 No. 7	0.721	395 500	34 390	.....	5 877	0.00931	530	0.365	0.375	0.385	0.390	0.433	0.490	0.549	0.573	0.247	0.473	0.567	0.250	0.1248	0.1040
21/32″	19 No. 8	0.642	313 700	28 380	.....	4 660	0.00829	470	0.460	0.470	0.480	0.485	0.546	0.608	0.672	0.698	0.253	0.485	0.582	0.258	0.1289	0.1074
9/16″	19 No. 9	0.572	248 800	23 390	.....	3 696	0.00739	410	0.580	0.590	0.600	0.605	0.688	0.756	0.826	0.853	0.259	0.496	0.595	0.266	0.1330	0.1108
5/8″	7 No. 4	0.613	292 200	22 310	.....	4 324	0.00792	470	0.492	0.500	0.508	0.512	0.584	0.624	0.664	0.680	0.255	0.489	0.587	0.261	0.1306	0.1088
9/16″	7 No. 5	0.546	231 700	18 510	.....	3 429	0.00705	410	0.620	0.628	0.636	0.640	0.736	0.780	0.843	0.840	0.261	0.501	0.601	0.269	0.1347	0.1122
1/2″	7 No. 6	0.486	183 800	15 330	.....	2 719	0.00628	350	0.782	0.790	0.798	0.802	0.928	0.975	1.021	1.040	0.267	0.513	0.615	0.278	0.1388	0.1157
7/16″	7 No. 7	0.433	145 700	12 670	.....	2 157	0.00559	310	0.986	0.994	1.002	1.006	1.170	1.220	1.271	1.291	0.273	0.524	0.629	0.286	0.1429	0.1191
3/8″	7 No. 8	0.385	115 600	10 460	.....	1 710	0.00497	270	1.244	1.252	1.260	1.264	1.476	1.530	1.584	1.606	0.279	0.536	0.644	0.294	0.1471	0.1226
11/32″	7 No. 9	0.343	91 650	8 616	.....	1 356	0.00443	230	1.568	1.576	1.584	1.588	1.861	1.919	1.978	2.00	0.285	0.548	0.658	0.303	0.1512	0.1260
5/16″	7 No. 10	0.306	72 680	7 121	.....	1 076	0.00395	200	1.978	1.986	1.994	1.998	2.35	2.41	2.47	2.50	0.291	0.559	0.671	0.311	0.1553	0.1294
3 No. 5	3 No. 5	0.392	99 310	8 373	.....	1 467	0.00621	250	1.445	1.450	1.455	1.457	1.714	1.738	1.762	1.772	0.269	0.514	0.617	0.293	0.1465	0.122
3 No. 6	3 No. 6	0.349	78 750	6 934	.....	1 163	0.00553	220	1.821	1.826	1.831	1.833	2.16	2.19	2.21	2.22	0.275	0.526	0.631	0.301	0.1506	0.1258
3 No. 7	3 No. 7	0.311	62 450	5 732	.....	922.4	0.00492	190	2.30	2.30	2.31	2.31	2.73	2.75	2.78	2.79	0.281	0.537	0.645	0.310	0.1547	0.1289
3 No. 8	3 No. 8	0.277	49 530	4 730	.....	731.5	0.00439	160	2.90	2.90	2.91	2.91	3.44	3.47	3.50	3.51	0.286	0.549	0.659	0.318	0.1589	0.1324
3 No. 9	3 No. 9	0.247	39 280	3 898	.....	580.1	0.00391	140	3.65	3.66	3.66	3.66	4.34	4.37	4.40	4.41	0.292	0.561	0.673	0.326	0.1629	0.1358
3 No. 10	3 No. 10	0.220	31 150	3 221	.....	460.0	0.00348	120	4.61	4.61	4.62	4.62	5.46	5.50	5.53	5.55	0.297	0.572	0.687	0.334	0.1671	0.1392
3 No. 12	3 No. 12	0.174	19 590	2 236	.....	289.3	0.00276	90	7.32	7.33	7.33	7.34	8.69	8.73	8.77	8.78	0.310	0.596	0.715	0.351	0.1754	0.1462

*Based on conductor temperature of 125°C. and an ambient of 25°C.
**Resistance at 75°C. total temperature, based on an ambient of 25°C. plus 50°C. rise due to heating effect of current.
The approximate magnitude of current necessary to produce the 50°C. rise is 75% of the "Approximate Current Carrying Capacity at 60 Cycles."

Table 5   Skin Effect Table

X	K	X	K	X	K	X	K
0.0	1.00000	1.0	1.00519	2.0	1.07816	3.0	1.31809
0.1	1.00000	1.1	1.00758	2.1	1.09375	3.1	1.35102
0.2	1.00001	1.2	1.01071	2.2	1.11126	3.2	1.38504
0.3	1.00004	1.3	1.01470	2.3	1.13069	3.3	1.41999
0.4	1.00013	1.4	1.01969	2.4	1.15207	3.4	1.45570
0.5	1.00032	1.5	1.02582	2.5	1.17538	3.5	1.49202
0.6	1.00067	1.6	1.03323	2.6	1.20056	3.6	1.52879
0.7	1.00124	1.7	1.04205	2.7	1.22753	3.7	1.56587
0.8	1.00212	1.8	1.05240	2.8	1.25620	3.8	1.60314
0.9	1.00340	1.9	1.06440	2.9	1.28644	3.9	1.64051

Table 6   Inductive Reactance Spacing Factor (x_d) Ohms per Conductor per Mile

Table 7   (Insert, bottom right) Zero-Sequence Resistance and Inductive Reactance Factors (r_e, x_e)*

25 CYCLES

SEPARATION — INCHES

Feet	0	1	2	3	4	5	6	7	8	9	10	11
0	–	−0.1256	−0.0906	−0.0701	−0.0555	−0.0443	−0.0350	−0.0273	−0.0205	−0.0145	−0.0092	−0.0044
1	0	0.0040	0.0078	0.0113	0.0145	0.0176	0.0205	0.0232	0.0258	0.0283	0.0306	0.0329
2	0.0350	0.0371	0.0391	0.0410	0.0428	0.0446	0.0463	0.0480	0.0496	0.0511	0.0527	0.0541
3	0.0555	0.0569	0.0583	0.0596	0.0609	0.0621	0.0633	0.0645	0.0657	0.0668	0.0679	0.0690
4	0.0701	0.0711	0.0722	0.0732	0.0741	0.0751	0.0760	0.0770	0.0779	0.0788	0.0797	0.0805
5	0.0814	0.0822	0.0830	0.0838	0.0846	0.0854	0.0862	0.0869	0.0877	0.0884	0.0892	0.0899
6	0.0906	0.0913	0.0920	0.0927	0.0933	0.0940	0.0946	0.0953	0.0959	0.0965	0.0972	0.0978
7	0.0984	0.0990	0.0996	0.1002	0.1007	0.1013	0.1019	0.1024	0.1030	0.1035	0.1041	0.1046

x_d at 25 cycles: $x_d = 0.1164 \log_{10} d$, d = separation, feet.

FUNDAMENTAL EQUATIONS

$$z_1 = z_2 = r_a + j(x_a + x_d)$$
$$z_0 = r_a + r_e + j(x_a + x_e - 2x_d)$$

50 CYCLES

SEPARATION — Inches

Feet	0	1	2	3	4	5	6	7	8	9	10	11
0	–	−0.2513	−0.1812	−0.1402	−0.1111	−0.0885	−0.0701	−0.0545	−0.0410	−0.0291	−0.0184	−0.0088
1	0	0.0081	0.0156	0.0226	0.0291	0.0352	0.0410	0.0465	0.0517	0.0566	0.0613	0.0658
2	0.0701	0.0742	0.0782	0.0820	0.0857	0.0892	0.0927	0.0960	0.0992	0.1023	0.1053	0.1082
3	0.1111	0.1139	0.1166	0.1192	0.1217	0.1242	0.1267	0.1291	0.1314	0.1337	0.1359	0.1380
4	0.1402	0.1423	0.1443	0.1463	0.1483	0.1502	0.1521	0.1539	0.1558	0.1576	0.1593	0.1610
5	0.1627	0.1644	0.1661	0.1677	0.1693	0.1708	0.1724	0.1739	0.1754	0.1769	0.1783	0.1798
6	0.1812	0.1826	0.1839	0.1853	0.1866	0.1880	0.1893	0.1906	0.1918	0.1931	0.1943	0.1956
7	0.1968	0.1980	0.1991	0.2003	0.2015	0.2026	0.2037	0.2049	0.2060	0.2071	0.2081	0.2092

x_d at 50 cycles: $x_d = 0.2328 \log_{10} d$, d = separation, feet.

60 CYCLES

SEPARATION — Inches

Feet	0	1	2	3	4	5	6	7	8	9	10	11
0	–	−0.3015	−0.2174	−0.1682	−0.1333	−0.1062	−0.0841	−0.0654	−0.0492	−0.0349	−0.0221	−0.0106
1	0	0.0097	0.0187	0.0271	0.0349	0.0423	0.0492	0.0558	0.0620	0.0679	0.0735	0.0789
2	0.0841	0.0891	0.0938	0.0984	0.1028	0.1071	0.1112	0.1152	0.1190	0.1227	0.1264	0.1299
3	0.1333	0.1366	0.1399	0.1430	0.1461	0.1491	0.1520	0.1549	0.1577	0.1604	0.1631	0.1657
4	0.1682	0.1707	0.1732	0.1756	0.1779	0.1802	0.1825	0.1847	0.1869	0.1891	0.1912	0.1933
5	0.1953	0.1973	0.1993	0.2012	0.2031	0.2050	0.2069	0.2087	0.2105	0.2123	0.2140	0.2157
6	0.2174	0.2191	0.2207	0.2224	0.2240	0.2256	0.2271	0.2287	0.2302	0.2317	0.2332	0.2347
7	0.2361	0.2376	0.2390	0.2404	0.2418	0.2431	0.2445	0.2458	0.2472	0.2485	0.2498	0.2511

x_d at 60 cycles: $x_d = 0.2794 \log_{10} d$, d = separation, feet.

x_d (Ohms per Conductor per Mile) for whole-foot separations

Feet	25 cycles	50 cycles	60 cycles
0	–	–	–
1	0	0	0
2	0.0350	0.0701	0.0841
3	0.0555	0.1111	0.1333
4	0.0701	0.1402	0.1682
5	0.0814	0.1627	0.1953
6	0.0906	0.1812	0.2174
7	0.0984	0.1968	0.2361
8	0.1051	0.2103	0.2523
9	0.1111	0.2222	0.2666
10	0.1164	0.2328	0.2794
11	0.1212	0.2425	0.2910
12	0.1256	0.2513	0.3015
13	0.1297	0.2594	0.3112
14	0.1334	0.2669	0.3202
15	0.1369	0.2738	0.3286
16	0.1402	0.2804	0.3364
17	0.1432	0.2865	0.3438
18	0.1461	0.2923	0.3507
19	0.1489	0.2977	0.3573
20	0.1515	0.3029	0.3635
21	0.1539	0.3079	0.3694
22	0.1563	0.3128	0.3751
23	0.1585	0.3170	0.3805
24	0.1607	0.3213	0.3856
25	0.1627	0.3255	0.3906
26	0.1647	0.3294	0.3953
27	0.1666	0.3333	0.3999
28	0.1685	0.3369	0.4043
29	0.1702	0.3405	0.4086
30	0.1720	0.3439	0.4127
31	0.1736	0.3472	0.4167
32	0.1752	0.3504	0.4205
33	0.1768	0.3536	0.4243
34	0.1783	0.3566	0.4279
35	0.1798	0.3595	0.4314
36	0.1812	0.3624	0.4348
37	0.1826	0.3651	0.4382
38	0.1839	0.3678	0.4414
39	0.1852	0.3704	0.4445
40	0.1865	0.3730	0.4476
41	0.1878	0.3755	0.4506
42	0.1890	0.3779	0.4535
43	0.1902	0.3803	0.4564
44	0.1913	0.3826	0.4592
45	0.1925	0.3849	0.4619
46	0.1936	0.3871	0.4646
47	0.1947	0.3893	0.4672
48	0.1957	0.3914	0.4697
49	0.1968	0.3935	0.4722

Table 7 — Zero-Sequence Resistance and Inductive Reactance Factors

	ρ Meter Ohm	FREQUENCY		
		25 Cycles	50 Cycles	60 Cycles
r_e	All	0.1192	0.2383	0.2860
x_e	1	0.921	1.736	2.050
	5	1.043	1.980	2.343
	10	1.095	2.085	2.469
	50	1.217	2.329	2.762
	100†	1.270	2.434	2.888
	500	1.392	2.679	3.181
	1000	1.444	2.784	3.307
	5000	1.566	3.028	3.600
	10 000	1.619	3.133	3.726

*From Formulas:

$$r_e = 0.004764f$$
$$x_e = 0.006985f \log_{10} 4\,665\,600 \frac{\rho}{f}$$

where f = frequency
ρ = Resistivity (meter-ohm)

†This is an average value which may be used in the absence of definite information.

Table 8   Shunt Capacitive Reactance Spacing Factor (x'_d) Megohms per Conductor per Mile

Table 9   (Insert, bottom right) Zero-Sequence Shunt Capacitive Reactance Factor

25 CYCLES

SEPARATION — INCHES

Feet	0	1	2	3	4	5	6	7	8	9	10	11
0	—	-0.1769	-0.1276	-0.0987	-0.0782	-0.0623	-0.0494	-0.0384	-0.0289	-0.0205	-0.0130	-0.0062
1	0	0.0057	0.0110	0.0159	0.0205	0.0248	0.0289	0.0327	0.0364	0.0398	0.0432	0.0463
2	0.0494	0.0523	0.0551	0.0577	0.0603	0.0628	0.0652	0.0676	0.0698	0.0720	0.0742	0.0762
3	0.0782	0.0802	0.0821	0.0839	0.0857	0.0875	0.0892	0.0909	0.0925	0.0941	0.0957	0.0972
4	0.0987	0.1002	0.1016	0.1030	0.1044	0.1058	0.1071	0.1084	0.1097	0.1109	0.1122	0.1134
5	0.1146	0.1158	0.1169	0.1181	0.1192	0.1203	0.1214	0.1225	0.1235	0.1246	0.1256	0.1266
6	0.1276	0.1286	0.1295	0.1305	0.1314	0.1324	0.1333	0.1342	0.1351	0.1360	0.1368	0.1377
7	0.1386	0.1394	0.1402	0.1411	0.1419	0.1427	0.1435	0.1443	0.1450	0.1458	0.1466	0.1473

x'_d at 25 cycles
$x'_d = .1640 \log_{10} d$
d = separation, feet.

FUNDAMENTAL EQUATIONS
$x'_1 = x'_2 = x'_a + x'_d$
$x'_o = x'_a + x'_e - 2x'_d$

25 cycles, x'_d at 0 inches (Feet 0–49):

Feet	x'_d	Feet	x'_d
0	—	25	0.2292
1	0	26	0.2320
2	0.0494	27	0.2347
3	0.0782	28	0.2373
4	0.0987	29	0.2398
5	0.1146	30	0.2422
6	0.1276	31	0.2445
7	0.1386	32	0.2468
8	0.1481	33	0.2490
9	0.1565	34	0.2511
10	0.1640	35	0.2532
11	0.1707	36	0.2552
12	0.1769	37	0.2571
13	0.1826	38	0.2590
14	0.1879	39	0.2609
15	0.1928	40	0.2627
16	0.1974	41	0.2644
17	0.2017	42	0.2661
18	0.2058	43	0.2678
19	0.2097	44	0.2695
20	0.2133	45	0.2711
21	0.2168	46	0.2726
22	0.2201	47	0.2742
23	0.2233	48	0.2756
24	0.2263	49	0.2771

50 CYCLES

SEPARATION — Inches

Feet	0	1	2	3	4	5	6	7	8	9	10	11
0	—	-0.0885	-0.0638	-0.0494	-0.0391	-0.0312	-0.0247	-0.0192	-0.0144	-0.0102	-0.0065	-0.0031
1	0	0.0028	0.0055	0.0079	0.0102	0.0124	0.0144	0.0164	0.0182	0.0199	0.0216	0.0232
2	0.0247	0.0261	0.0275	0.0289	0.0302	0.0314	0.0326	0.0338	0.0349	0.0360	0.0371	0.0381
3	0.0391	0.0401	0.0410	0.0420	0.0429	0.0437	0.0446	0.0454	0.0463	0.0471	0.0478	0.0486
4	0.0494	0.0501	0.0508	0.0515	0.0522	0.0529	0.0535	0.0542	0.0548	0.0555	0.0561	0.0567
5	0.0573	0.0579	0.0585	0.0590	0.0596	0.0601	0.0607	0.0612	0.0618	0.0623	0.0628	0.0633
6	0.0638	0.0643	0.0648	0.0652	0.0657	0.0662	0.0666	0.0671	0.0675	0.0680	0.0684	0.0689
7	0.0693	0.0697	0.0701	0.0705	0.0709	0.0713	0.0717	0.0721	0.0725	0.0729	0.0733	0.0737

x'_d at 50 cycles
$x'_d = 0.08198 \log_{10} d$
d = separation, feet.

50 cycles, x'_d at 0 inches (Feet 0–49):

Feet	x'_d	Feet	x'_d
0	—	25	0.1146
1	0	26	0.1160
2	0.0247	27	0.1173
3	0.0391	28	0.1186
4	0.0494	29	0.1199
5	0.0573	30	0.1211
6	0.0638	31	0.1223
7	0.0693	32	0.1234
8	0.0740	33	0.1245
9	0.0782	34	0.1255
10	0.0820	35	0.1266
11	0.0854	36	0.1276
12	0.0885	37	0.1286
13	0.0913	38	0.1295
14	0.0940	39	0.1304
15	0.0964	40	0.1313
16	0.0987	41	0.1322
17	0.1009	42	0.1331
18	0.1029	43	0.1339
19	0.1048	44	0.1347
20	0.1067	45	0.1355
21	0.1084	46	0.1363
22	0.1100	47	0.1371
23	0.1116	48	0.1378
24	0.1131	49	0.1386

60 CYCLES

SEPARATION — Inches

Feet	0	1	2	3	4	5	6	7	8	9	10	11
0	—	-0.0737	-0.0532	-0.0411	-0.0326	-0.0260	-0.0206	-0.0160	-0.0120	-0.0085	-0.0054	-0.0026
1	0	0.0024	0.0046	0.0066	0.0085	0.0103	0.0120	0.0136	0.0152	0.0166	0.0180	0.0193
2	0.0206	0.0218	0.0229	0.0241	0.0251	0.0262	0.0272	0.0282	0.0291	0.0300	0.0309	0.0318
3	0.0326	0.0334	0.0342	0.0350	0.0357	0.0365	0.0372	0.0379	0.0385	0.0392	0.0399	0.0405
4	0.0411	0.0417	0.0423	0.0429	0.0435	0.0441	0.0446	0.0452	0.0457	0.0462	0.0467	0.0473
5	0.0478	0.0482	0.0487	0.0492	0.0497	0.0501	0.0506	0.0510	0.0515	0.0519	0.0523	0.0527
6	0.0532	0.0536	0.0540	0.0544	0.0548	0.0552	0.0555	0.0559	0.0563	0.0567	0.0570	0.0574
7	0.0577	0.0581	0.0584	0.0588	0.0591	0.0594	0.0598	0.0601	0.0604	0.0608	0.0611	0.0614

x'_d at 60 cycles
$x'_d = 0.06831 \log_{10} d$
d = separation, feet.

60 cycles, x'_d at 0 inches (Feet 0–49):

Feet	x'_d	Feet	x'_d
0	—	25	0.0955
1	0	26	0.0967
2	0.0206	27	0.0978
3	0.0326	28	0.0989
4	0.0411	29	0.0999
5	0.0478	30	0.1009
6	0.0532	31	0.1019
7	0.0577	32	0.1028
8	0.0617	33	0.1037
9	0.0652	34	0.1046
10	0.0683	35	0.1055
11	0.0711	36	0.1063
12	0.0737	37	0.1071
13	0.0761	38	0.1079
14	0.0783	39	0.1087
15	0.0803	40	0.1094
16	0.0823	41	0.1102
17	0.0841	42	0.1109
18	0.0858	43	0.1116
19	0.0874	44	0.1123
20	0.0889	45	0.1129
21	0.0903	46	0.1136
22	0.0917	47	0.1142
23	0.0930	48	0.1149
24	0.0943	49	0.1155

Table 9

Conductor Height Above Ground Feet	FREQUENCY		
	25 Cycles	50 Cycles	60 Cycles
10	0.640	0.320	0.267
15	0.727	0.363	0.303
20	0.788	0.394	0.328
25	0.836	0.418	0.348
30	0.875	0.437	0.364
40	0.936	0.468	0.390
50	0.984	0.492	0.410
60	1.023	0.511	0.426
70	1.056	0.528	0.440
80	1.084	0.542	0.452
90	1.109	0.555	0.462
100	1.132	0.566	0.472

$x'_e = \dfrac{12.30}{f} \log_{10} 2h$
where h = height above ground.
f = frequency.

Table 10 Typical Constants of Three-Phase Synchronous Machines

	1 x_d (unsat)	2 x_q rated current	3 x_d' rated voltage	4 x_d'' rated voltage	5 x_2 rated current	6 (*) x_0 rated current	7 x_p	8 (†) r_2	9 (‡) r_1	10 (‡) r_a	11 T_{do}'	12 T_d'	13 T_d''	14 T_a	15 H
2-Pole turbine generators	$\dfrac{1.20}{0.95\text{-}1.45}$	$\dfrac{1.16}{0.92\text{-}1.42}$	$\dfrac{0.15}{0.12\text{-}0.21}$	$\dfrac{0.09}{0.07\text{-}0.14}$	$=x_d''$	$\dfrac{0.03}{0.01\text{-}0.08}$	$\dfrac{0.10}{0.07\text{-}0.17}$	0.025-0.04	0.004-0.011	0.001-0.007	5.0	0.6	$\dfrac{0.035}{0.02\text{-}0.05}$	$\dfrac{0.13}{0.04\text{-}0.24}$	
4-Pole turbine generators	$\dfrac{1.20}{1.00\text{-}1.45}$	$\dfrac{1.16}{0.92\text{-}1.42}$	$\dfrac{0.23}{0.20\text{-}0.28}$	$\dfrac{0.14}{0.12\text{-}0.17}$	$=x_d''$	$\dfrac{0.08}{0.015\text{-}0.14}$	$\dfrac{0.17}{0.12\text{-}0.24}$	0.03-0.045	0.003-0.008	0.001-0.005	8.0	1.0	$\dfrac{0.035}{0.02\text{-}0.05}$	$\dfrac{0.20}{0.15\text{-}0.35}$	
Salient-pole generators and motors (with dampers)	$\dfrac{1.25}{0.60\text{-}1.50}$	$\dfrac{0.70}{0.40\text{-}0.80}$	$\dfrac{0.30}{0.20\text{-}0.50}$(*)	$\dfrac{0.20}{0.13\text{-}0.32}$(*)	$\dfrac{0.20}{0.13\text{-}0.32}$(*)	$\dfrac{0.18}{0.03\text{-}0.23}$	$\dfrac{0.28}{0.17\text{-}0.40}$	0.012-0.020	0.005-0.020	0.003-0.015	$\dfrac{3.0\text{-}5.0}{1.5\text{-}10}$	$\dfrac{1.5}{0.5\text{-}3.3}$	$\dfrac{0.035}{0.01\text{-}0.05}$	$\dfrac{0.15}{0.03\text{-}0.25}$	
Salient-pole generators (without dampers)	$\dfrac{1.25}{0.60\text{-}1.50}$	$\dfrac{0.70}{0.40\text{-}0.80}$	$\dfrac{0.30}{0.20\text{-}0.50}$(*)	$\dfrac{0.30}{0.20\text{-}0.50}$(*)	$\dfrac{0.48}{0.35\text{-}0.65}$	$\dfrac{0.19}{0.03\text{-}0.24}$	$\dfrac{0.28}{0.17\text{-}0.40}$	0.03-0.045	0.005-0.020	0.003-0.015	$\dfrac{3.0\text{-}5.0}{1.5\text{-}10}$	$\dfrac{1.5}{0.5\text{-}3.3}$		$\dfrac{0.30}{0.10\text{-}0.50}$	
Condensers air cooled	$\dfrac{1.85}{1.25\text{-}2.20}$	$\dfrac{1.15}{0.95\text{-}1.30}$	$\dfrac{0.40}{0.30\text{-}0.50}$	$\dfrac{0.27}{0.19\text{-}0.30}$	$\dfrac{0.26}{0.18\text{-}0.40}$	$\dfrac{0.12}{0.025\text{-}0.15}$	$\dfrac{0.25}{0.20\text{-}0.35}$	0.025-0.07	$\dfrac{0.0065}{0.004\text{-}0.009}$	$\dfrac{0.0035}{0.0025\text{-}0.008}$	$\dfrac{9.0}{6.0\text{-}14.0}$	$\dfrac{2.0}{1.2\text{-}2.8}$	$\dfrac{0.035}{0.02\text{-}0.04}$	$\dfrac{0.17}{0.1\text{-}0.3}$	Large 2.4 Small 1.0
Condensers hydrogen cooled at ½ psi kva rating	$\dfrac{2.20}{1.50\text{-}2.65}$	$\dfrac{1.35}{1.10\text{-}1.55}$	$\dfrac{0.48}{0.36\text{-}0.60}$	$\dfrac{0.32}{0.23\text{-}0.36}$	$\dfrac{0.31}{0.22\text{-}0.48}$	$\dfrac{0.14}{0.030\text{-}0.18}$	$\dfrac{0.27}{0.22\text{-}0.37}$	0.025-0.07	$\dfrac{0.0065}{0.005\text{-}0.009}$	$\dfrac{0.0035}{0.0025\text{-}0.005}$	$\dfrac{9.0}{6.0\text{-}14.0}$	$\dfrac{2.0}{1.2\text{-}2.8}$	$\dfrac{0.035}{0.02\text{-}0.04}$	$\dfrac{0.20}{0.15\text{-}0.3}$	Large 2.0 Small 1.10

(*) High speed units tend to have low reactance and low speed units high reactance.
(*) X_0 varies so critically with armature winding pitch that an average value can hardly be given.
(*) Variation is from 0.1 to 0.7 of x_d''. Low limit is for 2/3 pitch windings.
(†) r_2 varies with damper resistance.
(‡) r_1 and r_a vary with machine rating, limiting values given are for about 50,000 kva and 500 kva.

461

Appendix B

Vectors and Matrices

B.1   DEFINITION OF A VECTOR

A vector $\mathbf{X}$ is defined as an ordered set of elements. The elements $x_1, x_2, \ldots,$ x_N may be real numbers, complex numbers, or functions of some dependent variable. An N-dimensional vector is shown in Eq. B.1.

$$\mathbf{X} = \begin{bmatrix} x_1 \\ x_2 \\ \vdots \\ x_N \end{bmatrix} = \begin{bmatrix} x_1(t) \\ x_2(t) \\ \vdots \\ x_N(t) \end{bmatrix} \tag{B.1}$$

B.2   VECTOR ALGEBRA

Equality of Vectors   Two vectors $\mathbf{X}$ and $\mathbf{Y}$ are said to be equal if and only if they have the same size and corresponding elements are equal; that is, $\mathbf{X} = \mathbf{Y}$ iff $x_i = y_i$ for all i.

Product of a Vector with a Scalar   To multiply a vector $\mathbf{X}$ by a scalar λ, multiply each element x_i by λ, that is,

$$\lambda \mathbf{X} = \mathbf{X} \lambda = \begin{bmatrix} \lambda x_1 \\ \lambda x_2 \\ \vdots \\ \lambda x_N \end{bmatrix} \tag{B.2}$$

Addition (Subtraction) of Vectors   The sum (difference) of two vectors $\mathbf{X}$ and $\mathbf{Y}$ results in a new vector $\mathbf{Z}$, which is obtained by adding (subtracting) corresponding elements as follows:

$$\mathbf{Z} = \mathbf{X} \pm \mathbf{Y} \begin{bmatrix} x_1 \pm y_1 \\ x_2 \pm y_2 \\ x_3 \pm y_3 \\ \vdots \\ x_N \pm y_N \end{bmatrix} \tag{B.3}$$

Inner (Dot) Product   The inner or dot product of two vectors $\mathbf{X}$ and $\mathbf{Y}$ of equal dimension or size is obtained as follows:

$$\mathbf{X} \cdot \mathbf{Y} = \langle \mathbf{X}, \mathbf{Y} \rangle = \Sigma_{k=1}^{N} x_k y_k = x_1 y_1 + x_2 y_2 + \cdots + x_N y_N \tag{B.4}$$

If the inner or dot product of two nonzero vectors (each having at least one nonzero element) is equal to zero, the vectors are said to be orthogonal.

B.3   DEFINITION OF A MATRIX

An $M \times N$ matrix $\mathbf{A}$ is defined as a rectangular array of MN elements as shown in Eq. B.5. The MN elements a_{ij} may be real numbers, complex numbers, or functions of some dependent variable. The double-subscript notation identifies the position of the element in the array; for example, element a_{ij} is found at the ith row and the jth column. The order (size) of the given matrix $\mathbf{A}$ is said to be $M \times N$ because the matrix contains M rows and N columns.

$$\mathbf{A}_{M \times N} \triangleq \begin{bmatrix} a_{11} & a_{12} & a_{13} & \dots & a_{1N} \\ a_{21} & a_{22} & a_{23} & \dots & a_{2N} \\ \vdots & \vdots & \vdots & \cdots & \vdots \\ a_{M1} & a_{M2} & a_{M3} & \cdots & a_{MN} \end{bmatrix} \triangleq [a_{ij}]_{M \times N} \tag{B.5}$$

A *square matrix* $\mathbf{A}$ is a matrix in which the number of rows is equal to the number of columns; that is, $M = N$. For a square matrix, the main (or principal) diagonal consists of the elements of the form a_{ii}.

A *diagonal matrix* is a square matrix in which all the off-diagonal elements are equal to zero; that is, $a_{ij} = 0$ for all $i \neq j$. The *identity* or *unit matrix* is a diagonal matrix whose elements in the principal diagonal are all equal to unity. The *null matrix* has elements that are all equal to zero.

The matrix $\mathbf{A}^T$ is called the *transpose* of $\mathbf{A}$ if the element a_{ij} in $\mathbf{A}$ is equal to element a_{ji} in $\mathbf{A}^T$ for all i and j. In general, $\mathbf{A}^T$ is formed by interchanging the rows and columns of $\mathbf{A}$.

A *symmetric matrix* has the property $a_{ij} = a_{ji}$ for all i and j; that is, it is symmetric about the principle diagonal. Therefore, it is equal to its own transpose matrix; thus, $\mathbf{A} = \mathbf{A}^T$.

B.4 MATRIX ALGEBRA

Matrix operations are defined for addition, subtraction, and multiplication. Division is not defined; however, it is replaced by matrix inversion.

Equality of Matrices   Two matrices $\mathbf{A} = [a_{ij}]$ and $\mathbf{B} = [b_{ij}]$ are said to be equal matrices if and only if (a) they have the same order (size) and (b) each element a_{ij} is equal to the corresponding b_{ij} for all i and j.

Addition (Subtraction) of Matrices   Two matrices $\mathbf{A} = [a_{ij}]$ and $\mathbf{B} = [b_{ij}]$ can be added (subtracted) if they are of the same order. The sum (difference) $\mathbf{C} = \mathbf{A} \pm \mathbf{B}$ is obtained by adding (subtracting) corresponding elements. Thus,

$$\mathbf{C}_{M \times N} = [c_{ij}]_{M \times N} = \mathbf{A}_{M \times N} \pm \mathbf{B}_{M \times N} = [a_{ij} \pm b_{ij}]_{M \times N} \qquad (B.6)$$

Product of a Matrix with a Scalar   A matrix is multiplied by a scalar λ by multiplying all mn elements by λ; that is,

$$\lambda \mathbf{A}_{M \times N} = \mathbf{A}_{MN} \lambda = \begin{bmatrix} \lambda a_{11} & \lambda a_{12} & \lambda a_{13} & \cdots & \lambda a_{1N} \\ \lambda a_{21} & \lambda a_{22} & \lambda a_{23} & \cdots & \lambda a_{2N} \\ \vdots & \vdots & \vdots & \cdots & \vdots \\ \lambda a_{M1} & \lambda a_{M2} & \lambda a_{M3} & \cdots & \lambda a_{MN} \end{bmatrix} \qquad (B.7)$$

Product of Matrices   Two matrices $\mathbf{A} = [a_{ij}]$ and $\mathbf{B} = [b_{ij}]$ can be multiplied in the order $\mathbf{AB}$ if and only if the number of columns of $\mathbf{A}$ is equal to the number of rows of $\mathbf{B}$. That is, if $\mathbf{A}$ is of order $(M \times P)$, then $\mathbf{B}$ has to be of order $(P \times N)$ where M and N are arbitrary. If the product matrix is denoted by $\mathbf{C} = \mathbf{AB}$, then $\mathbf{C}$ is of order $(M \times N)$. Its elements c_{ij} are given by

$$c_{ij} = \Sigma_{k=1}^P a_{ik} b_{kj} \quad \text{for } i = 1, 2, \cdots, M, j = 1, 2, \cdots, N \qquad (B.8)$$

EXAMPLE B.1

For the following two matrices $\mathbf{A}$ and $\mathbf{B}$, find the matrix product $\mathbf{C} = \mathbf{AB}$.

$$\mathbf{A} = \begin{bmatrix} 1 & 4 \\ 2 & 5 \\ 3 & 6 \end{bmatrix}_{3 \times 2} \qquad \mathbf{B} = \begin{bmatrix} 7 & 8 \\ 9 & 0 \end{bmatrix}_{2 \times 2}$$

Solution

$$\mathbf{C} = \begin{bmatrix} 1 & 4 \\ 2 & 5 \\ 3 & 6 \end{bmatrix}_{3 \times 2} \begin{bmatrix} 7 & 8 \\ 9 & 0 \end{bmatrix}_{2 \times 2}$$

$$= \begin{bmatrix} (1 \times 7 + 4 \times 9) & (1 \times 8 + 4 \times 0) \\ (2 \times 7 + 5 \times 9) & (2 \times 8 + 5 \times 0) \\ (3 \times 7 + 6 \times 9) & (3 \times 8 + 6 \times 0) \end{bmatrix}_{3 \times 2} = \begin{bmatrix} 43 & 0 \\ 59 & 16 \\ 79 & 24 \end{bmatrix}_{3 \times 2}$$

In general, the matrix multiplication is not commutative; that is, $\mathbf{AB} \neq \mathbf{BA}$ even if $\mathbf{AB}$ is defined. Indeed, the product $\mathbf{BA}$ may not even be defined, as in Example B.1.

B.5   INVERSE OF A MATRIX

Consider two N-square matrices $\mathbf{C}$ and $\mathbf{D}$. If $\mathbf{CD} = \mathbf{DC} = \mathbf{I}$, then $\mathbf{C}$ is called the inverse of $\mathbf{D}$. Conversely, $\mathbf{D}$ is called the inverse of $\mathbf{C}$. The inverse matrix is written as $\mathbf{C}^{-1}$ or $\mathbf{D}^{-1}$.

An Application of Matrix Inversion The inverse matrix may be used to solve a system of linear algebraic equations. Consider the following matrix equation.

$$\begin{bmatrix} a_{11} & a_{12} & a_{13} & \cdots & a_{1N} \\ a_{21} & a_{22} & a_{23} & \cdots & a_{2N} \\ \vdots & \vdots & \vdots & \cdots & \vdots \\ a_{N1} & a_{N2} & a_{N3} & \cdots & a_{NN} \end{bmatrix} \begin{bmatrix} x_1 \\ x_2 \\ \vdots \\ x_N \end{bmatrix} = \begin{bmatrix} b_1 \\ b_2 \\ \vdots \\ b_N \end{bmatrix} \qquad (\text{B.9})$$

where x_i represent the unknowns, a_{ij} are the coefficients, and b_i are constants. These N equations can be written in compact form as

$$\mathbf{AX} = \mathbf{B} \qquad (\text{B.10})$$

If the inverse of $\mathbf{A}$ exists, then multiplying Eq. B.10 by $\mathbf{A}^{-1}$ yields

$$\mathbf{A}^{-1}\mathbf{A}\mathbf{X} = \mathbf{A}^{-1}\mathbf{B} \quad \text{or} \quad \mathbf{X} = \mathbf{A}^{-1}\mathbf{B} \qquad (B.11)$$

Gauss–Jordan Method of Finding the Inverse Matrix Form the augmented matrix $\mathbf{A}^{\text{aug}} = [\mathbf{A}\,|\,\mathbf{I}]$. Apply elementary row operations on $\mathbf{A}^{\text{aug}}$ to transform the given matrix $\mathbf{A}$ into a unit matrix; simultaneously, the unit matrix $\mathbf{I}$ is transformed into the inverse matrix $\mathbf{A}^{-1}$.

The elementary row operations used are the following:

1. Multiplication of a row by any nonzero scalar
2. Interchange of any two rows
3. Addition of any multiple of one row to another row

EXAMPLE B.2

Solve the following system of equations.

$$\begin{bmatrix} 1 & 2 & -1 \\ 2 & 3 & -1 \\ 1 & 1 & 2 \end{bmatrix} \begin{bmatrix} x_1 \\ x_2 \\ x_3 \end{bmatrix} = \begin{bmatrix} 2 \\ 3 \\ -3 \end{bmatrix}$$

Solution To find the solution of the system of equations, the inverse of the coefficient matrix $\mathbf{A}$ is derived as follows:

Step 1 The augmented matrix is first formed.

$$\mathbf{A}^{\text{aug}} = \begin{bmatrix} 1 & 2 & 1 & | & 1 & 0 & 0 \\ 2 & 3 & -1 & | & 0 & 1 & 0 \\ 1 & 1 & 2 & | & 0 & 0 & 1 \end{bmatrix}$$

Step 2 $(-2R_1 + R_2); (-R_1 + R_3)$

$$\mathbf{A}^{\text{aug}} = \begin{bmatrix} 1 & 2 & -1 & | & 1 & 0 & 0 \\ 0 & -1 & 1 & | & -2 & 1 & 0 \\ 0 & -1 & 3 & | & -1 & 0 & 1 \end{bmatrix}$$

Step 3 $-R_2$

$$\mathbf{A}^{\text{aug}} = \begin{bmatrix} 1 & 2 & -1 & | & 1 & 0 & 0 \\ 0 & 1 & -1 & | & 2 & -1 & 0 \\ 0 & -1 & 3 & | & -1 & 0 & 1 \end{bmatrix}$$

Step 4 $(-2R_2 + R_1); (R_2 + R_3)$

$$\mathbf{A}^{\text{aug}} = \begin{bmatrix} 1 & 0 & 1 & | & -3 & 2 & 0 \\ 0 & 1 & -1 & | & 2 & -1 & 0 \\ 0 & 0 & 2 & | & 1 & -1 & 1 \end{bmatrix}$$

Step 5 $\frac{1}{2}R_3$

$$\mathbf{A}^{\text{aug}} = \begin{bmatrix} 1 & 0 & 1 & | & -3 & 2 & 0 \\ 0 & 1 & -1 & | & 2 & -1 & 0 \\ 0 & 0 & 1 & | & \frac{1}{2} & -\frac{1}{2} & \frac{1}{2} \end{bmatrix}$$

Step 6 $(-R_3 + R_1); (R_3 + R_2)$

$$\mathbf{A}^{\text{aug}} = \begin{bmatrix} 1 & 0 & 0 & | & -\frac{7}{2} & \frac{5}{2} & -\frac{1}{2} \\ 0 & 1 & 0 & | & \frac{5}{2} & -\frac{3}{2} & \frac{1}{2} \\ 0 & 0 & 1 & | & \frac{1}{2} & -\frac{1}{2} & \frac{1}{2} \end{bmatrix}$$

Thus, $\mathbf{A}^{-1}$ is found to be

$$\mathbf{A}^{-1} = \begin{bmatrix} -\frac{7}{2} & \frac{5}{2} & -\frac{1}{2} \\ \frac{5}{2} & -\frac{3}{2} & \frac{1}{2} \\ \frac{1}{2} & -\frac{1}{2} & \frac{1}{2} \end{bmatrix}$$

Therefore, the solution vector $\mathbf{X} = \mathbf{A}^{-1}\mathbf{B}$ is found as follows:

$$\mathbf{X} = \begin{bmatrix} x_1 \\ x_2 \\ x_3 \end{bmatrix} = \begin{bmatrix} -\frac{7}{2} & \frac{5}{2} & -\frac{1}{2} \\ \frac{5}{2} & -\frac{3}{2} & \frac{1}{2} \\ \frac{1}{2} & -\frac{1}{2} & \frac{1}{2} \end{bmatrix} \begin{bmatrix} 2 \\ 3 \\ -3 \end{bmatrix} = \begin{bmatrix} 2 \\ -1 \\ -2 \end{bmatrix}$$

B.6   PARTITIONING OF MATRICES

A large matrix can be subdivided into several submatrices of smaller dimension. Consider an $(M \times P)$ matrix $\mathbf{A}$ and a $(P \times N)$ matrix $\mathbf{B}$. $\mathbf{A}$ and $\mathbf{B}$ may be partitioned as follows:

$$\mathbf{A} = \left[\begin{array}{c|c|c} \mathbf{A}_{11} & \mathbf{A}_{12} & \mathbf{A}_{13} \\ \hline \mathbf{A}_{21} & \mathbf{A}_{22} & \mathbf{A}_{23} \end{array} \right] \tag{B.12}$$

$$
\mathbf{B} = \begin{bmatrix} \mathbf{B}_{11} & | & \mathbf{B}_{12} \\ ---+--- \\ \mathbf{B}_{21} & | & \mathbf{B}_{22} \\ ---+--- \\ \mathbf{B}_{31} & | & \mathbf{B}_{32} \end{bmatrix} \qquad \text{(B.13)}
$$

Product of Partitioned Matrices   If the submatrices $\mathbf{A}_{ij}$ and $\mathbf{B}_{ij}$ are chosen such that the number of columns of $\mathbf{A}_{ij}$ is equal to the number of rows of $\mathbf{B}_{ji}$ and such that the numbers of columns of $\mathbf{A}_{ij}$ and $\mathbf{A}_{(i+1)j}$ are equal, for all i and j, then the matrix product $\mathbf{C} = \mathbf{AB}$ is found as

$$
\mathbf{C} = \mathbf{AB} = \begin{bmatrix} \mathbf{C}_{11} & | & \mathbf{C}_{12} \\ ---+--- \\ \mathbf{C}_{21} & | & \mathbf{C}_{22} \end{bmatrix} \qquad \text{(B.14)}
$$

where

$$\mathbf{C}_{11} = \mathbf{A}_{11}\mathbf{B}_{11} + \mathbf{A}_{12}\mathbf{B}_{21} + \mathbf{A}_{13}\mathbf{B}_{31} \qquad \mathbf{C}_{12} = \mathbf{A}_{11}\mathbf{B}_{12} + \mathbf{A}_{12}\mathbf{B}_{22} + \mathbf{A}_{13}\mathbf{B}_{32}$$
$$\mathbf{C}_{21} = \mathbf{A}_{21}\mathbf{B}_{11} + \mathbf{A}_{22}\mathbf{B}_{21} + \mathbf{A}_{23}\mathbf{B}_{31} \qquad \mathbf{C}_{22} = \mathbf{A}_{21}\mathbf{B}_{12} + \mathbf{A}_{22}\mathbf{B}_{22} + \mathbf{A}_{23}\mathbf{B}_{32}$$

EXAMPLE B.3

Solve the following system of linear equations.

$$
\begin{bmatrix} a_{11} & a_{12} & | & a_{13} & a_{14} \\ a_{21} & a_{22} & | & a_{23} & a_{24} \\ ----+---- \\ a_{31} & a_{32} & | & a_{33} & a_{34} \\ a_{41} & a_{42} & | & a_{43} & a_{44} \end{bmatrix} \begin{bmatrix} x_1 \\ x_2 \\ -- \\ x_3 \\ x_4 \end{bmatrix} = \begin{bmatrix} b_1 \\ b_2 \\ -- \\ 0 \\ 0 \end{bmatrix}
$$

Solution   The system of equations can be written as follows:

$$
\begin{bmatrix} \mathbf{A}_{11} & | & \mathbf{A}_{12} \\ --+-- \\ \mathbf{A}_{21} & | & \mathbf{A}_{22} \end{bmatrix} \begin{bmatrix} \mathbf{X}_1 \\ -- \\ \mathbf{X}_2 \end{bmatrix} = \begin{bmatrix} \mathbf{B}_1 \\ -- \\ \mathbf{0} \end{bmatrix}
$$

Expanding,

$$\mathbf{A}_{11}\mathbf{X}_1 + \mathbf{A}_{12}\mathbf{X}_2 = \mathbf{B}_1$$
$$\mathbf{A}_{21}\mathbf{X}_1 + \mathbf{A}_{22}\mathbf{X}_2 = \mathbf{0}$$

By using the latter equation, the subvector X_2 can be expressed (provided the inverse of A_{22} exists) in terms of X_1 as follows:

$$X_2 = -A_{22}^{-1}A_{21}X_1$$

This last expression is substituted for X_2 in the other equation, and the expression for X_1 is derived as

$$X_1 = (A_{11} - A_{12}A_{22}^{-1}A_{21})^{-1}B_1$$

provided the inverse of the quantity inside the parentheses exists.

Appendix C

Solutions of Linear Equations

There are several ways to obtain the solution of a system of N simultaneous linear equations in N unknowns. Among these are

1. Matrix inversion
2. Gaussian elimination
3. Triangular factorization

Matrix inversion was discussed in Section B.5. In the following sections, the other two solution techniques are described.

C.1 GAUSSIAN ELIMINATION

Consider the following system of linear algebraic equations.

$$a_{11}x_1 + a_{12}x_2 + a_{13}x_3 = b_1 \qquad (C.1)$$

$$a_{21}x_1 + a_{22}x_2 + a_{23}x_3 = b_2 \qquad (C.2)$$

$$a_{31}x_1 + a_{32}x_2 + a_{33}x_3 = b_3 \qquad (C.3)$$

The Gaussian elimination technique proceeds as follows:

Step 1   Normalize the first of the equations; that is, divide Eq. C.1 by a_{11}. Thus,

$$x_1 + U_{12}x_2 + U_{13}x_3 = z_1 \qquad (C.4)$$

where

$$U_{12} = a_{12}/a_{11}$$
$$U_{13} = a_{13}/a_{11}$$
$$z_1 = b_1/a_{11}$$

Step 2   Multiply Eq. C.4 by a_{21}, and subtract the result from Eq. C.2 eliminating x_1. This yields

$$a'_{22}x_2 + U'_{23}x_3 = z'_2 \tag{C.5}$$

where

$$a'_{22} = a_{22} - a_{21}U_{12}$$
$$U'_{23} = a_{23} - a_{21}U_{13}$$
$$z'_2 = b_2 - a_{21}z_1$$

Step 3   Multiply Eq. C.4 by a_{31}, and subtract the result from Eq. C.3 eliminating x_1. This yields

$$a'_{32}x_2 + U'_{33}x_3 = z'_3 \tag{C.6}$$

where

$$a'_{32} = a_{32} - a_{31}U_{12}$$
$$a'_{33} = a_{33} - a_{31}U_{13}$$
$$z'_3 = b_3 - a_{31}z_1$$

Step 4   Repeat the general procedure of steps 1–3 on Eqs. C.5 and C.6 to eliminate x_2. Thus, dividing Eq. C.5 by a'_{22} yields

$$x_2 + U_{23}x_3 = z_2 \tag{C.7}$$

where

$$U_{23} = U'_{23}/a'_{22}$$
$$z_2 = z'_2/a'_{22}$$

Multiply Eq. C.7 by a'_{32} and subtract from Eq. C.6; thus,

$$a''_{33}x_3 = b''_3 \tag{C.8}$$

where

$$a''_{33} = a'_{33} - a'_{32}U_{23}$$
$$z''_3 = z'_3 - a'_{32}z_2$$

Dividing Eq. C.8 by a'_{33}, the value of x_3 is found as follows:

$$x_3 = z_3 \tag{C.9}$$

where $z_3 = z''_3 / a''_{33}$.

Step 5 Substitute the value of x_3 back into Eq. C.7 to solve for the value of x_2.

$$x_2 = z_2 - U_{23}x_3 \tag{C.10}$$

Next, substitute the values of x_2 and x_3 into Eq. C.4 to obtain the value of x_1.

$$x_1 = z_1 - U_{12}x_2 - U_{13}x_3 \tag{C.11}$$

This final process is referred to as back-substitution.

EXAMPLE C.1

Solve the following system by using the Gaussian elimination technique.

$$\begin{bmatrix} 1 & 2 & -1 \\ 2 & 3 & -1 \\ 1 & 1 & 2 \end{bmatrix} \begin{bmatrix} x_1 \\ x_2 \\ x_3 \end{bmatrix} = \begin{bmatrix} 2 \\ 3 \\ -3 \end{bmatrix}$$

Solution First, augment the constants to the coefficient matrix. The Gaussian elimination procedure is followed.

$$\mathbf{A}^{aug} = \begin{bmatrix} 1 & 2 & -1 & | & 2 \\ 2 & 3 & -1 & | & 3 \\ 1 & 1 & 2 & | & -3 \end{bmatrix}$$

Step 1 Since the first coefficient of the first row is already unity, continue with step 2.

Step 2 $-2R_1 + R_2$

$$\mathbf{A}^{aug} = \begin{bmatrix} 1 & 2 & -1 & | & 2 \\ 0 & -1 & 1 & | & -1 \\ 1 & 1 & 2 & | & -3 \end{bmatrix}$$

Step 3 $-R_1 + R_3$

$$\mathbf{A}^{\text{aug}} = \begin{bmatrix} 1 & 2 & -1 & | & 2 \\ 0 & -1 & 1 & | & -1 \\ 0 & -1 & 3 & | & -5 \end{bmatrix}$$

Step 4a $-R_2$

$$\mathbf{A}^{\text{aug}} = \begin{bmatrix} 1 & 2 & -1 & | & 2 \\ 0 & 1 & -1 & | & 1 \\ 0 & -1 & 3 & | & -5 \end{bmatrix}$$

Step 4b $R_2 + R_3$

$$\mathbf{A}^{\text{aug}} = \begin{bmatrix} 1 & 2 & -1 & | & 2 \\ 0 & 1 & -1 & | & 1 \\ 0 & 0 & 2 & | & -4 \end{bmatrix}$$

Step 4c $\frac{1}{2}R_3$

$$\mathbf{A}^{\text{aug}} = \begin{bmatrix} 1 & 2 & -1 & | & 2 \\ 0 & 1 & -1 & | & 1 \\ 0 & 0 & 1 & | & -2 \end{bmatrix}$$

The value of x_3 is read off as the last element in the last column of the final augmented matrix $\mathbf{A}^{\text{aug}}$. Therefore,

$$x_3 = -2$$

Step 5 Substitute the value of x_3 back into the second of the equations resulting from step 4a to solve for the value of x_2.

$$x_2 = 1 - (-1)(-2) = -1$$

Next, substitute the values of x_2 and x_3 into the first of the equations in step 1 to obtain the value of x_1.

$$x_1 = 2 - (2)(-1) - (-1)(-2) = 2$$

Thus, the back-substitution process yields the solution, which is presented in vector form as

$$\mathbf{X} = \begin{bmatrix} x_1 \\ x_2 \\ x_3 \end{bmatrix} = \begin{bmatrix} 2 \\ -1 \\ -2 \end{bmatrix}$$

C.2 TRIANGULAR FACTORIZATION

Triangular factorization is a modification of the Gaussian elimination technique. It is better adapted to computer use, particularly for a repeat solution of the system of equations with a new vector of right-hand side constants.

Consider the following system of three simultaneous linear algebraic equations in terms of the three unknown variables x_1, x_2, and x_3, where a_{ij} are constants for all i and j.

$$a_{11}x_1 + a_{12}x_2 + a_{13}x_3 = b_1 \tag{C.12}$$

$$a_{21}x_1 + a_{22}x_2 + a_{23}x_3 = b_2 \tag{C.13}$$

$$a_{31}x_1 + a_{32}x_2 + a_{33}x_3 = b_3 \tag{C.14}$$

These equations can also be written as follows:

$$\begin{bmatrix} a_{11} & a_{12} & a_{13} \\ a_{21} & a_{22} & a_{23} \\ a_{31} & a_{32} & a_{33} \end{bmatrix} \begin{bmatrix} x_1 \\ x_2 \\ x_3 \end{bmatrix} = \begin{bmatrix} b_1 \\ b_2 \\ b_3 \end{bmatrix} \tag{C.15}$$

In matrix notation, Eq. C.15 can be expressed as

$$\mathbf{AX} = \mathbf{B} \tag{C.16}$$

Assume that the coefficient matrix $\mathbf{A}$ can be written as the product of two matrices:

$$\mathbf{A} = \mathbf{LU} \tag{C.17}$$

Let $\mathbf{L}$ be a lower triangular matrix whose elements above the main or principal diagonal are all equal to zero. Also, let $\mathbf{U}$ be an upper triangular matrix whose elements on the principal diagonal are all unity and elements below the principal diagonal are all equal to zero. The matrices $\mathbf{L}$ and $\mathbf{U}$ are identified as the triangular factors of the coefficient matrix $\mathbf{A}$, and their standard forms are shown in Eqs. C.18 and C.19.

$$\mathbf{L} = \begin{bmatrix} L_{11} & 0 & 0 & \cdots & 0 \\ L_{21} & L_{22} & 0 & \cdots & 0 \\ L_{31} & L_{32} & L_{33} & \cdots & 0 \\ \vdots & \vdots & \vdots & \vdots & \vdots \\ L_{N1} & L_{N2} & L_{N3} & \cdots & L_{NN} \end{bmatrix} \tag{C.18}$$

$$\mathbf{U} = \begin{bmatrix} 1 & U_{12} & U_{13} & \cdots & U_{1N} \\ 0 & 1 & U_{23} & \cdots & U_{2N} \\ 0 & 0 & 1 & \cdots & U_{3N} \\ \vdots & \vdots & \vdots & \vdots & \vdots \\ 0 & 0 & 0 & \cdots & 1 \end{bmatrix} \qquad (C.19)$$

Equation C.16 can, therefore, be written as follows:

$$\mathbf{AX} = \mathbf{LUX} = \mathbf{B} \qquad (C.20)$$

Equation C.20 may also be written as

$$\mathbf{LZ} = \mathbf{B} \qquad (C.21)$$

where

$$\mathbf{UX} = \mathbf{Z} \qquad (C.22)$$

The vector $\mathbf{Z}$ is found by solving Eq. C.21; this process is called the forward pass. Then the solution vector $\mathbf{X}$ is found by solving Eq. C.22 using the previously determined vector $\mathbf{Z}$; this is the backward substitution process.

The matrix triangular factors $\mathbf{L}$ and $\mathbf{U}$ are found by multiplying Eqs. C.18 and C.19 and equating the product to the coefficient matrix $\mathbf{A}$. The following relationships are derived.

$$L_{ij} = a_{ij} - \sum_{k=1}^{j-1} L_{ij} U_{kj} \quad \text{for } i \geq j \qquad (C.23)$$

$$U_{ij} = \frac{1}{L_{ii}} \left(a_{ij} - \sum_{k=1}^{i-1} L_{ik} U_{kj} \right) \quad \text{for } i < j \qquad (C.24)$$

Equations C.23 and C.24 are used alternately to evaluate the elements of the triangular factors. The first column of $\mathbf{L}$ is initially determined by using Eq. C.23, followed by the first row of $\mathbf{U}$ by using Eq. C.24. Next, the second column of $\mathbf{L}$ is computed, followed by the second row of $\mathbf{U}$, and so forth.

Once the triangular factors $\mathbf{L}$ and $\mathbf{U}$ are found, they are used for repeat solutions of the system of equations. That is, for any vector of constants $\mathbf{B}$, there is no need to recalculate the elements of the vectors $\mathbf{L}$ and $\mathbf{U}$. The solutions of Eqs. C.21 and C.22 may be implemented by using the following relations.

$$z_i = \frac{1}{L_{ii}}\left(b_i - \sum_{k=1}^{i-1} L_{ik}z_k\right) \quad \text{for } i = 1, 2, \ldots, N \qquad (C.25)$$

$$x_i = z_i - \sum_{k=i+1}^{N} U_{ik}x_k \quad \text{for } i = N, N-1, \ldots, 2, 1 \qquad (C.26)$$

EXAMPLE C.2

Solve Example C.1 by using the triangular factorization technique.

$$\begin{bmatrix} 1 & 2 & -1 \\ 2 & 3 & -1 \\ 1 & 1 & 2 \end{bmatrix}\begin{bmatrix} x_1 \\ x_2 \\ x_3 \end{bmatrix} = \begin{bmatrix} 2 \\ 3 \\ -3 \end{bmatrix}$$

Solution By using Eqs. C.23 and C.24, the triangular factors are found as follows:

$$L_{11} = a_{11} = 1$$

$$L_{21} = a_{21} = 2$$

$$L_{31} = a_{31} = 1$$

$$U_{12} = \frac{1}{L_{11}}a_{12} = \frac{1}{1}(2) = 2$$

$$U_{13} = \frac{1}{L_{11}}a_{13} = \frac{1}{1}(-1) = -1$$

$$L_{22} = a_{22} - L_{21}U_{12} = 3 - (2)(2) = -1$$

$$L_{32} = a_{32} - L_{31}U_{12} = 1 - (1)(2) = -1$$

$$U_{23} = \frac{1}{L_{22}}(a_{23} - L_{21}U_{13}) = \frac{1}{-1}[-1 - (2)(-1)] = -1$$

$$L_{33} = a_{33} - L_{31}U_{13} - L_{32}U_{23}$$
$$= 2 - (1)(-1) - (-1)(-1) = 2$$

The forward pass is performed by using Eq. C.25 to solve for the **Z** vector as follows:

$$z_1 = \frac{1}{L_{11}}b_1 = \frac{1}{1}(2) = 2$$

$$z_2 = \frac{1}{L_{22}}(b_2 - L_{21}z_1) = \frac{1}{-1}[3 - (2)(2)] = 1$$

$$z_3 = \frac{1}{L_{33}}(b_3 - L_{31}z_1 - L_{32}z_2)$$
$$= \tfrac{1}{2}[-3 - (1)(2) - (-1)(1)] = -2$$

By using Eq. C.26, the back substitution process yields

$$x_3 = z_3 = -2$$
$$x_2 = z_2 - U_{23}z_3 = 1 - (-1)(-2) = -1$$
$$x_1 = z_1 - U_{12}x_2 - U_{13}x_3 = 2 - (2)(-1) - (-1)(-2) = 2$$

Appendix D

Maxwell's Equations

Maxwell's equations are a collection, and are generalizations, of various phenomena and laws previously described by different scientists on the relationships among currents, charges, electric fields, and magnetic fields. These fields have both magnitudes and directions; thus, they are represented as vectors.

Maxwell's equations can be written in either integral or differential form. They are first presented in integral form in the next section, and the differential form is given in the following section.

D.1   INTEGRAL FORM OF MAXWELL'S EQUATIONS

Maxwell's equations in integral form are used to analyze electromagnetic systems that exhibit symmetry (e.g., rectangular, cylindrical) with respect to one or more dimensions that are usually found in electromechanical conversion devices and systems. These equations describe the relationships among electric fields, magnetic fields, charge densities, and current densities over a specified area or volume in space. The following notation is used.

$\mathbf{E}$ = electric field intensity (volts/meter)

$\mathbf{H}$ = magnetic field intensity (amperes/meter)

$\mathbf{D}$ = electric flux density (coulombs/square meter)

$\mathbf{B}$ = magnetic flux density (webers/square meter)

$\mathbf{M}_i$ = applied magnetic current density (volts/square meter)

$\mathbf{J}_i$ = applied electric current density (amperes/square meter)

ρ_e = electric charge density (coulombs/cubic meter)

ρ_m = magnetic charge density (webers/cubic meter)

478

The first of Maxwell's equations in integral form can be written as

$$\oint_C \mathbf{E} \cdot \mathbf{dl} = -\int\int_S \mathbf{M_i} \cdot \mathbf{dS} - \frac{\partial}{\partial t}\int\int_S \mathbf{B} \cdot \mathbf{dS} \qquad (D.1)$$

where $\mathbf{dS}$ is the normal vector to the surface $\mathbf{S}$ which is bounded by the contour $\mathbf{C}$. When there is no applied magnetic current density ($\mathbf{M_i} = 0$), Eq. D.1 reduces to Faraday's law, which states that the electromagnetic force (emf) induced across the open-circuited terminals of a coil is equal to the time rate of change of magnetic flux linking the coil.

The second of Maxwell's equations in integral form can be written as

$$\oint_C \mathbf{H} \cdot \mathbf{dl} = -\int\int_S \mathbf{J_i} \cdot \mathbf{dS} - \frac{\partial}{\partial t}\int\int_S \mathbf{D} \cdot \mathbf{dS} \qquad (D.2)$$

In the absence of the applied electric current density ($\mathbf{J_i} = 0$), Eq. D.2 reduces to Ampère's law, which states that the line integral of the electric field about any closed path is equal to the current enclosed by that path.

The next two Maxwell's equations in integral form can be written as follows:

$$\oint\oint_S \mathbf{D} \cdot \mathbf{dS} = \int\int\int_V \rho_e \, dv = Q_e \qquad (D.3)$$

$$\oint\oint_S \mathbf{B} \cdot \mathbf{dS} = \int\int\int_V \rho_m \, dv = 0 \qquad (D.4)$$

Equations D.3 and D.4 are Gauss's law for electric fields and magnetic fields, respectively. Gauss's law for electric fields states that the total electric flux passing through a closed surface is equal to the total charge enclosed by that surface. Since no magnetic source for magnetic flux lines has ever been discovered, the right-hand side of Gauss's law for magnetic fields is identically equal to zero.

D.2 DIFFERENTIAL FORM OF MAXWELL'S EQUATIONS

Maxwell's equations in differential form are used to describe the relationships among electric fields, magnetic fields, current densities, and charge densities at any point in space at any time. It is assumed that the field vectors are analytic functions of both position and time with continuous derivatives. Electromagnetic fields exhibit these characteristics except where there are abrupt changes in charge and current densities, which usually occur when there are changes in the properties of the medium or flux path. At these discontinuities, the associated boundary conditions need to be specified.

Maxwell's equations in differential form can be written as

$$\nabla \times \mathbf{E} = -\mathbf{M}_i - \frac{\partial \mathbf{B}}{\partial t} \tag{D.5}$$

$$\nabla \times \mathbf{H} = \mathbf{J}_i + \frac{\partial \mathbf{D}}{\partial t} \tag{D.6}$$

$$\nabla \cdot \mathbf{D} = \rho_e \tag{D.7}$$

$$\nabla \cdot \mathbf{B} = 0 \tag{D.8}$$

Appendix E

Constants and Conversion Factors

E.1 CONSTANTS

Permeability of free space	$\mu_0 = 4\pi \times 10^{-7}$ H/m
Permittivity of free space	$\epsilon_0 = 8.854 \times 10^{-12}$ F/m
Resistivity of annealed copper	$\rho = 1.72 \times 10^{-8}$ Ω-m
Acceleration due to gravity	$g = 9.807$ m/s^2

E.2 CONVERSION FACTORS

Length	1 meter (m)	= 3.281 feet (ft)
		= 39.36 inches (in)
	1 mile (mi)	= 1.609 kilometers (km)
Mass	1 kilogram (kg)	= 0.0685 slug
		= 2.205 pounds (lb)
Time	1 second (s)	= 1/60 minute (min)
		= 1/3600 hour (h)
Force	1 Newton (N)	= 0.02248 lb force (lbf)
		= 0.102 kg force
Torque	1 N-m	= 0.7376 lbf-ft
Moment of inertia	1 kg-m^2	= 23.7 lb-ft^2

Power	1 watt (W)	= 0.7376 ft-lbf/s
	1 horsepower (hp)	= 746 W
Energy	1 joule (J)	= 1 W-s
		= 0.7376 ft-lbf
		= 2.778×10^{-7} kWh
Current	1 ampere (A)	= 1 coulomb/s
Voltage	1 volt (V)	= 1 watt/ampere
Magnetic flux	1 weber (Wb)	= 10^8 maxwells or lines
Magnetic flux density	1 tesla (T)	= 1 Wb/m^2
		= 64,500 lines/in^2
Magnetic field intensity	1 A-turn/m	= 0.0254 A-turn/in
Flux linkage	1 Wb-turn	
Resistance	1 ohm (Ω)	
Inductance	1 henry (H)	
Capacitance	1 farad (F)	

Appendix F

Trigonometric Identities

a. $\sin^2 \theta + \cos^2 \theta = 1$

b. $\cos 2\theta = \cos^2 \theta - \sin^2 \theta = 2\cos^2 \theta - 1 = 1 - 2\sin^2 \theta$

c. $\sin 2\theta = 2\sin \theta \cos \theta$

d. $\cos^2 \theta = \frac{1}{2}(1 + \cos 2\theta)$

e. $\sin^2 \theta = \frac{1}{2}(1 - \cos 2\theta)$

f. $\cos(\alpha - \beta) = \cos \alpha \cos \beta + \sin \alpha \sin \beta$
$\cos(\alpha + \beta) = \cos \alpha \cos \beta - \sin \alpha \sin \beta$

g. $\sin(\alpha + \beta) = \sin \alpha \cos \beta + \cos \alpha \sin \beta$
$\sin(\alpha - \beta) = \sin \alpha \cos \beta - \cos \alpha \sin \beta$

h. $\cos \alpha \cos \beta = \frac{1}{2}[\cos(\alpha - \beta) + \cos(\alpha + \beta)]$

$\sin \alpha \sin \beta = \frac{1}{2}[\cos(\alpha - \beta) - \cos(\alpha + \beta)]$

$\sin \alpha \cos \beta = \frac{1}{2}[\sin(\alpha + \beta) + \sin(\alpha - \beta)]$

i. $\cos \theta + \cos(\theta - 120°) + \cos(\theta - 240°) = 0$

j. $\sin \theta + \sin(\theta - 120°) + \sin(\theta - 240°) = 0$

k. $\cos^2 \theta + \cos^2(\theta - 120°) + \cos^2(\theta - 240°) = \frac{3}{2}$

l. $\sin^2 \theta + \sin^2(\theta - 120°) + \sin^2(\theta - 240°) = \frac{3}{2}$

m. $\sin \theta \cos \theta + \sin(\theta - 120°)\cos(\theta - 120°) + \sin(\theta - 240°)\cos(\theta - 240°) = 0$

n. $\tan \theta = \sin \theta / \cos \theta$

o. $\cot \theta = \cos \theta / \sin \theta$

p. $\sec \theta = 1/\cos \theta$

q. $\csc \theta = 1/\sin \theta$

r. $d(\sin\theta) = \cos\theta\, d\theta$

s. $d(\cos\theta) = -\sin\theta\, d\theta$

t. $d(\tan\theta) = \sec^2\theta\, d\theta$

u. $d(\cot\theta) = -\csc^2\theta\, d\theta$

v. $d(\sec\theta) = \sec\theta\tan\theta\, d\theta$

w. $d(\csc\theta) = -\csc\theta\cot\theta\, d\theta$

Appendix G

Glossary

The key terms as well as some of the most commonly used terms that have been highlighted in this book are defined in the following pages. Most of the definitions given in this glossary are based on the *IEEE Standard Dictionary of Electrical and Electronics Terms,* 4th ed, IEEE, New York, 1988.

Air-gap power   The power transferred across the air gap from the stator to the rotor.

Aluminum cable steel-reinforced (ACSR)   A composite conductor made up of a combination of aluminum wires surrounding the steel.

Ampère's law   The magnetic field strength, at any point in the neighborhood of a circuit in which there is a current i, is equal to the vector sum of the contributions from all the differential elements of the circuit. The contribution, $d\mathbf{H}$, caused by a current i in an element $\mathbf{ds}$ at a distance $\mathbf{r}$ from a point P is given by

$$d\mathbf{H} = \frac{i[\mathbf{r} \times \mathbf{ds}]}{r^2}$$

An-bn-cn or abc sequence   The order in which the successive members of the set reach their positive maximum values. Phase a is followed by phase b and then by phase c.

Apparent power   The product of the root-mean-square voltage and the root-mean-square current.

Armature winding   The winding in which alternating voltage is generated by virtue of relative motion with respect to a magnetic flux field.

Autotransformer   A transformer in which at least two windings have a common section.

Average power   The time average of the instantaneous power, the average being taken over one period.

Back emf.   *See* Counter emf.

Balanced set   A set of phasor currents (or phasor voltages) that have equal magnitudes and are separated from each other by equal phase angles.

Blocked-rotor test   A test applied to an induction motor, in which the rotor is blocked so it cannot rotate.

Bundled conductors   An assembly of two or more conductors used as a single conductor and employing spacers to maintain a predetermined configuration. The individual conductors of this assembly are called subconductors.

Bus   A conductor, or group of conductors, that serve as a common connection for two or more circuits.

Bus admittance matrix   A matrix whose elements have the dimension of admittance and, when multiplied into the vector of bus voltages, gives the vector of bus currents.

Capacitive reactance at 1-ft spacing, x_a'   Capacitive reactance of one conductor to neutral of a circuit consisting of two conductors 1 foot apart.

Capacitive reactance spacing factor, x_d'   Capacitive reactance of one conductor of a circuit consisting of two conductors separated by a distance D expressed in feet.

Circuit breaker   A switching device capable of making, carrying, and breaking currents under normal circuit conditions and also making, carrying for a specified time, and breaking currents under specified abnormal conditions such as those of a short circuit.

Circular mil   A unit of area equal to $\pi/4$ of a square mil (0.7854 square mil). The cross-sectional area of a circle in circular mils is therefore equal to the square of its diameter in mils.

Coenergy   A function associated with the field energy function such that the sum of the energy and coenergy functions is equal to the sum of the products of the flux linkage of a coil multiplied by the corresponding current flowing through it.

Complex power   The product of the phasor voltage multiplied by the complex conjugate of the phasor current.

Core losses   The power dissipated in a magnetic core subjected to a time-varying magnetizing force.

Counter emf   The effective electromotive force within the system that opposes the passage of current in a specified direction.

Critical resistance The value of field circuit resistance of a DC shunt generator above which the generator fails to build up.

DC machine An electromechanical conversion device whose armature terminal voltage and current, as well as field voltage and excitation current, are all DC.

Developed power The power converted from electrical to mechanical or from mechanical to electrical.

Diagonal matrix A square matrix in which all the elements not in the principal diagonal are equal to zero; that is, $a_{ij} = 0$ for all $i \neq j$.

Direct-axis synchronous reactance, X_d The ratio of the sustained fundamental component of armature voltage that is produced by the total direct-axis flux due to direct-axis armature current to the fundamental component of this current, the machine running at rated speed.

Economic dispatch The distribution of total generation requirements among alternative sources for optimum system economy, with due consideration of both incremental generating costs and incremental transmission losses.

Effective value. *See* Root-mean-square value.

Efficiency The ratio of the useful power output to the total power input.

Electrical load Electric power used by devices connected to an electrical generating system.

Electromotive force (emf) A voltage produced in a closed path or circuit by the relative motion of the circuit or its parts with respect to magnetic flux.

Equivalent radius The conductor radius r multiplied by $e^{-1/4}$ (i.e., $re^{-1/4}$) where e is the base of the natural logarithm.

Faraday's law The electromotive force induced is proportional to the time rate of change of magnetic flux linked with the circuit.

Fast-decoupled method An iterative technique for solving the nonlinear power flow equations by separately and alternatively solving for the voltage angles and voltage magnitudes.

Fault A physical condition that causes a device, a component, or an element to fail to perform in a required manner, for example, a short circuit, a broken wire, or an intermittent connection.

Field energy The energy stored in the magnetic field of an electromagnetic system.

Field winding A winding on either the stationary or the rotating part of a machine whose sole purpose is the production of the main electromagnetic field of the machine.

Flux linkages The sum of the fluxes linking the turns forming the coil.

Gauss–Seidel method An iterative technique for solving a set of nonlinear equations (e.g., the power flow equations), which uses the latest values of the variables in seeking improved values of the other variables.

Gauss's law The integral over any closed surface of the normal component of the electric flux density is equal in a rationalized system to the electric charge Q_0 within the surface.

Geometric mean distance The mean of (n) distances produced by taking the nth root of their product.

Geometric mean radius The equivalent radius of a multistrand conductor.

Ground wire A conductor having grounding connections at intervals, which is suspended usually above but not necessarily over the line conductor to provide a degree of protection against lightning discharges.

Ideal transformer A transformer characterized by no winding resistance, no leakage flux, and a lossless and infinitely permeable magnetic core.

Identity, or unit, matrix A diagonal matrix whose elements in the principal diagonal are all equal to unity and all elements not in the principal diagonal are equal to zero.

Induction machine An asynchronous AC machine that comprises a magnetic circuit interlinked with two electric circuits, rotating with respect to each other, and in which power is transferred from one circuit to another by electromagnetic induction.

Inductive reactance at 1-ft spacing, x_a The inductive reactance of one conductor of a circuit consisting of two conductors 1 foot apart.

Inductive reactance spacing factor, x_d The inductive reactance of one conductor of a circuit consisting of two conductors separated by a distance D expressed in feet.

Iron losses. *See* Core losses.

Lagging power factor An operating power factor condition such that the phasor current lags the phasor voltage.

Leading power factor An operating power factor condition such that the phasor current leads the phasor voltage.

Linear electromagnetic system An electromagnetic system whose flux linkages are expressed as linear combinations of the currents in terms of the self-inductance of each winding and mutual inductances between the windings.

Load characteristic. *See* Terminal characteristic.

Load curve A curve of power versus time showing the value of a specific load for each unit of the period covered.

Magnetic circuit The region containing essentially all the flux, such as the core of a transformer.

Magnetic flux The surface integral of the normal component of the magnetic induction over the area.

Magnetic flux density Flux per unit area through an element normal to the direction of flux.

Magnetization curve. *See* Saturation curve.

Magnetomotive force (mmf) The line integral of the magnetizing force around the path.

Maximum power The maximum output that an electric machine is capable of developing at rated voltage and speed.

Mutual inductance The common property of two electric circuits whereby an electromotive force is induced in one circuit by a change of current in the other circuit.

Negative-sequence impedance The quotient of that component of negative-sequence sinusoidal voltage that is due to the negative-sequence component of the current, divided by the negative-sequence component of the current at the same frequency.

Negative-sequence network The equivalent representation of a power system constructed by using only the negative-sequence impedances of the various components.

Newton–Raphson method An iterative technique for solving a set of non-linear equations (e.g., the power flow equations) by solving a succession of linearized equations to derive improvements to the latest estimate of the solution.

Node. *See* Bus.

No-load characteristic The saturation curve of a machine at no load.

No-load test A test applied to a machine on no load at rated voltage and frequency.

Nominal π circuit A network composed of three branches connected in series with each other to form a mesh, the three junction points forming an input terminal, an output terminal, and a common input and output terminal, respectively.

Nonideal or actual transformer A transformer having winding resistance and leakage flux, and a magnetic core having finite permeability and core losses.

Nonsalient, round, or cylindrical The part of a core, usually circular, that by virtue of DC excitation of a winding embedded in slots and distributed over the interpolar space acts as a pole.

Normally excited The operating condition of a synchronous machine at unity power factor.

490 APPENDIX G GLOSSARY

Null matrix   A matrix whose elements are all equal to zero.

One-line, or single-line, diagram   A diagram that shows, by means of single lines and graphic symbols, the course of an electric circuit or system of circuits and the component devices or parts used therein.

Open-circuit characteristic (OCC)   The saturation curve of a machine with an open-circuited armature winding.

Open-circuit test   A test in which the machine is run as a generator with its terminals open-circuited.

Overexcited   The operating condition of a synchronous machine delivering reactive power.

Pickup   The action of a relay as it makes designated responses to a progressive increase of input.

Pickup value   The minimum input that will cause a device to complete contact operation or similar designated action.

Positive-sequence impedance   The quotient of that component of positive-sequence sinusoidal voltage that is due to the positive-sequence component of current, divided by the positive-sequence component of the current at the same frequency.

Positive-sequence network   The equivalent representation of a power system constructed by using only the positive-sequence impedances of the various components.

Power angle   The phase angle between the generated voltage phasor and the terminal voltage phasor.

Power-angle characteristic   The expression for the real power developed by a synchronous machine in terms of its generated voltage, terminal voltage, synchronous reactance, and power angle.

Power-angle curve   The plot of the power-angle characteristic of a synchronous machine.

Power factor   The ratio of the average power in watts to the root-mean-square (RMS) volt-amperes.

Power factor angle   The angle whose cosine is the power factor.

Power flow equations   The system of nonlinear algebraic equations relating the phasor bus voltages to the complex power injections into the buses of the power system.

Protective relay   A device whose function is to detect defective lines or apparatus or other power system conditions of an abnormal or dangerous nature and to initiate appropriate control action.

Pull-out torque   The maximum sustained torque that the synchronous machine will develop at synchronous speed with rated voltage applied at rated frequency and with normal excitation.

Quadrature-axis synchronous reactance, X_q   The ratio of the fundamental component of reactive armature voltage, due to the fundamental quadrature-axis component of armature current, to this component of current under steady-state conditions and at rated frequency.

Reactive power   The product of voltage and out-of-phase components of alternating current.

Real power   The average power, or active power, or the real part of the complex power.

Reluctance   The ratio of the magnetomotive force to the magnetic flux through any cross section of the magnetic circuit.

Reluctance power   The component of the power delivered by a synchronous generator representing the effects of generator saliency.

Reset   The action of a relay as it makes designated responses to decreases in input.

Reset value   The maximum value of an input quantity reached by progressive decreases that will permit the relay to reach the state of complete reset from pickup.

Residual voltage   The generated voltage due to the residual flux in the field poles even when the field circuit remains unexcited.

Root-mean-square value   The square root of the average of the square of the value of the function taken throughout one period.

Rotor   The rotating member of a machine, with shaft.

Salient, or projecting, pole   A field pole that projects from the yoke or hub toward the primary winding core.

Saturation curve   A characteristic curve that expresses the degree of magnetic saturation as a function of some property of the magnetic excitation.

Self-inductance   The property of an electric circuit whereby an electromotive force is induced in that circuit by a change of current in the circuit.

Short-circuit characteristic (SCC)   The relationship between the current in the short-circuited armature winding and the field current.

Short-circuit test   A test applied to a transformer with one winding short-circuited and reduced voltage applied to the other winding such that rated current flows in the windings.

Single-phase transformer   A device consisting of two or more windings coupled by a magnetic core that is used to transform a single-phase voltage.

Skin effect   The tendency of an alternating current to concentrate in the areas of lowest impedance.

Slip rpm   The difference between the synchronous speed and the actual speed of the rotor, expressed in revolutions per minute.

Slip s The quotient of the difference between the synchronous speed and the actual speed of a rotor, to the synchronous speed, expressed as a ratio or as a percentage.

Speed regulation   The relationship between the speed and the load of a motor under specified conditions.

Square matrix   A matrix in which the number of rows is equal to the number of columns.

Squirrel-cage rotor   A rotor core assembly consisting of a number of conducting bars having their extremities connected by metal rings or plates at each end, like a squirrel's cage.

Stator   The portion of the rotating machine that includes and supports the stationary active parts.

Stranded conductor   A conductor composed of a group of wires or of any combination of groups of wires.

Swing equation   The differential equation used to describe the dynamic motion of a synchronous machine.

Symmetric matrix   A matrix whose elements are symmetric about the principal diagonal, that is, $a_{ij} = a_{ji}$, for all i and j. Therefore, it is equal to its own transpose matrix; thus, $\mathbf{A} = \mathbf{A}^T$.

Synchronization   The process whereby a synchronous machine, with its voltage and phase suitably adjusted, is paralleled with another synchronous machine or system.

Synchronous machine   A machine in which the average speed of normal operation is exactly proportional to the frequency of the system to which it is connected.

Synchronous reactance   The steady-state reactance of a generator during fault conditions used to calculate the steady-state fault current.

Synchronous speed   The speed of rotation of the magnetic flux, produced by or linking the primary winding.

Terminal characteristic   A plot of the terminal voltage versus load current.

Three-phase transformer   A transformer consisting of three pairs of windings used to transform a balanced set of three-phase voltages from one voltage level to another.

Transpose   The matrix $\mathbf{A}^T$ whose element a_{ji} is equal to the element a_{ij} of matrix $\mathbf{A}$ for all i and j. In general, $\mathbf{A}^T$ is formed by interchanging the rows and columns of $\mathbf{A}$.

Turns ratio   The ratio of the primary winding turns to the secondary winding turns.

Underexcited   The operating condition of a synchronous machine absorbing reactive power.

V-curve   The characteristic of a synchronous motor showing the variation of the stator current versus the field current.

Voltage regulation   (1) In a transformer, the change in output (secondary) voltage that occurs when the load (at a specified power factor) is reduced from rated value to zero, with the primary impressed terminal voltage maintained constant. (2) In a DC generator, the final change in voltage with constant field-rheostat setting when the specified load is reduced gradually to zero, expressed as a percent of rated-load voltage, the speed being kept constant. (3) In a synchronous generator, the rise in voltage with constant field current, when, with the synchronous generator operated at rated voltage and rated speed, the specified load at the specified power factor is reduced to zero, expressed as a percent of rated voltage.

Winding factor   The product of the distribution factor and the pitch factor.

Wound rotor   A rotor core assembly having a winding made up of individually insulated wires.

Zero-sequence impedance   The quotient of the zero-sequence component of the voltage, assumed to be sinusoidal, supplied to a synchronous machine, and the zero-sequence component of the current at the same frequency.

Zero-sequence network   The equivalent representation of a power system constructed by using only the zero-sequence impedances of the various components.

Index

nominal π circuit, 337, 343
propagation constant, 342
resistance, 310–311
surge impedance, *see* characteristic
  impedance
transposition, 317
voltage regulation, 335–338
Transmission line transients, *see* Power
  system analysis
Transpose, 492. *See also* Matrix
Transposition, *see* Transmission lines
Triangular factorization, *see* Linear
  equations, solutions of
Turns ratio, 492. *See also* Transformers

Underexcited, 492. *See also*
  Synchronous generators
Unit matrix, *see* Identity matrix

V-curve, 493. *See also* Synchronous
  motors

Vector, 462
Voltage buildup in a shunt generator, *see*
  DC generators
Voltage regulation:
  in a DC generator, 181
  in a synchronous generator, 230
  in a transformer, 89
  in a transmission line, 335–338

Wave winding, 169
Winding factor, 138, 493
Wound rotor, 493. *See also* Induction
  machines
Wye-connected load, 44–45

Zero-sequence component, *see*
  Symmetrical components
Zero-sequence impedance, 493. *See
  also* Sequence impedances
Zero-sequence network, 493. *See also*
  Sequence networks
Zone of protection, 413